T0292843

Monographs on Statistics and Applied Probability 146

Semialgebraic Statistics and Latent Tree Models

Piotr Zwiernik
Università degli Studi di Genova, Italy

CRC Press
Taylor & Francis Group
Boca Raton   London   New York

CRC Press is an imprint of the
Taylor & Francis Group, an **informa** business

A CHAPMAN & HALL BOOK

MONOGRAPHS ON STATISTICS AND APPLIED PROBABILITY

General Editors

F. Bunea, V. Isham, N. Keiding, T. Louis, R. L. Smith, and H. Tong

CRC Press
Taylor & Francis Group
6000 Broken Sound Parkway NW, Suite 300
Boca Raton, FL 33487-2742

© 2016 by Taylor & Francis Group, LLC
CRC Press is an imprint of Taylor & Francis Group, an Informa business

No claim to original U.S. Government works

Printed on acid-free paper
Version Date: 20150617

International Standard Book Number-13: 978-1-4665-7621-6 (Hardback)

This book contains information obtained from authentic and highly regarded sources. Reasonable efforts have been made to publish reliable data and information, but the author and publisher cannot assume responsibility for the validity of all materials or the consequences of their use. The authors and publishers have attempted to trace the copyright holders of all material reproduced in this publication and apologize to copyright holders if permission to publish in this form has not been obtained. If any copyright material has not been acknowledged please write and let us know so we may rectify in any future reprint.

Except as permitted under U.S. Copyright Law, no part of this book may be reprinted, reproduced, transmitted, or utilized in any form by any electronic, mechanical, or other means, now known or hereafter invented, including photocopying, microfilming, and recording, or in any information storage or retrieval system, without written permission from the publishers.

For permission to photocopy or use material electronically from this work, please access www.copyright.com (http://www.copyright.com/) or contact the Copyright Clearance Center, Inc. (CCC), 222 Rosewood Drive, Danvers, MA 01923, 978-750-8400. CCC is a not-for-profit organization that provides licenses and registration for a variety of users. For organizations that have been granted a photocopy license by the CCC, a separate system of payment has been arranged.

Trademark Notice: Product or corporate names may be trademarks or registered trademarks, and are used only for identification and explanation without intent to infringe.

**Visit the Taylor & Francis Web site at
http://www.taylorandfrancis.com**

**and the CRC Press Web site at
http://www.crcpress.com**

To Natalia and Julian

Contents

Preface

Algebraic tools have been used in statistical research since the very beginning of the field. In recent years, the interaction between statistics and pure mathematics has intensified and many new directions are now explored. One instance of that is a growing interest in *algebraic statistics*, an emerging field aimed at solving statistical inference problems using concepts from algebraic geometry as well as related computational and combinatorial techniques. As a result, algebraic statistics provides a dictionary that enables mathematicians and statisticians to work together on statistically relevant problems.

There is a natural question: Is algebraic statistics useful or is it just a way to redefine statistical objects in a more algebraic language with no real benefit to the field? More concretely, are there some real statistical problems which were previously unsolved before applying the algebraic machinery? There are many such problems that have been solved only after translating them into algebraic geometry. There are even more problems that are naturally formulated in terms of algebraic geometry but statisticians did not have the tools to solve them from that perspective. Some examples of both types include designing procedures to sample exactly from a given conditional distribution for discrete exponential families (see Diaconis and Sturmfels [1998]), using commutative algebra to study the structure of conditional independence models (see [Drton et al., 2009, Chapter 4]), and identifiability of statistical models with or without hidden variables (see Allman et al. [2009], Drton et al. [2011]).

Algebro-geometric methods are very important in the study of statistical models with hidden variables, which are generally poorly understood within statistics. Lazarsfeld and Henry [1968] studied algebraic constraints on probability distributions in the mixture model. More recently, algebraic geometry has been used to establish the identifiability of many models with hidden variables, which amounts to studying the parameterization of those models. In addition to that, the work of Watanabe [2009] uses algebraic geometry to study the asymptotic behavior of empirical processes arising in the context of hidden data, where the classical asymptotic theory does not apply.

The focus of this book is on statistical models with hidden variables. The first part is a general introduction to some important concepts in algebraic statistics with an emphasis on methods that are helpful in the study of models with hidden variables. The three main ideas behind this approach are

(i) Using tensor geometry as a natural language to deal with multivariate probability distributions,

(ii) Developing new combinatorial tools that can be used to study models with hidden data, and

(iii) Focusing on the *semialgebraic* structure of statistical models.

The last item is the most important here. In algebraic statistics the geometry of statistical models is typically studied over the complex numbers. This approach is based on the fact that many statistical models correspond, from the parameterization point of view, to geometric objects described by polynomial *equations*. This is useful in a wide variety of contexts, but it gives unsatisfactory results in the case of models with hidden variables where the geometry over the real numbers is always much richer and it admits additional polynomial *inequalities*.

An important example of models with hidden variables is given by the latent tree models studied in the second part of this book. Models of this type are widely used in biology and they generalize various popular models used in machine learning like the hidden Markov model, the naive Bayes model, and various state-space models. A general statistical understanding of these models is very limited and typically estimation is done using fragile numerical procedures with no guarantees of convergence to the global maximum. This book shows how combinatorics and algebraic geometry can give a better understanding of these models. It contains many results on the geometry of these models, which includes a very detailed analysis of identifiability and the defining polynomial constraints.

I would like to thank all of the collaborators who worked with me on the material presented in this book. I am grateful to Jan Draisma, Mathias Drton, Diane Maclagan, Mateusz Michałek, John Rhodes, and Jim Q. Smith for many helpful discussions. This book also benefited from the comments of Elizabeth Allman, Steffen Klaere, and anonymous referees. I am especially thankful to Bernd Sturmfels for being a great mentor, collaborator, and for reading an early version of this book. Kaie Kubjas helped me to get rid of many typos. This book was written partly during my Marie Skłodowska-Curie Fellowship at the University of California, Berkeley and Università degli Studi di Genova. I thank Eva Riccomagno and both institutions for hosting me, and the European Commission for funding my research.

Genoa, Italy

List of Figures

List of Tables

List of Symbols

$\dashrightarrow$	rational mapping
$\emptyset$	the empty set
$\hat{0}$	minimal element of a lattice
$\hat{1}$	maximal element of a lattice
$\mathcal{A}(u)$	$\{1,\ldots,1,\ldots,m,\ldots,m\}$, where i is repeated u_i times
$\mathcal{A}(\mathcal{X})$	$\{\mathcal{A}(x) : x \in \mathcal{X}\}$
$\mathbb{C}$	the complex numbers
$\mathbb{C}^*$	$\mathbb{C} \setminus \{0\}$
$\mathrm{corr}(X,Y)$	correlation between X and Y
$\mathrm{cov}(X,Y)$	covariance between X and Y
$\Delta_{\mathcal{X}}^0$	the interior of $\Delta_{\mathcal{X}}$
Det	hyperdeterminant
$\det$	determinant
$\Delta_{\mathcal{X}}$	probability simplex of distributions of $X \in \mathcal{X}$
$\mathbb{E}$	expectation
$E(\mathcal{T})$	edges of a semi-labeled tree $\mathcal{T}$
$\mathcal{F}$	semi-labeled forest
$\mathcal{F}^{\pi}$	see (5.5)
$\mathcal{F}_{\pi}$	see (5.6)
$\overline{ij}$	the set of edges on the path between i and j
$\mathrm{ind}(\alpha)$	$\mathrm{ind}(\alpha)_i = 1$ if $\alpha_i \neq 0$ and is zero otherwise
k	a field, typically $k = \mathbb{R}, \mathbb{C}$ or $\mathbb{Q}$
$\boldsymbol{M}_{++}(\mathcal{T},2)$	the positive part of $\boldsymbol{M}(\mathcal{T},2)$
$\boldsymbol{M}_{+}(\mathcal{T},2)$	the nonnegative part of $\boldsymbol{M}(\mathcal{T},2)$
mat	mat operator
$\mathfrak{m}$	Möbius function
$\boldsymbol{M}(\mathcal{T})$	latent tree model on $\mathcal{T}$
$\boldsymbol{M}(\mathcal{T},k)$	general Markov model for state space of size k
μ	moment

μ'	central moment
$\mathbb{N}$	$1, 2, 3, \ldots$
$\boldsymbol{N}_+(T, 2)$	nonnegative part of $\boldsymbol{N}(T, 2)$
$\mathbb{N}_0$	$0, 1, 2, 3, \ldots$
$\mathbf{N}(T)$	Markov process on T
PD_m	the set of $m \times m$ matrices that are symmetric and positive definite
$\boldsymbol{\Pi}(\mathcal{T})$	lattice of tree partitions
$\mathbb{P}^m$	projective space of dimension m
$\mathbb{Q}$	the rational numbers
$\mathbb{R}$	the real numbers
$\mathbb{R}_+$	$\{x \in \mathbb{R} : x \geq 0\}$
ρ	standardized moment, correlation
$\mathbb{RP}^m$	real projective space of dimension m
σ_i	the variance of X_i
$\mathcal{S}(\mathcal{T})$	set of tree splits of $\mathcal{T}$
$\mathcal{T}$	semi-labeled tree
T^r	rooted tree
t	tree cumulant
$\mathrm{Tuff}(m)$	the Tuffley poset
$\overline{uv}$	a path between two vertices u, v
$\mathrm{var}(X)$	variance of the random variable X
vec	vec operator
$\mathcal{V}_k(I)$	the variety given by the ideal I over the field k
$V(\mathcal{T})$	vertices of a semi-labeled tree $\mathcal{T}$
$\mathcal{X}$	$\{0, \ldots, r_1\} \times \cdots \times \{0, \ldots, r_m\}$

Chapter 1

Introduction

[]

1.1 A statistical model as a geometric object

Let X be a discrete random variable with values in a finite set $\mathcal{X}$. If $\mathcal{X}$ has m elements, then without loss of generality, we assume that $\mathcal{X} = \{1, \ldots, m\}$ and we identify the probability distribution of X with a point $p = (p_1, \ldots, p_m) \in \mathbb{R}^m$ such that $p_x \geq 0$ for every $x \in \mathcal{X}$ and $\sum_{x \in \mathcal{X}} p_x = 1$. The *probability simplex* is the set of all such points

$$\Delta_{\mathcal{X}} \quad := \quad \{p \in \mathbb{R}^m : p_x \geq 0, \sum_{x \in \mathcal{X}} p_x = 1\}. \tag{1.1}$$

Any statistical model for $\mathcal{X}$ is by definition a family of probability distributions and hence a family of points in $\Delta_{\mathcal{X}}$. This gives a basic identification of discrete statistical models with geometric objects. In this book we study only parametric models and hence we are always given a parameter space $\Theta \subseteq \mathbb{R}^d$ and a map $p : \Theta \to \Delta_{\mathcal{X}}$ such that the model is equal to the image of Θ under p. The coordinates of this map are typically denoted by $p_x(\theta)$ or $p(x; \theta)$ for $x \in \mathcal{X}$ and $\theta \in \Theta$.

For most interesting models, possibly after some reparameterization, we can assume that all $p_x(\theta)$ are polynomials in $\theta = (\theta_1, \ldots, \theta_d)$. Important examples are the discrete exponential families and their mixtures.

Example 1.1. Let $X \in \mathcal{X} = \{0, 1, 2\}$, then the probability distribution for X lies in $\Delta_{\mathcal{X}}$, which is a triangle in $\mathbb{R}^3$ with vertices given by $(1, 0, 0)$, $(0, 1, 0)$, and $(0, 0, 1)$. Assume that X follows the binomial $\text{Bin}(2, \theta)$ distribution for some $\theta \in [0, 1]$. Then $p_x(\theta) = \binom{2}{x}(1-\theta)^{2-x}\theta^x$ and hence

$$p(\theta) = (p_0(\theta), p_1(\theta), p_2(\theta)) = ((1-\theta)^2, 2\theta(1-\theta), \theta^2).$$

The model is part of a curve contained in the probability simplex depicted in Figure 1.1, which is parameterized by $\theta \in [0, 1]$. For example, $p(0) = (1, 0, 0)$, $p(0.3) = (0.49, 0.42, 0.09)$, $p(0.6) = (0.16, 0.48, 0.36)$, and $p(1) = (0, 0, 1)$. Alternatively, the model is given as a subset of $\Delta_{\mathcal{X}}$ defined by a single equation $p_1^2 - 4p_0p_2 = 0$, namely,

$$\text{Bin}(2, \theta) \quad = \quad \{p \in \Delta_{\mathcal{X}} : p_1^2 - 4p_0p_2 = 0\}.$$

The curve containing this model is the intersection of two hypersurfaces in $\mathbb{R}^3$ given by $p_0 + p_1 + p_2 - 1 = 0$ and $p_1^2 - 4p_0p_2 = 0$ presented in Figure 1.2.

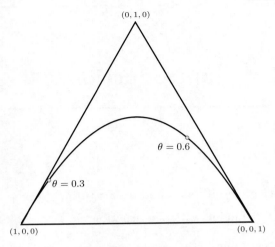

Figure 1.1 *Graphical representation of* Bin$(2, \theta)$ *model in* $\Delta_\mathcal{X}$ *projected on a plane.*

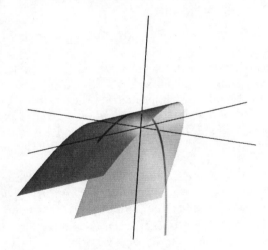

Figure 1.2: *Graphical representation of* Bin$(2, \theta)$ *model in* $\mathbb{R}^3$.

1.2 Algebraic statistics

The geometric representation of statistical models has been known for a long time. The most classical example is the linear model; see Herr [1980] for the historical sketch of how geometric ideas developed in this context. There are, however, many other examples of how geometric representation of models became important in understanding their statistical properties. These include probability (see Diaconis [1977]), graphical models (see Geiger et al. [2001], Mond et al. [2003]), contingency tables (see Fienberg and Gilbert [1970]), phylogenetics (see Chor et al. [2000], Kim [2000]), information theory (see

Amari [1985]), and asymptotic theory (see Kass and Vos [1997], Watanabe [2009]).

In this book, we focus on geometric aspects of statistical models that can be systematically studied from the algebraic geometry point of view. A seminal paper by Diaconis and Sturmfels [1998] began intensified research on applying computational algebraic geometry in various parts of statistics (see also Drton et al. [2009], Pachter and Sturmfels [2005], Pistone et al. [2001]). A historical sketch of these developments was given by Riccomagno [2009]. The geometric analysis of statistical models is focused on understanding the structure of the model in order to understand and improve existing inference procedures. This involves three closely related concepts:

1. understanding the geometry of the parameterization defining the model,

2. providing explicit description of all probability distributions in the model, and

3. understanding the geometry of the likelihood function and various pseudo-likelihoods.

Understanding the geometry of the parameterization defining the model is necessary to address the issue of model identifiability and multimodality of the likelihood function. A statistical model is identifiable if and only if the parameterization is injective, that is $p(\theta) = p(\theta')$ implies $\theta = \theta'$. For many interesting classes of models, identifiability does not hold, and then for a given θ it is important to understand the structure of all $\theta' \in \Theta$ such that $p(\theta) = p(\theta')$. A typical question is whether this set is finite. If it is finite, then one may ask if there is a natural group acting on this set so that modulo this group action, the model is identifiable; see Allman et al. [2009].

Providing explicit description of all probability distributions in the model is important for a number of reasons. The most basic is in the construction of simple diagnostic tests. Suppose that we are given two models M, M' and sample proportions $\hat{p}$ for a random variable $X \in \mathcal{X} = \{1, \dots, m\}$. Since both models and $\hat{p}$ lie in the probability simplex $\Delta_{\mathcal{X}}$, then we can compare $\min_{p \in M} d(p, \hat{p})$ and $\min_{p \in M'} d(p, \hat{p})$ for a suitable distance d in $\Delta_{\mathcal{X}}$ and pick the model with the minimum distance.

The complete description of a model as a subset of the probability simplex gives also a potentially better understanding of the behavior of the likelihood function and other inference procedures. In case of the likelihood function we may, for example, want to understand how many local maxima it has, how many of them are critical points, and how many lie on the boundary of the model. An example of such an analysis will be given later in Section 7.4.

A typical approach to address all three aspects above is to pass to the geometry over complex numbers, possibly in the projective space. In this setting, algebraic problems are typically much easier to study. However, none of the three aspects above can be studied in full detail from this purely algebraic point of view, which is why recently more researchers become interested in semialgebraic statistics.

1.3   Toward semialgebraic statistics

The algebraic structure of a statistical model typically involves providing all polynomial equations defining the model inside the simplex Δ_X. For instance in Example 1.1 there is only one equation given by $p_1^2 - 4p_0p_2 = 0$. However, for many interesting statistical models, the defining equations make only part of the whole story. Geometric analysis typically becomes harder when some of the variables in the system are not observed.

Example 1.2. Let $X \in \{0, 1, 2\}$ be a random variable with a distribution described as follows. Consider two biased coins, a θ_1-coin and a θ_2-coin. We pick one of the coins so that $\mathbb{P}(C = 1) = \pi = 1 - \mathbb{P}(C = 2)$, where $\pi \in [0, 1]$. After we pick a coin, we toss it twice and record the number of heads X. The distribution of X is a mixture of $\mathrm{Bin}(2, \theta_1)$ and $\mathrm{Bin}(2, \theta_2)$. For each $x = 0, 1, 2$ we have

$$\mathbb{P}(X = x) = \mathbb{P}(C = 1)\,\mathbb{P}(X = x|C = 1) + \mathbb{P}(C = 2)\,\mathbb{P}(X = x|C = 2)$$

and hence the possible points in the model are parameterized by $\theta = (\pi, \theta_1, \theta_2)$ as

$$p(\theta) = \pi \cdot \left((1 - \theta_1)^2, 2\theta_1(1 - \theta_1), \theta_1^2\right) + (1 - \pi) \cdot \left((1 - \theta_2)^2, 2\theta_2(1 - \theta_2), \theta_2^2\right),$$

where $\theta_1, \theta_2 \in [0, 1]$. The image of this map is a mixture model denoted by $\mathcal{M}_{\mathrm{mix}}$. Every point in $\mathcal{M}_{\mathrm{mix}}$ can be described (non-uniquely!) as a convex combination of two points from $\mathrm{Bin}(2, \theta)$ as described in Example 1.1. The model is depicted in Figure 1.3. For example $p((0.2, 0.3, 0.6)) = (0.226, 0.468, 0.306)$. The model has dimension two and hence there are no

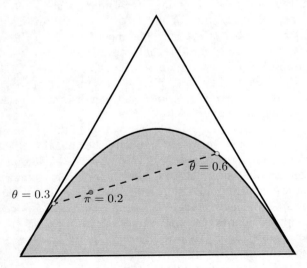

Figure 1.3: *The mixture of two* $\mathrm{Bin}(2, \theta)$ *distributions.*

non-trivial equations defining it as a subset of $\Delta_{\mathcal{X}}$. In fact the model is given by a single non-trivial inequality $p_1^2 - 4p_0p_2 \leq 0$.

What is *not* illustrated by the above example is that the likelihood function for models with hidden variables is usually multimodal and the inference has to be supported with fragile numerical algorithms. This has been pointed out in many places; see for example the discussion of Dempster et al. [1977], Lindsay [1995], Chickering and Heckerman [1997], Bartholomew et al. [2011]. The resulting optima will often lie on the boundary of the parameter space and they will not be critical points of the likelihood function, which gives another example of why inequality constraints are relevant.

Classical algebraic geometry is usually studied over the complex numbers which makes the analysis much easier due to the Fundamental Theorem of Algebra. In this case, we usually embed the original problem into the complex domain by neglecting the inequality constraints on $\Theta \subseteq \mathbb{R}^d$ and $\Delta_{\mathcal{X}} \subseteq \mathbb{R}^m$. Thus we consider the map $p_{\mathbb{C}} : \mathbb{C}^d \to \mathbb{C}^m$ given by exactly the same polynomials. In this case we have some efficient algorithms to obtain equations defining the image of $p_{\mathbb{C}}$. However, as illustrated by Example 1.2, in general, there is no hope to obtain in this way the full description of the model.

Example 1.3. Let $p_{\mathbb{C}} : \mathbb{C} \to \mathbb{C}^3$ given by $\theta \mapsto ((1-\theta)^2, 2\theta(1-\theta), \theta^2)$ as in Example 1.1, then the image of $p_{\mathbb{C}}$ is

$$p_{\mathbb{C}}(\mathbb{C}) = \{p \in \mathbb{C}^3 : p_1^2 - 4p_0p_2 = 0, \ p_0 + p_1 + p_2 = 1\}.$$

In this case, $p_{\mathbb{C}}(\mathbb{C}) \cap \Delta_{\mathcal{X}} = \mathcal{M}_{\text{Bin}(2,\theta)}$ and hence considering the complex parameterization still allows us to recover the original model by constraining to the probability simplex $\Delta_{\mathcal{X}}$. However, for the map $p_{\mathbb{C}} : \mathbb{C}^3 \to \mathbb{C}^3$ given in Example 1.2, we have

$$p_{\mathbb{C}}(\mathbb{C}^3) = \{p \in \mathbb{C}^3 : p_0 + p_1 + p_2 = 1\}.$$

For example, the point $(0, 1, 0)$, which is in $\Delta_{\mathcal{X}}$ but not in $\mathcal{M}_{\text{mix}}$, can be obtained as an image of a point in $\mathbb{R}^3$ given by $\pi = 2$, $\theta_1 = \frac{2+\sqrt{2}}{4}$ and $\theta_2 = \frac{1+\sqrt{2}}{2}$. Therefore, $p_{\mathbb{C}}(\mathbb{C}^3) \cap \Delta_{\mathcal{X}} = \Delta_{\mathcal{X}} \neq \mathcal{M}_{\text{mix}}$ and hence the equations fail to provide the complete description of the model.

Semialgebraic statistics is a new approach within algebraic statistics that aims at studying models in their natural habitat of real numbers. The focus is on understanding the full structure of models under consideration. Such a description can be used to better understand the statistical inference in non-regular settings — for example, where the only tractable way to do inference is by the EM algorithm and the likelihood function is highly complicated with a lot of local maxima.

1.4 Latent tree models

Latent tree models are graphical models defined as Bayesian networks on rooted trees such that each inner vertex of the tree is assumed to represent a hidden random variable; for a formal definition see Section 5.2. They seem to have been first defined by Pearl [1986, 1988], Pearl and Dechter [1989] as tree-decomposable distributions. Latent tree models or their submodels are widely used in phylogenetic analysis (see Semple and Steel [2003], Yang [2006]), in causal modeling (see Pearl and Tarsi [1986]) and in machine learning (see Choi et al. [2011], Zhang [2003/04]). They also contain other well-known classes of models like hidden Markov models, naive Bayes models, and many popular models used in phylogenetics. For example, the models of Cavender [1978, 1979], Jukes and Cantor [1969], Kimura [1980], among others, are all latent tree models with some additional constraints on the parameters. Latent tree models for continuous data were also considered Pearl and Xu [1987], Choi et al. [2011].

The algebraic structure of latent tree models has been studied extensively over the last two decades. This focused on the study of phylogenetic invariants; see for example Cavender and Felsenstein [1987], Allman and Rhodes [2003], Sturmfels and Sullivant [2005], Casanellas and Fernández-Sánchez [2007], Allman and Rhodes [2007]. However, the semialgebraic structure of the model has been largely neglected. The main reason is that this problem is generally considered to be hard, which was pointed out, for example, by Drton and Sullivant [2007], Garcia et al. [2005], and Settimi and Smith [2000]. Only very recently more work has been done to understand the full geometric structure of these models; see Matsen [2009], Zwiernik and Smith [2011], Allman et al. [2014], Klaere and Liebscher [2012], Allman et al. [2015].

To realize how important it may be to include inequalities in the model description, consider a simple model of three binary random variables, which are conditionally independent given a hidden binary variable; for details, see Section 6.1.1. The algebraic structure of this model is trivial because there are no equations on the probability distributions. Consequently, *every* probability distribution on three binary variables satisfies the equality constraints. However, basic Monte Carlo simulation reveals that the model covers only about 8% of the total volume of the probability simplex. This shows that using only the algebraic description may lead to invalid conclusions.

The complicated structure of models with hidden variables usually leads to difficulties in establishing the identifiability of their parameters; see e.g. Allman et al. [2009]. The models are not regular in the classical statistical sense, which in certain cases requires the singular learning theory of Watanabe [2009]. Consequently, the behavior of inferential procedures for models with hidden variables is often unexpected and hard to handle. In particular, the likelihood function typically has many maxima, which lie on the boundary of the parameter space and thus need not correspond to the critical points. This book gives a geometric perspective that allows us to study these questions in a more systematic way.

One of the main concepts developed in this book is the link between latent tree models and various tree spaces like the space of tree metrics or the space of phylogenetic oranges. This link is underlying most of the learning algorithms created for this model class. We believe that developing a good intuition behind how these algorithms are constructed is more important that a detailed analysis of any of them.

1.5   Structure of the book

The book is divided in two parts. The first part contains preliminaries for the second part and some standard results in algebraic statistics. Chapter 2 introduces basic concepts in algebraic geometry, real algebraic geometry, and other closely related notions. We introduce there also the tensor notation, which is going to be used throughout the book. In Chapter 3 we introduce results that allow us to study discrete statistical models from the geometric viewpoint. In Chapter 4 we discuss some less standard results on tensors in statistics. The concepts of L-cumulants is one of the most important developments of this book.

The second part of this book focuses on latent tree models. In Chapter 5 we present some standard results defining trees, their underlying models and various related combinatorial concepts. In Chapter 6 we introduce a change of coordinates in which the latent tree models are easily studied and then use it to analyze in detail their local geometry. Chapter 7 extends these results to the global geometry. In particular, we provide the complete semialgebraic description of latent tree models. In Chapter 8 we extend the results of previous chapters to Gaussian latent tree models.

Part I

Semialgebraic statistics

Chapter 2

Algebraic and analytic geometry

Even very basic algebraic geometry provides a powerful set of tools to study discrete statistical models. In this chapter we provide some elementary introduction to polynomial algebra and algebraic and analytic geometry. This material is by no means complete, but may help to focus on the study of more specialized textbooks like Cox et al. [2007] and Smith et al. [2000]. In addition, we provide a basic account of real algebraic geometry and tensor algebra. See the bibliographical notes at the end of this chapter for further references.

In this book we typically work with the field of real numbers $\mathbb{R}$. However, sometimes we may also consider other *fields of numbers*: the field of *complex numbers* $\mathbb{C}$ and the field of *rational numbers* $\mathbb{Q}$. Whenever we do not want to specify the field, we let k denote any of $\mathbb{R}$, $\mathbb{C}$, or $\mathbb{Q}$. The choice of the field plays an important role in this book. An applied statistician works over the rational numbers and a theoretical statistician works over the real numbers. On the other hand, an algebraic statistician often works with complex numbers, in which case the geometric analysis is typically *much* simpler.

2.1 Basic concepts

2.1.1 Polynomials and varieties

Let $f(\mathbf{x})$ be a function in n *indeterminates* $\mathbf{x} = (x_1, \ldots, x_n)$ such that $\mathbf{x} \in \mathbb{R}^n$, where $\mathbb{R}^n$ denotes the real n-dimensional space. If $n \leq 3$, the indeterminates are often denoted by x, y, z. Also in the statistical context we frequently denote unknowns with other letters like p, μ, κ.

Polynomials   A *monomial* in indeterminates $\mathbf{x} = (x_1, \ldots, x_n)$ is a product of the form

$$\mathbf{x}^\alpha = x_1^{\alpha_1} x_2^{\alpha_2} \cdots x_n^{\alpha_n}, \tag{2.1}$$

where $\alpha = (\alpha_1, \ldots, \alpha_n) \in \mathbb{N}_0^n$. The *degree* of monomial $\mathbf{x}^\alpha$ is the sum $|\alpha| := \alpha_1 + \cdots + \alpha_n$. For example, x^2yz^4 is a monomial in x, y, z of degree 7. A *polynomial* is any sum of the form:

$$f(\mathbf{x}) = \sum_{\alpha_1=0}^{\infty} \cdots \sum_{\alpha_n=0}^{\infty} c_{\alpha_1 \cdots \alpha_n} x_1^{\alpha_1} \cdots x_n^{\alpha_n}, \qquad c_{\alpha_1 \cdots \alpha_n} \in \mathbb{R},$$

such that only a finite number of $c_\alpha = c_{\alpha_1 \cdots \alpha_n}$ is non-zero. This is written in a shorter way as $f(\mathbf{x}) = \sum_\alpha c_\alpha \mathbf{x}^\alpha$. We call c_α the *coefficient* of $\mathbf{x}^\alpha$. If

$c_\alpha \neq 0$, then we call $c_\alpha \mathbf{x}^\alpha$ a *term* of f. The *degree* of f, denoted $\deg(f)$, is the maximum $|\alpha|$ such that the coefficient c_α is non-zero. We denote the set of all polynomials in $x_1, \ldots, x_n$ with coefficients in the field k by $k[\mathbf{x}]$ or $k[x_1, \ldots, x_n]$. This set forms a *commutative ring* with standard addition and multiplication of polynomials, with 0 denoting the zero polynomial.

Algebraic sets Every polynomial $f = \sum_\alpha c_\alpha \mathbf{x}^\alpha$ defines a function $f : \mathbb{R}^n \to \mathbb{R}$, which is called a *polynomial function*. We distinguish between polynomials and the maps they define. Thus $f = 0$ will mean that f is a *zero polynomial*, and $f(x_1, \ldots, x_n) = 0$ denotes that a polynomial function takes value zero. Whenever we say that a polynomial vanishes at some point, we implicitly mean that the corresponding polynomial function evaluated to zero at this point. The distinction between a polynomial and the corresponding polynomial function allows us to link algebra and geometry by relating a polynomial f with the set of zeros of the function it defines.

Definition 2.1. Let $f_1, \ldots, f_s$ be polynomials in $k[x_1, \ldots, x_n]$. The *affine algebraic variety* defined by $f_1, \ldots, f_s$ is

$$\mathcal{V}_k(f_1, \ldots, f_s) := \{(a_1, \ldots, a_n) \in k^n : f_i(a_1, \ldots, a_n) = 0 \text{ for all } 1 \leq i \leq s\}.$$

Thus an affine algebraic variety $\mathcal{V}_k(f_1, \ldots, f_s)$ is the set of all solutions in k^n to the system of equations

$$f_1(x_1, \ldots, x_n) \quad = \quad \cdots \quad = \quad f_s(x_1, \ldots, x_n) \quad = \quad 0. \qquad (2.2)$$

For example, the variety $\mathcal{V}_\mathbb{R}(x^2 + y^2 - 1)$ is the circle of radius 1 in $\mathbb{R}^2$ and $\mathcal{V}_\mathbb{R}(x^2 + y^2 + 1)$ gives the empty set. In Example 1.1 the affine algebraic variety $\mathcal{V}_\mathbb{R}(p_0 + p_1 + p_2 - 1, p_1^2 - 4p_0 p_1)$ is the smallest (with respect to inclusion) affine algebraic variety in $\mathbb{R}^3$ containing the binomial model.

Remark 2.2. Over the real numbers, every affine algebraic variety can be represented by vanishing of a single polynomial. In particular, the set of equations (2.2) is equivalent to

$$f_1(x_1, \ldots, x_n)^2 + \cdots + f_s(x_1, \ldots, x_n)^2 \quad = \quad 0.$$

For example, $x_1^2 + x_2^2 = 0$ defined a point in $\mathbb{R}^2$ and a curve in $\mathbb{C}^2$. This curve contains the real point $(0, 0)$ but also the point $(1, i)$, whose second coordinate is imaginary.

An affine algebraic variety is *irreducible* if it cannot be written as a union of two proper algebraic varieties. For example, the set $\mathcal{V}_\mathbb{R}(x_1 x_2)$ is reducible because in $\mathbb{R}^2$ it forms a union of two coordinate lines $\mathcal{V}_\mathbb{R}(x_1)$ and $\mathcal{V}_\mathbb{R}(x_2)$. Note that irreducibility depends on the field; for example, the polynomial

$$f(x) = x^3 - x^2 + x - 1 = (x - 1)(x^2 + 1)$$

defines an irreducible variety over $\mathbb{R}$ and a reducible variety over the complex numbers.

Zariski closure and Zariski topology   Another important concept is that of Zariski closure. The simplest way to define the Zariski closure $\overline{S}$ of a set $S \subseteq k^n$ is as the smallest (with respect to inclusion) algebraic variety in k^n containing S. We now present a more elegant way to obtain this definition by introducing the concept of the *Zariski topology*.

Definition 2.3. A *topological space* is a set M together with a collection of subsets of M called the *closed sets* satisfying the following:

(i) Both M and the empty set $\emptyset$ are closed.

(ii) Any intersection of a collection of closed sets is closed.

(iii) Any union of finitely many closed sets is closed.

An *open set* is a complement of a closed set. A *topology* is any collection of open sets such that their complements are closed sets satisfying (i)–(iii).

We now show that k^n forms a topological space with closed sets given by affine algebraic varieties. Indeed, the empty set is the set of zeros of a constant non-zero polynomial function and therefore it forms an affine algebraic variety. Similarly, the whole space k^n is given as the set of zeros of the zero polynomial function. The *intersection* of any number of affine algebraic varieties in k^n is an algebraic variety, because it is defined by the union of the sets of polynomials defining the given varieties. The *union* of two affine algebraic varieties in k^n is an affine algebraic variety. This union is defined by the set of all pairwise products of the polynomials defining the original varieties:

$$\mathcal{V}_k(\{f_i\}_{i \in I}) \ \cup \ \mathcal{V}_k(\{g_i\}_{i \in J}) \quad = \quad \mathcal{V}_k(\{f_i g_j\}_{(i,j) \in I \times J}).$$

This implies by induction that the union of a finite number of affine algebraic varieties is an affine algebraic variety.

We have verified that the empty set, the whole k^n, the intersection of arbitrarily many affine algebraic varieties, and the union of finitely many affine algebraic varieties are all affine algebraic varieties. It follows that the set of all affine algebraic varieties satisfies the three axioms in Definition 2.3. The corresponding topology is called the *Zariski topology* on k^n.

Similarly, as in the classical (Euclidean) topology, we define the closure $\overline{S}$ of the set $S \subseteq k^n$ as an intersection of all closed sets in k^n containing S. In the Zariski topology this procedure gives that the Zariski closure of S is the smallest algebraic variety containing S, so it is consistent with the definition given earlier. For instance, the model in Example 1.2 is not an affine algebraic variety. Its closure $\overline{\mathcal{M}}$ in $\mathbb{R}^3$ is the whole affine subspace given by $p_0 + p_1 + p_2 = 1$.

2.1.2   Ideals and morphisms

Ideals   Consider the set of all polynomials vanishing at a point $a \in k^n$. For any two such polynomials, their sum is a polynomial which also vanishes at a. Moreover, every such polynomial multiplied by an arbitrary polynomial also vanishes at a. In addition, the zero polynomial $0 \in k[x_1, \ldots, x_n]$, which vanishes everywhere, vanishes at a as well. This motivates the following definition.

Definition 2.4. A subset $I \subseteq k[x_1, \ldots, x_n]$ is an *ideal* if it satisfies:

(i) $0 \in I$.

(ii) If $f, g \in I$, then $f + g \in I$.

(iii) If $f \in I$ and $h \in k[x_1, \ldots, x_n]$, then $hf \in I$.

The first very important example of an ideal is the ideal generated by a finite number of polynomials.

Definition 2.5. Let $f_1, \ldots, f_s$ be polynomials in $k[x_1, \ldots, x_n]$. Then we set

$$\langle f_1, \ldots, f_s \rangle := \left\{ \sum_{i=1}^{s} h_i f_i : h_1, \ldots, h_s \in k[x_1, \ldots, x_n] \right\}.$$

In other words, $\langle f_1, \ldots, f_s \rangle$ is the set of all polynomial combinations of $f_1, \ldots, f_s$. It is easily checked that $\langle f_1, \ldots, f_s \rangle$ is an ideal. We call it the *ideal generated by* $f_1, \ldots, f_s$.

Example 2.6. Let $I = \langle x_1^2 - x_2, x_2 x_3 \rangle \subseteq k[x_1, x_2, x_3]$. The monomial $x_1^2 x_3$ lies in I. To see this, we write $x_1^2 x_3$ as a polynomial combination of the generators of I:

$$x_1^2 x_3 = x_3(x_1^2 - x_2) + x_2 x_3.$$

We say that I is *finitely generated* if there exist $f_1, \ldots, f_s \in k[x_1, \ldots, x_n]$ such that $I = \langle f_1, \ldots, f_s \rangle$. When this is so, we say that $f_1, \ldots, f_s$ form a *basis* of I. By the following important theorem, finitely generated polynomials as in Definition 2.5 are the only ones we need to consider.

Theorem 2.7 (Hilbert Basis Theorem). *Every ideal in $k[x_1, \ldots, x_n]$ is finitely generated.*

There is a close relation between affine algebraic varieties and ideals. More specifically, a variety depends only on the ideal generated by its defining equations.

Lemma 2.8. *If $f_1, \ldots, f_s$ and $g_1, \ldots, g_t$ are two bases of the same ideal in $k[x_1, \ldots, x_n]$, so that $\langle f_1, \ldots, f_s \rangle = \langle g_1, \ldots, g_t \rangle$, then $\mathcal{V}_k(f_1, \ldots, f_s) = \mathcal{V}_k(g_1, \ldots, g_t)$.*

This Lemma allows us to define $\mathcal{V}_k(I)$ for any ideal I in $k[x_1, \ldots, x_n]$.

Definition 2.9. Let $S \subset k^n$ be any subset. Define *the ideal of S* by

$$\mathcal{I}(S) = \{f \in k[x_1, \ldots, x_n] : f(a_1, \ldots, a_n) = 0 \text{ for all } (a_1, \ldots, a_n) \in S\}.$$

We leave it as an exercise to check that $\mathcal{I}(S)$ is an ideal.

Proposition 2.10. *If $S \subset k^n$, the affine algebraic variety $\mathcal{V}_k(\mathcal{I}(S))$ is the Zariski closure $\overline{S}$.*

Morphisms The definition of a polynomial function can be generalized to maps between arbitrary affine or projective algebraic varieties.

Definition 2.11. We say that a function $\phi : V \to W$ is a polynomial mapping from $V \subset k^m$ to $W \subset k^n$ if there exist polynomials $f_1, \ldots, f_n$ in $k[x_1, \ldots, x_m]$ such that

$$(f_1(a_1, \ldots, a_m), \ldots, f_n(a_1, \ldots, a_m)) \in W$$

for all $(a_1, \ldots, a_m) \in V$. Moreover, a polynomial mapping $\phi : V \to W$ is a *polynomial isomorphism* if there exists a polynomial mapping $\psi : W \to V$ such that $\psi \circ \phi = \mathrm{id}_V$ and $\phi \circ \psi = \mathrm{id}_W$, where id_V denotes the *identity map on V* defined by

$$\mathrm{id}_V(a_1, \ldots, a_m) = (a_1, \ldots, a_m) \qquad \text{for all } (a_1, \ldots, a_m) \in V.$$

The identity map id_W is defined in the same way.

If $\psi : V \to W$ is a polynomial mapping, then for any subset $U \subseteq V$ the set

$$\psi(U) := \{\psi(a) : a \in U\}$$

is a subset of W called the *image* of U under ψ. We say that ψ is *surjective* (or *onto*) if $\psi(V) = W$, and we say that ψ is *injective* if for any $a, a' \in V$ we have $\psi(a) = \psi(a')$ only if $a = a'$. For example, the identity map $\mathrm{id}_V : V \to V$ is both surjective and injective. Note that every polynomial isomorphism is necessarily both surjective and injective.

An important case of a polynomial mapping is when $W = k$, in which case ϕ simply becomes a *polynomial function on V*. We usually specify a polynomial function by giving an explicit polynomial representative which is rarely uniquely defined. For example, consider the variety $V = \mathcal{V}_{\mathbb{R}}(p_0 + p_1 + p_2 - 1, p_1^2 - 4p_0p_2) \subset \mathbb{R}^3$ in Example 1.1. The polynomial $f = p_0^3 + p_1^3$ represents a polynomial function on V. However, for instance,

$$g = p_0^3 + p_1^3 + (p_0 + p_1 + p_2 - 1)^2 + (p_1^2 - 4p_0p_2)$$

defines the same polynomial function on V. Indeed, we have $p_0 + p_1 + p_2 - 1 = 0$ and $p_1^2 - 4_0p_2 = 0$ on V and hence $f - g$ is zero on V.

In general let $V \subset k^m$ be an affine algebraic variety. Then f and g in $k[x_1, \ldots, x_m]$ define the same polynomial function on V if and only if $f - g \in \mathcal{I}(V)$. We denote by $k[V]$ the collection of polynomial functions $\phi : V \to k$. It is easy to check that $k[V]$ forms a commutative ring with the usual pointwise operations on functions. We sometimes call $k[V]$ the *coordinate ring* of V.

To study coordinate rings we need to understand the quotient of $k[x_1, \ldots, x_m]$ by an ideal I, which is a partition of $k[x_1, \ldots, x_m]$ into equivalence classes of polynomials which define the same polynomial functions on $\mathcal{V}(I)$. Let $I \subset k[x_1, \ldots, x_m]$ be an ideal, and let $f, g \in k[x_1, \ldots, x_m]$. We say that f and g are *congruent modulo I*, written $f \equiv g \bmod I$, if $f - g \in I$. The congruence modulo I is an equivalence relation on $k[x_1, \ldots, x_m]$. Hence it partitions $k[x_1, \ldots, x_m]$ into equivalence classes. For any $f \in k[x_1, \ldots, x_m]$, the *class* of f is the set

$$[f] = \{g \in k[x_1, \ldots, x_m] : g \equiv f \bmod I\}.$$

Definition 2.12. The *quotient* of $k[x_1, \ldots, x_m]$ modulo I, written $k[x_1, \ldots, x_m]/I$, is the set of equivalence classes for congruence modulo I:

$$k[x_1, \ldots, x_m]/I = \{[f] : f \in k[x_1, \ldots, x_m]\}.$$

It is a standard result in algebra (see for example [Cox et al., 2007, Theorem 6, Section 5.2]) that $k[x_1, \ldots, x_m]/I$ is a commutative ring, where the algebraic operations are defined in a natural way

$$\begin{aligned} [f] + [g] &= [f + g] \\ [f] \cdot [g] &= [f \cdot g]. \end{aligned} \tag{2.3}$$

Moreover there is a one-to-one correspondence between $k[x_1, \ldots, x_m]/\mathcal{I}(V)$ and $k[V]$.

2.1.3 Projective space and projective varieties

The *projective line*, denoted by $\mathbb{P}^1$, is the set of all lines in k^2 through the origin. Formally, we form an equivalence class in $k^2 \setminus (0,0)$ such that $(x, y) \sim (x', y')$ if there exists $\lambda \in k$ such that $(x, y) = \lambda(x', y')$ and hence when (x, y) and (x', y') lie on the same line through the origin in $\mathbb{R}^2$. We denote this equivalence class by $(x : y)$. In the same way, we define the *projective space* $\mathbb{P}^m$ for $m \geq 1$ as a set of lines through the origin in k^{m+1}. The equivalence class in $k^{m+1} \setminus \{0\}$ is constructed in the same way as for $\mathbb{P}^1$ and we write $(x_0 : \ldots : x_m)$ to denote a point in $\mathbb{P}^m$.

An important fact is that $\mathbb{P}^m$ can be *covered* by $m + 1$ copies of k^m. Indeed, let U_i for $i = 0, \ldots, m$ be a subset of $\mathbb{P}^m$ defined by $x_i \neq 0$. Note that in U_i

$$(x_0 : \cdots : x_i : \cdots : x_m) \sim \left(\frac{x_0}{x_i} : \cdots : 1 : \cdots : \frac{x_m}{x_i} \right).$$

Thus, if we denote $y_j = \frac{x_j}{x_i}$ we can identify U_i with k^n. Moreover, every point $x \in \mathbb{P}^m$ lies in at least one of U_i so that these sets really cover $\mathbb{P}^m$.

Definition 2.13. We say that a polynomial $f \in k[x_0, \ldots, x_m]$ is *homogeneous of degree d* if all its terms have degree d or equivalently $f(\lambda \boldsymbol{x}) = \lambda^d f(\boldsymbol{x})$ for any $\lambda \in k$.

Note that the notion of a polynomial mapping $\mathbb{P}^m \to k$ is not well defined because for an arbitrary function typically $f(\boldsymbol{x}) \neq f(\lambda \boldsymbol{x})$ even though $\boldsymbol{x}$ and $\lambda \boldsymbol{x}$ define the same point in $\mathbb{P}^m$. However, if $f \in k[x_0, \ldots, x_m]$ is homogeneous, then at least the condition $f(x_0, \ldots, x_m) = 0$ is well defined. Moreover, if f, g are homogeneous of the same degree, then f/g forms a well-defined function on $\mathbb{P}^m$.

Definition 2.14 (Projective variety). An ideal is said to be *homogeneous* if it is generated by homogeneous polynomials. Let $I \subseteq k[x_0, \ldots, x_n]$ be a homogeneous ideal. We define a *projective variety* as the set of zeros of the elements of I.

If $f \in k[x_1, \ldots, x_m]$, then $f(\boldsymbol{x}) = 0$ describes an affine algebraic variety V in $U_0 = k^m$. By multiplying each term of f by an appropriate power of x_0, we obtain a homogeneous polynomial, which defines a projective variety. This process is called *homogenization*. For example, take

$$f(x_1, x_2) = x_1^3 x_2 - x_2^2 + 1;$$

this can be extended to a homogeneous polynomial given by

$$f(x_0, x_1, x_2) = x_1^3 x_2 - x_0^2 x_2^2 + x_0^4.$$

The equation $f(x_0, x_1, x_2) = 0$ defines a projective variety in $\mathbb{P}^2$, which on the open subset U_0 is equal to the original affine algebraic variety.

2.1.4 Parametric representation and toric varieties

Both in statistics and in algebraic geometry it is often convenient to define a variety as a family of points parameterized by a simpler set. A typical example is when a single parameter $\theta \in k$ describes a curve in the affine space k^n as in Example 1.1.

Definition 2.15. Let $V = \mathcal{V}_k(f_1, \ldots, f_s) \subset k^n$. Then a *polynomial parametric representation* $\psi : k^d \to k^n$ of V consists of polynomial functions $\psi_1, \ldots, \psi_n \in k[t_1, \ldots, t_d]$ such that the points given by

$$x_i = \psi_i(t_1, \ldots, t_d) \quad \text{for all } i = 1, \ldots, n$$

lie in V. We also require that $V = \overline{\psi(k^d)}$. The original defining equations $f_1 = \cdots = f_s = 0$ of V are called the *implicit representation* of V.

For instance, in Example 1.1 we considered $V = \mathcal{V}_\mathbb{R}(p_1^2 - 4p_0 p_2, p_0 + p_1 + p_2 - 1)$. The polynomial parametric representation of this affine algebraic variety is given precisely by $\theta \mapsto ((1-\theta)^2, 2\theta(1-\theta), \theta^2)$, where $\theta \in \mathbb{R}$. Note, however, that not every affine algebraic variety has a polynomial parametric representation.

Example 2.16. Let $n = 4$ and index the indeterminates with the set $\{0,1\}^2$. Consider the affine algebraic variety defined by $x_{11} x_{00} - x_{01} x_{10} = 0$ and $x_{00} + x_{01} + x_{10} + x_{11} - 1 = 0$. It is an exercise to verify that this variety is parameterized by

$$x_{00} = (1-t_1)(1-t_2) \quad x_{01} = (1-t_1)t_2$$
$$x_{10} = t_1(1-t_2) \quad x_{11} = t_1 t_2.$$

Statisticians refer to this example as the independence model of two binary random variables. In this probabilistic setting we additionally assume that $x_{ij} \geq 0$ for all $i, j \in \{0, 1\}$, or equivalently, $t_1, t_2 \in [0, 1]$ and x_{ij} is interpreted as the probability that the first random variable takes value i and the second random variable takes value j.

Definition 2.17. With any parameterization, we associate the *Jacobian matrix*, which is the following matrix of partial derivatives

$$\frac{\partial \psi}{\partial \mathbf{t}}(\mathbf{t}) = \begin{bmatrix} \frac{\partial \psi_1}{\partial t_1}(\mathbf{t}) & \cdots & \frac{\partial \psi_1}{\partial t_d}(\mathbf{t}) \\ \vdots & & \vdots \\ \frac{\partial \psi_n}{\partial t_1}(\mathbf{t}) & \cdots & \frac{\partial \psi_n}{\partial t_d}(\mathbf{t}) \end{bmatrix},$$

where $\mathbf{t} = (t_1, \ldots, t_d)$. If $d = n$, we define the *Jacobian* of g as the determinant of the Jacobian matrix $\frac{\partial \psi}{\partial \mathbf{t}}(\mathbf{t})$.

An important notion related to affine algebraic varieties is the *dimension*, denoted by $\dim(V)$. In full generality, the definition of the dimension of an affine algebraic variety is subtle and can be defined in many ways usually related to some invariants of the coordinate ring. In this book we focus on real algebraic varieties that are given in a parametric form. In this case, the dimension can be defined by studying the rank of the Jacobian matrix of the corresponding parameterization. It turns out that this rank is constant outside of a measure zero set and the dimension of V is equal to that constant. The *codimension* of an affine algebraic variety $V \subseteq k^n$ is equal to $n - \dim(V)$.

The main distinction between the complex and the real numbers is that over the complex numbers we have the *Fundamental Theorem of Algebra* which states that every non-constant polynomial $f \in \mathbb{C}[x]$ has a root in $\mathbb{C}$. Of course, over the real numbers, it is usually not true as, for example, $x^2 + 1 = 0$ has no real solutions. An important consequence is that over the complex numbers, parameterization always very closely approximates the Zariski closure of its image.

Theorem 2.18. *Let* $\psi : \mathbb{C}^d \to \mathbb{C}^n$ *be a polynomial parametric representation of* $V = \overline{\psi(\mathbb{C}^d)}$. *Then*

$$\dim(V \setminus \psi(\mathbb{C}^d)) \quad < \quad \dim(\psi(\mathbb{C}^d)).$$

This theorem does not hold over the real numbers. For example, if $\psi : \mathbb{R} \to \mathbb{R}$ is given by $\psi(x) = x^2$, then $\psi(\mathbb{R}) = [0, \infty)$ but $\mathcal{V}_{\mathbb{R}}(\mathcal{I}(\psi(\mathbb{R}^d))) = \mathbb{R}$ and hence the difference has full dimension. The real case is discussed in more detail in Section 2.2.

Ideals and parameterizations   Given a parametric representation $\psi : k^d \to k^n$ of a variety V given by

$$\psi = (\psi_1(t_1, \ldots, t_d), \ldots, \psi_n(t_1, \ldots, t_d)),$$

we may be interested in its implicit representation in the form of the ideal $\mathcal{I}(V)$. In order to obtain this, we define the ring homomorphism

$\psi^* : k[x_1, \ldots, x_n] \to k[t_1, \ldots, t_d]$ given by $\psi^*(x_i) = \psi_i(t_1, \ldots, t_d)$. Hence ψ^* is a map such that

$$\psi^*(1) = 1, \quad \psi^*(fg) = \psi^*(f)\psi^*(g) \quad \text{and} \quad \psi^*(f + g) = \psi^*(f) + \psi^*(g)$$

for every $f, g \in k[x_1, \ldots, x_n]$. Note that in particular $\psi^*(cx^\alpha) = c\prod_{i=1}^{n} \psi_i(t_1, \ldots, t_d)^{\alpha_i}$ for every $c \in k$ and $\alpha \in \mathbb{N}_0^n$. The *kernel* of ψ^* is defined as

$$\ker(\psi^*) = \{f \in k[x_1, \ldots, x_n] : \psi^*(f) = 0\}.$$

A polynomial $f \in k[x_1, \ldots, x_n]$ lies in the kernel of ψ^* if and only if it evaluates to zero for every fixed value of $t_1, \ldots, t_d$. This observation gives the following result.

Proposition 2.19. *The kernel* $\ker(\psi^*)$ *forms an ideal in* $k[x_1, \ldots, x_n]$. *Moreover,* $\mathcal{V}_k(\ker(\psi^*))$ *defines the Zariski closure of the image of* $\psi : k^d \to k^n$.

The computations of the kernel of a polynomial ring map are done typically with computer algebra software like CoCoA, Macaulay2, Singular; see CoCoATeam, Grayson and Stillman, Decker et al. [2011]. In Example 2.16, the corresponding kernel is generated by $x_{11}x_{00} - x_{01}x_{10} = 0$ and $x_{00} + x_{01} + x_{10} + x_{11} - 1 = 0$.

Toric varieties   One of the simplest instances of a parametric family is when its parameterization is given by monomials. *Toric varieties* form an important part of algebraic geometry and statistics. We define toric varieties as subsets of the complex projective space $\mathbb{P}^n$, that is, always with a concrete embedding into the projective space.

Definition 2.20. Let $\mathcal{A} = \{m_0, \ldots, m_n\} \subset \mathbb{Z}^d$. A *projective toric variety* $Y_\mathcal{A}$ is given as the Zariski closure of the image of a monomial parameterization

$$\psi_\mathcal{A} : (\mathbb{C}^*)^d \to \mathbb{P}^n \tag{2.4}$$

$$\mathbf{t} = (t_1, \ldots, t_d) \mapsto [\mathbf{t}^{m_0}, \ldots, \mathbf{t}^{m_n}], \tag{2.5}$$

where $m_0, \ldots, m_n \in \mathbb{N}^d$ are some non-negative exponents and $\mathbb{C}^* = \mathbb{C} \setminus 0$.

To describe the set parameterized by $\psi_\mathcal{A}$, we need to compute the Zariski closure of its image. Equivalently, we need to describe the ideal of polynomials in $\mathbb{C}[x_0, \ldots, x_n]$ vanishing on the image of $\psi_\mathcal{A}$.

Definition 2.21. The *toric ideal* $I_\mathcal{A}$ is the homogeneous ideal of polynomials whose vanishing defines the projective toric variety $Y_\mathcal{A}$. Equivalently, $I_\mathcal{A}$ is the ideal of all the homogeneous polynomials vanishing on $\psi_\mathcal{A}((\mathbb{C}^*)^d)$.

Denote by A the $d \times (n+1)$-matrix obtained by putting elements of $\mathcal{A}$ as columns. Here we assume that the vector $(1, \ldots, 1)$ lies in the rowspan of A. For a vector $u \in \mathbb{Z}^{n+1}$, we write $u = u^+ - u^-$, where $u^+, u^- \in \mathbb{N}_0^{n+1}$ and they have disjoint supports. We have the following result.

Theorem 2.22. *The toric ideal* $I_\mathcal{A}$ *is the linear span of all homogeneous binomials* $x^u - x^v$ *with* $Au = Av$. *Even more,*

$$I_\mathcal{A} = \langle x^{u^+} - x^{u^-} : u \in \ker(A) \rangle.$$

The fact that this ideal is homogeneous follows from the fact that the vector of ones lies in the rowspan of A.

2.2   Real algebraic and analytic geometry

The geometry of algebraic varieties and their relation to ideals is substantially different in the complex than in the real case. For example, the equation $x^2 + y^2 + 1 = 0$ defines a curve in the complex plane and the empty set in $\mathbb{R}^2$. Moreover, a projection of a real affine algebraic variety on one of the axes need not be an affine algebraic variety anymore.

Example 2.23. Consider the parabola $\mathcal{V}_{\mathbb{R}}(y - x^2) \subseteq \mathbb{R}^2$. Its projection on the y axis is the half-line $\mathbb{R}_{\geq 0}$. This is not an affine algebraic variety. The smallest affine algebraic variety containing a half-line is the whole real line $\mathbb{R}$.

Another difference between the complex and the real algebraic varieties is that every real affine algebraic variety is given by a single polynomial equation since $\mathcal{V}_{\mathbb{R}}(f_1, \ldots, f_s) = \mathcal{V}_{\mathbb{R}}(f_1^2 + \cdots + f_s^2)$. There are also rational functions which are defined everywhere but they are not polynomial. An example is $(1 + x^2)^{-1}$.

2.2.1   Real analytic manifolds

A *power series* is any sum of the form:

$$f(\mathbf{x}) = \sum_{\alpha_1=0}^{\infty} \cdots \sum_{\alpha_n=0}^{\infty} c_{\alpha_1 \cdots \alpha_n}(x_1 - b_1)^{\alpha_1} \cdots (x_n - b_n)^{\alpha_n},$$

where $c_{\alpha_1 \cdots \alpha_n} \in \mathbb{R}$, $b_i \in \mathbb{R}$; this is written in a shorter way as $f(\mathbf{x}) = \sum_{\alpha} c_{\alpha}(\mathbf{x} - b)^{\alpha}$. If there exists an open set $U \subset \mathbb{R}^n$, which contains b such that for every $\mathbf{x} \in U$

$$\sum_{\alpha} |c_{\alpha}||\mathbf{x} - b|^{\alpha} \quad < \quad \infty$$

then $f(\mathbf{x})$ is called an *absolutely convergent power series*. If $f(\mathbf{x})$ is an absolutely convergent power series then it uniquely defines a function $f : U \to \mathbb{R}$. This function is called a *real analytic function*.

If $f(\mathbf{x})$ is a real analytic function, then the coefficient of the Taylor series expansion around b satisfies

$$c_{\alpha} = \frac{1}{\alpha!} \frac{\partial^{\alpha} f}{\partial \mathbf{x}^{\alpha}}(b),$$

where $\alpha! = \alpha_1! \cdots \alpha_n!$ and

$$\frac{\partial^{\alpha}}{\partial \mathbf{x}^{\alpha}} \quad := \quad \frac{\partial^{\alpha_1}}{\partial x_1^{\alpha_1}} \cdots \frac{\partial^{\alpha_n}}{\partial x_n^{\alpha_n}}.$$

The concept of a polynomial ideal can be extended to ideals in the ring of analytic functions. Although, we will not use this fact explicitly in this book, in statistics the concept of an analytic ideal helps in singular learning theory to understand the asymptotic behavior of some empirical processes ; see for example Watanabe [2009], Lin [2011].

Definition 2.24 (Function of class C^r). Let $U \subset \mathbb{R}^n$ be an open set. A map $f : U \to \mathbb{R}^m$ is said to be *of class C^r in U*, which we denote by $f \in C^r(U)$, if partial derivatives

$$\frac{\partial^\alpha f}{\partial \mathbf{x}^\alpha}(\mathbf{x})$$

are well defined and continuous for all $\alpha \in \mathbb{N}_0^n$ such that $|\alpha| \leq r$. If $f \in C^r(U)$ for all $r \geq 0$, we say it is $C^\infty(U)$. If a function f is real analytic in U, we say it is $C^\omega(U)$.

The following definition generalizes the polynomial isomorphism.

Definition 2.25 (C^r isomorphism). Let U, V be two open subsets of $\mathbb{R}^d$. If there exists a one-to-one map $f : U \to V$ such that $f \in C^r(U)$ and $f^{-1} \in C^r(V)$, then U is said to be C^r *isomorphic* to V, and f is called a C^r *isomorphism*. If both f and f^{-1} are analytic functions, then U is said to be *analytically isomorphic* to V and f is called an analytic isomorphism.

A topological space M with the topology of open sets $\mathcal{U}$ is called a *Hausdorff space* if, for arbitrary $x, y \in M$ ($x \neq y$), there exist open sets $U, V \in \mathcal{U}$ such that $x \in U$, $y \in V$ and $U \cap V = \emptyset$.

Definition 2.26 (System of local coordinates). Let M be a Hausdorff space with topology $\mathcal{U} = (U_\alpha)$ such that the open sets in $\mathcal{U}$ cover M (every point $x \in M$ lies in some element of $\mathcal{U}$). Suppose that for each U_α there exists a map $\phi_\alpha : U_\alpha \to \mathbb{R}^d$. The pair $\{U_\alpha, \phi_\alpha\}$ is called a system of local coordinates if for every α and any open set U in U_α, ϕ_α is a continuous and one-to-one map from U to $\phi_\alpha(U)$, whose inverse is also continuous.

Definition 2.27 (Manifold). A Hausdorff space that has a system of local coordinates is called a *manifold*. For each $r = 0, 1, \ldots, \infty, \omega$, a manifold M is said to be of class C^r if it has a system of local coordinates such that, if $U^* = U_1 \cap U_2 \neq \emptyset$, both of the maps

$$\phi_1(U^*) \ni x \mapsto \phi_2(\phi_1^{-1}(x)) \in \phi_2(U^*),$$
$$\phi_2(U^*) \ni x \mapsto \phi_1(\phi_2^{-1}(x)) \in \phi_1(U^*)$$

are of class C^r. If M is a manifold of class C^ω, it is called a real analytic manifold.

The concepts of the affine algebraic variety and the real analytic manifold are related. Roughly speaking, M is a real analytic manifold if one can cover M with small sets (*charts*) such that M constrained to each of the charts looks like a set of zeros of a real analytic map.

2.2.2 *Semialgebraic sets*

Real affine algebraic varieties and real analytic manifolds cannot be studied without also considering inequality constraints. In this section we provide a very basic introduction to semialgebraic geometry.

Definition 2.28. A *basic semialgebraic set* is a subset of $\mathbb{R}^n$ given by polynomial equations and inequalities. More formally, it is a set of the form

$$\bigcap_{i=1}^{r} \{\boldsymbol{x} \in \mathbb{R}^n : f_i *_i 0\},$$

where $f_i \in \mathbb{R}[x_1, \ldots, x_n]$ and $*_i$ is either $<$ or $=$, for $i = 1, \ldots, r$. A *semialgebraic* subset of $\mathbb{R}^n$ is a finite union of basic semialgebraic sets.

Remark 2.29. Semialgebraic subsets of $\mathbb{R}^n$ form the smallest family of subsets containing all sets of the form

$$\{\boldsymbol{x} \in \mathbb{R}^n : f(\boldsymbol{x}) > 0\}, \quad \text{where } f \in \mathbb{R}[x_1, \ldots, x_n],$$

and closed under taking finite intersections, finite unions, and complements.

The following result shows that the family of semialgebraic sets is stable under certain transformations (c.f. Example 2.23).

Theorem 2.30. *If S is a semialgebraic subset of $\mathbb{R}^n$, and $\pi : \mathbb{R}^n \to \mathbb{R}^m$ is the projection on the first m coordinates, then $\pi(S)$ is a semialgebraic subset of $\mathbb{R}^n$. Moreover, the closure and the interior of S are semialgebraic subsets of $\mathbb{R}^n$.*

Note that the closure of a basic semialgebraic set is not always obtained just by relaxing the strict inequalities describing it. Consider the following example.

Example 2.31. The closure of the set

$$A = \{(x, y) \in \mathbb{R}^2 : x^3 - x^2 - y^2 > 0\}$$

is not the set

$$B = \{(x, y) \in \mathbb{R}^2 : x^3 - x^2 - y^2 \geq 0\}.$$

The closure of A is obtained by removing the point $(0,0)$ from B, and can be described as

$$\mathrm{cl}(A) = \{(x, y) \in \mathbb{R}^2 : x^3 - x^2 - y^2 \geq 0, \ x \geq 1\}.$$

This situation is depicted in Figure 2.1.

Let $y = f(\mathbf{x})$ be a function in n indeterminates $\mathbf{x} = (x_1, \ldots, x_n) \in \mathbb{R}^n$. The *graph* of this function is the set in $\mathbb{R}^{n+1}$ defined by

$$\{(\mathbf{x}, f(\mathbf{x})) : \mathbf{x} \in \mathbb{R}^n\} = \{(\mathbf{x}, y) : y - f(x) = 0\}.$$

In the same way we define a graph of any polynomial map $f : \mathbb{R}^n \to \mathbb{R}^m$.

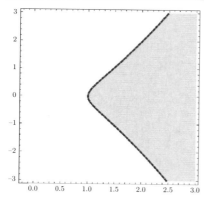 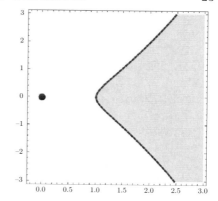

Figure 2.1 *The closure of a semialgebraic set given by $x^3 - x^2 - y^2 > 0$ (left) is not equal to the set given by $x^3 - x^2 - y^2 \geq 0$ (right).*

Definition 2.32. Let $A \subset \mathbb{R}^m$ and $B \subset \mathbb{R}^n$ be two semialgebraic sets. A mapping $f : A \to B$ is semialgebraic if its graph is a semialgebraic set in $\mathbb{R}^{m+n}$.

Semialgebraic sets are stabilized by semialgebraic maps in the following sense.

Proposition 2.33. *Let $f : A \to B$ be a semialgebraic mapping. If $S \subset A$ is semialgebraic, then its image $f(S)$ is semialgebraic. If $T \subset B$ is semialgebraic, then the inverse image $f^{-1}(T)$ is semialgebraic.*

Checking if a map is semialgebraic may be complicated. In our case, however, we work with polynomial maps which are semialgebraic by the following proposition.

Proposition 2.34. *The composition of two semialgebraic mappings is semialgebraic. Moreover, any polynomial mapping is semialgebraic.*

The importance of this result follows from the fact that in statistics many interesting discrete models are parameterized by polynomials with the parameter space being a semialgebraic subset of $\mathbb{R}^d$. In this case, the resulting model can be identified with a set of points that forms a semialgebraic subset. All statistical models considered in this book will correspond to semialgebraic sets. In particular, since polynomial maps are semialgebraic, both $\mathrm{Bin}(2, \theta)$ in Example 1.1 and $\mathcal{M}_{\mathrm{mix}}$ in Example 1.2 are semialgebraic.

To understand the structure of a semialgebraic set $S \subseteq \mathbb{R}^n$, in algebraic statistics we often first want to find the Zariski closure $\overline{S}$. This already gives us some insight into the geometry of S. For example, the dimension of S is equal to the dimension of $\overline{S}$. The following definition lists some aspects of the geometry of S that cannot be studied without taking into account the underlying inequalities of S.

Definition 2.35. The *relative interior* of S, denoted by $\mathrm{int}(S)$, is the set of points a in S such that a sufficiently small neighborhood of a in $\overline{S}$ is contained in S. Moreover, the *boundary* of S is $\mathrm{bd}(S) = S \setminus \mathrm{int}(S)$ and the *algebraic boundary* is the Zariski closure of $\mathrm{bd}(S)$.

If S has the full dimension (codimension zero), then the relative interior and the interior of S coincide. We now illustrate these concepts in a basic example.

Example 2.36. Consider the set S described in $\mathbb{R}^3$ by $z = x^2 + y^2$ and $z \leq 1$; see Figure 2.2. Its Zariski closure is given by $z = x^2 + y^2$. The boundary is given by the circle $x^2 + y^2 = z = 1$ and the relative interior is given by $z = x^2 + y^2$ and $z < 1$. The dimension of S is 2, and S is isomorphic to the disc $x^2 + y^2 \leq 1$ in $\mathbb{R}^2$, which can be realized by projecting down onto the $z = 0$ plane.

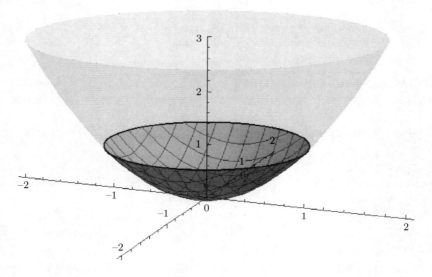

Figure 2.2: *The semialgebraic set given by $z = x^2 + y^2$ and $z \leq 1$.*

2.2.3   Real toric varieties, toric cubes, and the moment map

Real toric varieties   Denote the *real projective space* by $\mathbb{RP}^m$ and by $\mathbb{RP}^m_{\geq 0}$ denote its *nonnegative part*, that is, the set of points $(x_0 : \ldots : x_m) \in \mathbb{RP}^m$ such that $x_i \geq 0$ for all i. The following Lemma shows that the probability simplex can be identified with $\mathbb{RP}^m_{\geq 0}$ and hence every discrete statistical model lies naturally in $\mathbb{RP}^m_{\geq 0}$ for some $m \in \mathbb{N}$; see Definition 3.1.

Lemma 2.37. *Let Δ_m be the probability simplex defined in (1.1). Then*

$$\Delta_m \quad \simeq \quad \mathbb{RP}^m_{\geq 0}.$$

Proof. If $(x_0 : \ldots : x_m) \in \mathbb{RP}^m_{\geq 0}$, then $\sum_{i=0}^{m} x_i > 0$, and hence we have

$$(x_0, \ldots, x_m) \sim \frac{1}{\sum_{i=0}^{m} x_i} (x_0, \ldots, x_m) = (x_0', \ldots, x_m'),$$

where $x'_i \geq 0$ and $\sum_{i=0}^{m} x'_i = 1$. It follows that every point in $\mathbb{RP}^m_{\geq}$ corresponds to a unique point in Δ_m. □

By definition the toric variety $Y_{\mathcal{A}}$ (see Section 2.1.4) is obtained as the Zariski closure of the image of the parameterization $(\mathbb{C}^*)^d \to \mathbb{P}^m$ defined by $\mathbf{t} \mapsto (\mathbf{t}^{a_1}, \ldots, \mathbf{t}^{a_m})$, where $\mathcal{A} = \{a_1, \ldots, a_m\}$. The *real part* $Y_{\mathcal{A}}(\mathbb{R})$ of the toric variety $Y_{\mathcal{A}}$ is the intersection of $Y_{\mathcal{A}}$ with the real projective space $\mathbb{RP}^m$. The following result can be found, for example, in [Sottile, 2003, Section 6].

Proposition 2.38. *The real part $Y_{\mathcal{A}}(\mathbb{R})$ of the toric variety $Y_{\mathcal{A}}$ can be defined by the three equivalent conditions:*

- *the intersection of $Y_{\mathcal{A}}$ with the real projective space $\mathbb{RP}^m$,*
- *the variety of points in $\mathbb{RP}^m$ described by the toric ideal $I_{\mathcal{A}}$*
- *the closure in $\mathbb{RP}^m$ of the parameterization $(\mathbb{R}^*)^d \to \mathbb{R}^m$ given by $\mathbf{t} \mapsto (\mathbf{t}^{a_1}, \ldots, \mathbf{t}^{a_m})$.*

The non-negative part $(Y_{\mathcal{A}})_{\geq}$ of a toric variety $Y_{\mathcal{A}}$ is the intersection of $Y_{\mathcal{A}}$ with $\mathbb{RP}^m_{\geq}$. Consider the *torus* $(\mathbb{R}^*)^d$, which forms a dense subset of $Y_{\mathcal{A}}(\mathbb{R})$. This torus has 2^d components called *orthants*, each identified by the sign vector $\epsilon \in \{\pm 1\}^d$ recording the signs of coordinates of points in that component. The *nonnegative component* is the orthant containing identity and it has sign vector $(1, \ldots, 1)$. Denote this component by $\mathbb{R}^d_>$. For the following result we again refer to [Sottile, 2003, Section 6].

Proposition 2.39. *The non-negative part $(Y_{\mathcal{A}})_{\geq}$ of a toric variety $Y_{\mathcal{A}}$ can be defined by any of the three equivalent conditions:*

- *the intersection of $Y_{\mathcal{A}}$ with $\mathbb{RP}^m_{\geq}$,*
- *the variety of points in $\mathbb{RP}^m_{\geq}$ described by the toric ideal $I_{\mathcal{A}}$,*
- *the closure in $Y_A(\mathbb{R})$ of the parameterization $(\mathbb{R}_>)^d \to \mathbb{R}^m$ given by $\mathbf{t} \mapsto (\mathbf{t}^{a_1}, \ldots, \mathbf{t}^{a_m})$.*

By the isomorphism between $\mathbb{RP}^m_{\geq 0}$ and the probability simplex Δ_m, the points in the non-negative part of $\bar{Y}_{\mathcal{A}}$ are exactly the points on $Y_{\mathcal{A}}$ corresponding to probability distributions.

Remark 2.40. By the equivalence of the three conditions in Proposition 2.39 it follows that to understand the semialgebraic description of $(Y_{\mathcal{A}})_{\geq}$ it suffices to understand the algebraic description of $Y_{\mathcal{A}}$ and then add the inequalities $x_i \geq 0$ defining $\mathbb{RP}^m_{\geq}$.

We could also consider the closures of other components of the torus $(\mathbb{R}^*)^d$, obtaining $2^d - 1$ other pieces analogous to the non-negative part $(Y_{\mathcal{A}})_{\geq}$. Since the component of $(\mathbb{R}^*)^d$ having sign vector ϵ is simply $\epsilon \mathbb{R}^d_>$, these other pieces are transformations of $(Y_{\mathcal{A}})_{\geq}$ by the appropriate sign vector, and hence are all isomorphic. Since $Y_{\mathcal{A}}(\mathbb{R})$ is the closure of $(\mathbb{R}^*)^d$ and each piece $\epsilon \cdot X_{\geq}$ is the closure of the orthant $\epsilon \cdot \mathbb{R}^d_{\geq}$, we obtain a concrete picture of $Y_{\mathcal{A}}(\mathbb{R})$: it is pieced together from 2^d copies of its non-negative part.

The moment map   The analysis of the boundary of $(Y_\mathcal{A})_\geq$ plays an important role in statistics in the study of discrete exponential families; see Section 3.2. For this study it is important to note that the non-negative part of any projective toric variety $Y_\mathcal{A}$ admits an identification with the convex hull $\mathcal{P}_\mathcal{A}$ of $\mathcal{A}$, which enables us to use tools from combinatorics and polyhedral geometry.

Proposition 2.41. *Let $Y_\mathcal{A} \subset \mathbb{P}^n$ be a projective toric variety given by a collection of exponent vectors $\mathcal{A} \subset \mathbb{R}^d$ with convex hull $\mathcal{P}_\mathcal{A}$. Define the algebraic moment map as:*

$$\alpha_\mathcal{A} : (Y_\mathcal{A})_\geq \quad \to \quad \mathbb{R}^d \tag{2.6}$$

$$x \quad \mapsto \quad \sum_{u \in \mathcal{A}} x_u \, u.$$

Then $\alpha_\mathcal{A}$ is a homeomorphism onto $\mathcal{P}_\mathcal{A}$.

Example 2.42. Let $\mathcal{A} = \{(1, 0, 1, 0), (1, 0, 0, 1), (0, 1, 1, 0), (0, 1, 0, 1)\}$. The convex hull of this set forms a square $\mathcal{P}_\mathcal{A} \subset \mathbb{R}^4$. This square lies in a plane given by $y_1 + y_2 = y_3 + y_4 = 1$.

Let $\mathbb{P}^3$ have coordinates $x_{00}, x_{01}, x_{10}, x_{11}$. Then $\mathcal{A}$ induces a parameterization $\psi_\mathcal{A} : (\mathbb{C}^*)^4 \to \mathbb{P}^3$ given by $x_{ij} = s_i t_j$ for $i, j = 0, 1$. The Zariski closure of its image gives a real toric variety $Y_\mathcal{A}$. Its intersection with $\mathbb{RP}^3_{\geq 0}$ is isomorphic to the independence model of two binary random variables; see Example 2.16. Note that this model lies in the hyperplane given by $x_{00} + x_{01} + x_{10} + x_{11} = 1$. This enables us to draw it as a two-dimensional object in a three-dimensional probability simplex; see Figure 3.1 in the next chapter. The vertices of this simplex correspond to setting one of the x_{ij}'s to one and the rest to zero. The algebraic moment map constitutes a homeomorphism between the independence model and a square.

Toric cubes   In applications we are often interested in a set which is strictly smaller than the nonnegative part of a toric variety. An important case is the semialgebraic set obtained as the image of the hypercube $[0, 1]^d$ under the toric map $\mathbf{t} \mapsto (\mathbf{t}^{a_1}, \ldots, \mathbf{t}^{a_m})$.

Definition 2.43. A *toric cube* in $[0, 1]^m$ is an image of a hypercube $[0, 1]^d$, for any positive integer d, under a monomial map. By Proposition 2.33 it is a semialgebraic set.

An interesting example of toric cubes is given by the *space of phylogenetic oranges* described in Chapter 7; see also Kim [2000], Moulton and Steel [2004]. Here we give the easiest example.

Example 2.44. Consider a map $f : [0, 1]^3 \to \mathbb{R}^3$ given by

$$x_{12} = t_1 t_2, \quad x_{13} = t_1 t_3, \quad x_{23} = t_2 t_3,$$

where x_{12}, x_{13}, x_{23} are coordinates of $\mathbb{R}^3$. Let $M = f([0, 1]^3)$. A point $x = (x_{12}, x_{13}, x_{23}) \in \mathbb{R}^3$ lies in M if and only if $x \in [0, 1]^3$ and either

(a) $\min\{x_{12}, x_{13}, x_{23}\} > 0$ and $\max\{\frac{x_{12} x_{13}}{x_{23}}, \frac{x_{12} x_{13}}{x_{23}}, \frac{x_{12} x_{13}}{x_{23}}\} \leq 1$, or

(b) $\min\{x_{12}, x_{13}, x_{23}\} = 0$ and the minimum is attained at least twice.

We leave it as an exercise to show that this set can be more compactly described as the set of points in $[0,1]^3$ satisfying

$$x_{12} \geq x_{13}x_{23}, \quad x_{13} \geq x_{12}x_{23}, \quad x_{23} \geq x_{12}x_{13}.$$

In particular, the points in (b) lie in the closure of (a). This toric cube is depicted in Figure 2.3.

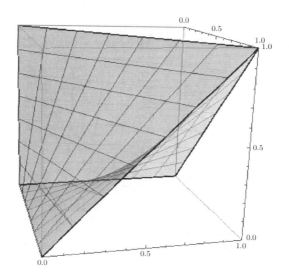

Figure 2.3: *The phylogenetic orange space defined in Example 2.44.*

In the above example the considered toric cube had a nice description in terms of binomial inequalities. Define a *toric precube* as any subset of $[0,1]^m$ defined by some binomial inequalities and hence constraints of the form

$$x^u \quad \leq \quad x^v \qquad \text{where } u, v \in \mathbb{N}_0^d. \tag{2.7}$$

By definition, toric precubes are basic semialgebraic sets. We say that a set $V \subseteq \mathbb{RP}_{\geq}^m$ is equal to the closure of its strictly positive part $V^+ := V \cap \mathbb{RP}_{>}^m$ if the closure of V^+ is equal to V. Here we mean the closure in the classical topology. The following has been proved in Engström et al. [2012].

Theorem 2.45. *Toric cubes correspond to toric precubes which are equal to the closure of its strictly positive points. In particular, toric cubes are basic semialgebraic sets.*

To see why toric cubes are toric precubes, let $\mathcal{C}$ be a toric cube given by monomial parameterization induced by $\mathcal{A} = \{a_1, \ldots, a_m\}$ like in Section 2.1.4. Denote by $A \in \mathbb{N}^{d \times m}$ the matrix with the a_i's as columns. The set $\mathcal{C}^+$ of strictly positive points of $\mathcal{C}$ has an extremely nice structure. Let

$\log \mathcal{C}^+$ denote the transformation of $\mathcal{C}^+$ by taking minus logarithm on each of the coordinates. For example, the transformed toric cube in Example 2.44 is depicted in Figure 2.4. Denote $y_i = -\log x_i$. The cone $\log \mathcal{C}^+$ is the image of the positive orthant $\log((0,1]^d) = \mathbb{R}^d_\geq$ under the linear map $A : \mathbb{R}^d \to \mathbb{R}^n$. By the Weyl–Minkowski Theorem (see for example Ziegler [1995]), every such cone can be written as a solution set of a finite system of linear inequalities of the form

$$u_1 y_1 + \cdots + u_n y_n \quad \geq \quad v_1 y_1 + \cdots + v_n y_n.$$

By applying the exponential map, conclude that $\mathcal{C}^+$ is defined as a subset of $(0,1]^n$ by a finite set of binomial inequalities. Since $\mathcal{C}$ is equal to the closure of $\mathcal{C}^+$, then also $\mathcal{C}$ is given by those inequalities. For details, see Engström et al. [2012].

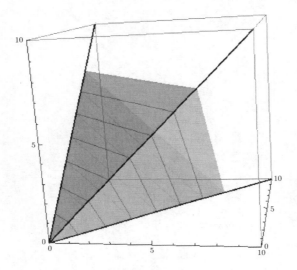

Figure 2.4 *The cone obtained from the strictly positive part of the toric cube in Figure 2.3 by taking minus logarithms.*

As we mentioned when discussing semialgebraic sets in Section 2.2.2, in applications, researchers are often interested in a very detailed description of a given semialgebraic set. For toric cubes, this structure is remarkably beautiful. It turns out that every toric cube can be subsequently decomposed into smaller toric cubes. To introduce this result we need the following definition.

Definition 2.46. An n-dimensional *CW-complex* is a topological subspace X^m of $\mathbb{R}^m$ that is constructed recursively in the following way:

(1) If $n = 0$, then X^0 is a discrete set of points.

(2) If $n > 1$, then X^n is given by:

    (i) an $(n-1)$-dimensional CW-complex X^{n-1} in $\mathbb{R}^m$,

    (ii) a partition $\amalg_{\alpha \in I} \sigma^n_\alpha$ of $X^n \setminus X^{n-1}$ into open n-cells,

(iii) for every index $\alpha \in I$, there is a *characteristic map* $\Phi_\alpha : D^n \to X^n$ such that $\Phi_\alpha(\partial D^n) \subseteq X^{n-1}$ and the restriction of Φ_α to the open cell $\mathring{D}^n$ is a homeomorphism with image σ_α^n.

The simplest case is when the image of each ∂D^n is homeomorphic to a sphere. Such a CW-complex is called *regular*. For example, the toric cube in Example 2.44 forms a regular CW-complex. It turns out that all toric cubes are regular CW-complexes.

Theorem 2.47. *Every toric cube can be realized as a CW-complex whose open cells are interiors of toric cubes. This CW-complex has the further property that the boundary of each open cell is a subcomplex.*

Manifold with corners   The concept if a toric cube can be extended in many ways. For example, the *manifold with corners* can be informally obtained by gluing in a smooth way nonnegative parts of toric varieties; see Kottke and Melrose [2013]. Formally a manifold with corners is a manifold (see Definition 2.27) which locally is diffeomorphic to $\mathbb{R}_{\geq}^k \times \mathbb{R}^{m-k}$ for some m and $k \leq m$.

In this book we consider a very concrete example which directly generalizes toric cubes (c.f. Section 6.3.3). In this case we consider the image in $\mathbb{R}^m$ of the hypercube $[-1, 1]^d$ under the toric map $\mathbf{t} \mapsto (\mathbf{t}^{a_1}, \dots, \mathbf{t}^{a_m})$. Modulo the sign this image is a toric cube. Thus it suffices to analyze which sign patterns are possible in the image and how these toric cubes are glued together.

Example 2.48. Consider the map f defined as in Example 2.44. The image M' of $[-1, 1]^3$ satisfies $x_{12}x_{23}x_{13} \geq 0$. In particular, M' consists of four copies of the toric cube $M = f([0, 1]^3)$ as given in Figure 2.5. In particular, each toric cube in M has dimension 3, and any two cubes are glued along an interval. The intersection of all four cubes is the origin $(0, 0, 0)$.

2.3   Tensors and flattenings

In this section we define tensors, multilinear transformations on tensors, and we set up the basic notation that will be used throughout the book.

2.3.1   Basic definitions

For every vector space in this section we fix a basis. With this, we can think about tensors as generalizations of matrices to higher dimensions and hence as n-dimensional arrays of numbers, functions, or polynomials.

Definition 2.49 (Tensor product $v \otimes w$). Let $V \simeq \mathbb{R}^{r+1}$ and $W \simeq \mathbb{R}^{s+1}$ be two linear spaces. The *tensor product* $v \otimes w$ of vectors $v = [v_i] \in V$ and $w = [w_j] \in W$ is the array

$$\alpha = [\alpha_{ij}] = [v_i \cdot w_j]$$

of products of the coordinates of v and w. Any tensor product of two vectors

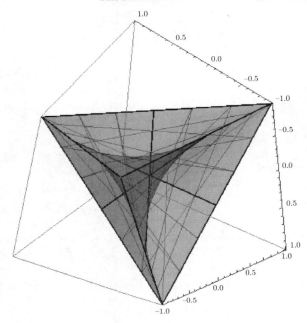

Figure 2.5: *The set $M = f([-1,1]^3)$ defined in Example 2.48.*

is called a *rank-one matrix*. By definition $\otimes$ is a bilinear operation, i.e.,

$$(v + v') \otimes w = v \otimes w + v' \otimes w$$
$$v \otimes (w + w') = v \otimes w + v \otimes w'$$
$$(cv) \otimes w = v \otimes (cw) = c(v \otimes w)$$

for all $v, v' \in V$, $w, w' \in W$ and $c \in \mathbb{R}$.

Definition 2.50 (Tensor product $V \otimes W$). Let $e_0, \ldots, e_r$ and $e'_0, \ldots, e'_s$ be the standard basis vectors for V and W. Then the *tensor product* $V \otimes W$ of V and W is the space of all linear combinations $\sum_{i,j} \alpha_{ij} \, e_i \otimes e'_j$. In particular, $V \otimes W \simeq \mathbb{R}^{(r+1) \times (s+1)}$.

Both definitions generalize to a bigger number of components. Thus a tensor product of tensors $v_1 \in V_1, \ldots, v_m \in V_m$ is the tensor

$$\alpha = [\alpha_{i_1 \cdots i_m}] = [v_{1 i_1} \cdots v_{m i_m}],$$

called a *rank-one tensor*. Similarly, the tensor product $V_1 \otimes \cdots \otimes V_m$ is the space of all linear combinations of rank-one tensors. If $V_i \simeq \mathbb{R}^{r_j+1}$, then $V_1 \otimes \cdots \otimes V_m \simeq \mathbb{R}^{(r_1+1) \times \cdots \times (r_m+1)}$.

The *order* of a tensor is the dimension of its array of coordinates. For instance, vectors are tensors of order one while matrices have order two. We say that a tensor α has *(real) rank* k if k is the minimal number such that α can be written as a sum of k rank-one tensors.

If V is a vector space, then its *dual* V^* is the vector space of linear maps $V \to \mathbb{R}$. If $e_0, \ldots, e_r$ is the basis of V, then the basis of V^* is $e^0, \ldots, e^r$ such that $e^i(e_j) = \delta_{ij}$, where δ_{ij} is the *Kronecker delta*

$$\delta_{ij} \quad = \quad \begin{cases} 1 & \text{if } i = j, \\ 0 & \text{otherwise.} \end{cases}$$

When working in a tensor product of vector spaces and their duals, we denote the coordinates by

$$v = (v^{j_1 \cdots j_n}_{i_1 \cdots i_m}) \ \in \ V_{a_1} \otimes \cdots \otimes V_{a_m} \otimes V^*_{b_1} \otimes \cdots \otimes V^*_{b_n},$$

where the superscripts and subscripts indicate the vector spaces involved. Thus the indices of dual bases are written as superscripts.

Given a tensor $\alpha \in V \otimes W$ of order two, we define its *inverse*

$$\alpha^{-1} \in V^* \otimes W^*$$

to be the unique (if exists) tensor that satisfies

$$\sum_j \alpha_{ij}(\alpha^{-1})^{kj} = \delta_{ik} \quad \text{and} \quad \sum_i \alpha_{ij}(\alpha^{-1})^{ik} = \delta_{kj}.$$

Given a tensor

$$\alpha = (\alpha_{i_1 \cdots i_n}) \in V^{\otimes n} = V \otimes \cdots \otimes V,$$

we say α is *symmetric* if $\alpha_{i_1 \cdots i_n} = \alpha_{i_{\sigma(1)} \cdots i_{\sigma(n)}}$ for all permutations σ on $\{1, \ldots, n\}$. If $v \in V$, then for a fixed $k \geq 1$, the diagonal diag_k is a tensor in $V^{\otimes k}$ given by

$$\mathrm{diag}_k(v) \quad = \quad \sum_i v_i e_i \otimes \cdots \otimes e_i \qquad (2.8)$$

and it is symmetric by definition. If $k = 2$ we often omit k in the notation.

2.3.2 Multilinear transformations and contractions

A linear transformation $A : V \to W$ is represented by a matrix, which is a tensor in $W \otimes V^*$. If $v \in V$, then $A \cdot v \in W$. Linear transformations are generalized to tensor products of vector spaces.

Definition 2.51. Consider tensor products $V_1 \otimes \cdots \otimes V_d$ and $W_1 \otimes \cdots \otimes W_d$ and linear transformations $A_i : V_i \to W_i$ on each of the components. Then the *multilinear transformation*

$$(A_1, \ldots, A_d) : \quad V_1 \otimes \cdots \otimes V_d \to W_1 \otimes \cdots \otimes W_d$$
$$\alpha \ \mapsto \ \tilde{\alpha} \ = \ (A_1, \ldots, A_d) \cdot \alpha$$

is defined for every tensor $\alpha \in V_1 \otimes \cdots \otimes V_d$ by

$$\tilde{\alpha}_{i_1 \cdots i_d} \quad = \quad \sum_{j_1, \ldots, j_d} (A_1)^{j_1}_{i_1} \cdots (A_d)^{j_d}_{i_d} \alpha_{j_1 \cdots j_d}.$$

Example 2.52. Let $d = 2$, $V_i = W_i = \mathbb{R}^2$. In this case, A_1, A_2 and α are represented by 2×2 matrices and we have

$$(A_1, A_2) \cdot \alpha \quad = \quad A_1 \alpha A_2^T.$$

Lastly, we come to an important class of linear transformations known as *contractions*. Contraction is an operation on tensors that comes from the natural pairing of a vector space and its dual. In its simplest form, a contraction $V \otimes V^* \to \mathbb{R}$ corresponds to the bilinear form $u \otimes v \mapsto \langle u, v \rangle = \sum_i u_i v^i$ defined for all rank-one tensors, which is then extended to all other elements of $V \otimes V^*$ by bilinearity of the tensor product.

Scalar products and matrix traces are instances of contractions. Matrix multiplication can also be expressed as a contraction. Given matrices $(\alpha_i^j) \in U \otimes V^*$ and $(\beta_j^k) \in V \otimes W^*$, the contraction

$$(\alpha_i^j) \otimes (\beta_j^k) \mapsto (\gamma_i^k), \quad \gamma_i^k = \sum_j \alpha_i^j \beta_j^k \tag{2.9}$$

of the tensor product $\alpha \otimes \beta$ gives us the matrix product γ.

2.3.3   *Multivariate notation and the operators* `vec` *and* `mat`

In this section, we introduce convenient index notation, which we are going to use throughout the book. Motivated by computer implementations, we also introduce two operations that map tensors to vectors and matrices.

Fix integers $r_1, \ldots, r_m \geq 1$ and consider the set

$$\mathcal{X} \quad = \quad \{0, \ldots, r_1\} \times \cdots \times \{0, \ldots, r_m\}. \tag{2.10}$$

For every subset $A = \{i_1, \ldots, i_d\} \subseteq [m]$ such that $i_1 < \cdots < i_d$, define

$$\mathcal{X}_A := \{0, \ldots, r_{i_1}\} \times \cdots \times \{0, \ldots, r_{i_d}\}.$$

In particular, $\mathcal{X}_i = \{0, \ldots, r_i\}$ for every $i \in [m]$. Consider the vector space $\mathbb{R}^{\mathcal{X}_i} = \mathbb{R}^{r_i+1}$ with the standard basis $e_0, \ldots, e_{r_i}$ for every $i = 1, \ldots, m$ and a tensor space

$$\mathbb{R}^{\mathcal{X}} \quad := \quad \mathbb{R}^{\mathcal{X}_1} \otimes \cdots \otimes \mathbb{R}^{\mathcal{X}_m}$$

with the basis $(e_x)_{x \in \mathcal{X}}$, where $x = (x_1, \ldots, x_m)$ and $e_x = e_{x_1} \otimes \cdots \otimes e_{x_m}$. If $v \in \mathbb{R}^{\mathcal{X}}$, then its coordinates are denoted by v_x. In a similar way we define $\mathbb{R}^{\mathcal{X}_A}$ for every $A \subseteq [m]$. By $\mathbb{C}^{\mathcal{X}}$ we denote the complex tensor space and by $\mathbb{P}^{\mathcal{X}}$ its projectivization.

There is a canonical total ordering $\prec$ of the basis elements of $\mathbb{R}^{\mathcal{X}}$, which comes from the lexicographic ordering of $\mathcal{X}$. For example, if $\mathcal{X} = \{0, 1\}^3$, the elements of $\mathcal{X}$ are ordered by

$$(0,0,0) \prec (0,0,1) \prec (0,1,0) \prec (0,1,1) \prec (1,0,0) \prec (1,0,1) \prec (1,1,0) \prec (1,1,1).$$

The vec operator acts on a tensor space and returns a vector. Thus vec : $\mathbb{R}^{\mathcal{X}} \to \mathbb{R}^{|\mathcal{X}|}$ takes a tensor $v = [v_x] \in \mathbb{R}^{\mathcal{X}}$ and stacks its elements in a vector in the order given by the total ordering $\prec$. The resulting vector is denoted by vec(v). We have the same definition on the tensor space $\mathbb{R}^{\mathcal{X}_A}$ for any $A \subseteq [m]$.

Definition 2.53. For two matrices $A = [a_{ij}] \in \mathbb{R}^{k \times l}$, $B = [b_{ij}] \in \mathbb{R}^{m \times n}$ we define their *Kronecker product* as the matrix $C \in \mathbb{R}^{km \times ln}$ given in a block form by

$$\begin{bmatrix} a_{11}B & \cdots & a_{1l}B \\ \vdots & \ddots & \vdots \\ a_{k1}B & \cdots & a_{kl}B \end{bmatrix}.$$

The Kronecker product of an m-tuple of matrices is also well defined because taking a Kronecker product is associative.

In the literature, the Kronecker product is typically denoted by the tensor product $\otimes$. Here, to avoid confusion, we refrain from doing that. Instead, we realize the Kronecker product of matrices as a special instance of the mat operator, which we now define. The vec and mat operators help to translate tensor operations to operations on vectors and matrices, which can then be used in concrete computations.

Definition 2.54. Let $V_A := \mathbb{R}^{\mathcal{X}_A}$. For any disjoint $A, B \subseteq [m]$, the mat operator takes a tensor in $V_A \otimes V_B$ and returns a matrix in $\mathbb{R}^{|\mathcal{X}_A| \times |\mathcal{X}_B|}$ whose rows and columns are ordered by the total ordering $\prec$ on $\mathcal{X}_A$ and $\mathcal{X}_B$, respectively.

Let $A \in V_1 \otimes W_1^*$, $B \in V_2 \otimes W_2^*$; then their tensor product is

$$A \otimes B \in (V_1 \otimes W_1^*) \otimes (V_2 \otimes W_2^*) = (V_1 \otimes V_2) \otimes (W_1 \otimes W_2)^*.$$

Now $\mathrm{mat}(A \otimes B)$ is exactly the Kronecker product of A and B. Similarly $\mathrm{mat}(A_1 \otimes \cdots \otimes A_n)$ for matrices $A_i \in V_i \otimes W_i^*$ is their Kronecker product. Now a multilinear transformation can be written in linear algebra terms as

$$\mathrm{vec}(\tilde{\alpha}) \quad = \quad \mathrm{mat}(A_1 \otimes \cdots \otimes A_n) \cdot \mathrm{vec}(\alpha). \qquad (2.11)$$

Definition 2.55. Let $A \subseteq [m]$; then we split $[m]$ into two sets: A and its complement $B := [m] \setminus A$. This split induces a map

$$F_{A|B} : \quad \mathbb{R}^{\mathcal{X}} \to V_A \otimes V_B^*$$
$$v = [v_{i_1 \cdots i_m}] \mapsto [v_{i_A}^{i_B}].$$

The matrix

$$v_{A;B} \quad := \quad \mathrm{mat}(F_{A|B}(v)) \qquad (2.12)$$

is called the *(matrix) flattening map* of the tensor v.

Example 2.56. Let $m = 4$ and $\mathcal{X} = \{0,1\}^4$. Consider the flattening map induced by $A = \{1,3\}$. Every element $v = [v_{ijkl}]$ is mapped to the matrix

$$v_{12;34} \quad = \quad \begin{bmatrix} v_{0000} & v_{0001} & v_{0100} & v_{0101} \\ v_{0010} & v_{0011} & v_{0110} & v_{0111} \\ v_{1000} & v_{1001} & v_{1100} & v_{1101} \\ v_{1010} & v_{1011} & v_{1110} & v_{1111} \end{bmatrix}.$$

2.4   Classical examples

In this section we present some classical algebraic varieties, which have some interesting links to statistics that will be developed in later chapters.

Segre variety Let $\mathcal{X}_1 = \mathcal{X}_2 = \{0, 1\}$ for $i = 1, 2$, $\mathcal{X} = \mathcal{X}_1 \times \mathcal{X}_2$ and consider two copies of the projective line $\mathbb{P}^{\mathcal{X}_i} \simeq \mathbb{P}^1$ with coordinates $(s_0 : s_1)$ and $(t_0 : t_1)$. Their Cartesian product $\mathbb{P}^{\mathcal{X}_1} \times \mathbb{P}^{\mathcal{X}_2}$ is a two-dimensional variety. To realize it as a projective variety, we need to embed $\mathbb{P}^{\mathcal{X}_1} \times \mathbb{P}^{\mathcal{X}_2}$ into a projective space. We now show that such an embedding is possible into $\mathbb{P}^{\mathcal{X}} \simeq \mathbb{P}^3$. The map $\mathbb{P}^{\mathcal{X}_1} \times \mathbb{P}^{\mathcal{X}_2} \to \mathbb{P}^{\mathcal{X}}$ is given by $(s, t) \mapsto y = s \otimes t$, or in coordinates $y_{ij} = s_i t_j$ for $i, j = 0, 1$. Denote this variety by $\mathrm{Seg}(1, 1)$. To show that this map is well defined, we need to show that at least one y_{ij} is nonzero for any choice of $s = (s_0 : s_1)$ and $t = (t_0 : t_1)$. This follows from the fact that at least one coordinate of both points $(s_0 : s_1)$ and $(t_0 : t_1)$ needs to be nonzero. For example, if s_1 and t_0 are nonzero, then y_{10} is nonzero in the image.

To show that $(s, t) \mapsto s \otimes t$ is an embedding, we need to show that for every point in the image we can uniquely identify the points $(s_0 : s_1)$ and $(t_0 : t_1)$ mapping to it. Consider a point $y \in \mathbb{P}^{\mathcal{X}}$ which lies in $\mathrm{Seg}(1, 1) \cap U_{00}$, where $U_{00} = \{y \in \mathbb{P}^{\mathcal{X}} : y_{00} \neq 0\}$. Then $s_0 t_0 \neq 0$, and hence we can assume $s_0 = t_0 = 1$ in $\mathbb{P}^{\mathcal{X}_1}$ and $\mathbb{P}^{\mathcal{X}_2}$, respectively. In this case $y = (1 : s_1 : t_1 : s_1 t_1)$ for some $s_1, t_1 \in \mathbb{C}$, which identifies s_1 and t_1. We easily check that the same holds for every open subset U_x covering $\mathbb{P}^{\mathcal{X}}$ and hence for every point on $\mathrm{Seg}(1, 1)$.

The Segre variety $\mathrm{Seg}(1, 1) \subseteq \mathbb{P}^{\mathcal{X}}$ is given by a single equation

$$y_{00} y_{11} - y_{01} y_{10} = 0,$$

which is the determinant of the flattening $V \mapsto V_1 \otimes V_2^*$. In other words, the points $y \in \mathbb{P}^{\mathcal{X}}$, which lie in the Segre embedding of $\mathbb{P}^{\mathcal{X}_1} \times \mathbb{P}^{\mathcal{X}_2}$, correspond to rank-one matrices.

More generally, let $r_1, \ldots, r_m \geq 1$ and consider m projective spaces $\mathbb{P}^{\mathcal{X}_1}, \ldots, \mathbb{P}^{\mathcal{X}_m}$, where $\mathcal{X}_i = \{0, \ldots, r_i\}$. By $t_i = (t_{i0} : \cdots : t_{ir_i})$, denote the coordinates on $\mathbb{P}^{\mathcal{X}_i}$ for $i = 1, \ldots, m$. The *Segre variety* $\mathrm{Seg}(r_1, \ldots, r_m)$ is the *embedding* of the product $\mathbb{P}^{\mathcal{X}_1} \times \cdots \times \mathbb{P}^{\mathcal{X}_m}$ into $\mathbb{P}^{\mathcal{X}}$ with coordinates (y_x), where $x \in \mathcal{X}$. This embedding is defined by

$$y_x \;\; = \;\; t_{1x_1} \cdots t_{mx_m} \qquad \text{for every } x \in \mathcal{X},$$

which in tensor notation can be written by

$$y = t_1 \otimes \cdots \otimes t_m. \tag{2.13}$$

This map is well defined by exactly the same argument as for $\mathrm{Seg}(1, 1)$. It describes the set of all rank-one tensors of $\mathbb{P}^{\mathcal{X}}$ denoted by $\mathrm{Seg}(r_1, \ldots, r_m)$.

Remark 2.57. Consider the parameterization of the Segre variety constrained to $\Delta_{\mathcal{X}_1} \times \cdots \times \Delta_{\mathcal{X}_m} \to \Delta_{\mathcal{X}}$. The image is given by all rank-one tensors y in $\Delta_{\mathcal{X}}$, where $\Delta_{\mathcal{X}_i}, \Delta_{\mathcal{X}}$ are probability simplices in $\mathbb{R}^{\mathcal{X}_i}$ and $\mathbb{R}^{\mathcal{X}}$, respectively. In particular, by Lemma 2.37, it is isomorphic to the real non-negative part of the Segre variety.

Veronese embedding   Let $r_1, \ldots, r_m, n \geq 1$ and let $\mathcal{X}$ be as before. Consider the set $\mathcal{U}$ of all tensors in $\mathbb{R}^{\mathcal{X}}$ with integer entries whose sum is equal to n

$$\mathcal{U} \quad := \quad \mathcal{U}(\mathcal{X}, n) \quad = \quad \{u = (u_x) \in \mathbb{N}_0^{\mathcal{X}} : \sum_{x \in \mathcal{X}} u_x = n\}. \tag{2.14}$$

Consider the projective space $\mathbb{P}^{\mathcal{X}}$ with coordinates $\mathbf{t} = (t_x)_{x \in \mathcal{X}}$ and $\mathbb{P}^{\mathcal{U}}$ with coordinates y_u for $u \in \mathcal{U}$.

Definition 2.58. The *Veronese embedding* of $\mathbb{P}^{\mathcal{X}}$ into $\mathbb{P}^{\mathcal{U}}$ is defined by

$$y_u \quad = \quad t^u := \prod_{x \in \mathcal{X}} t_x^{u_x} \qquad \text{for } u \in \mathcal{U}(\mathcal{X}, n).$$

The *multinomial embedding* is given by

$$y_u \quad = \quad \frac{n!}{\prod u_x!} \mathbf{t}^u \qquad \text{for } u \in \mathcal{U}(\mathcal{X}, n).$$

The fact that both maps are embeddings is easy to verify.

For example, if $n = 2$, $m = 1$, $r_1 = 1$, then the Veronese embedding is defined by

$$y_{20} \quad = \quad t_0^2, \quad y_{11} \quad = \quad t_0 t_1, \quad y_{02} \quad = \quad t_1^2.$$

The multinomial embedding is $y_{20} = t_0^2$, $y_{11} = 2t_0 t_1$, $y_{02} = t_1^2$ and it corresponds to the model defined in Example 1.1 by setting $t_1 = 1 - t_0 = \theta$.

Secants and mixtures   Consider the Segre embedding $\mathrm{Seg}(r_1, \ldots, r_m) \subseteq \mathbb{P}^{\mathcal{X}}$. For any two points in $\mathrm{Seg}(r_1, \ldots, r_m)$, consider the line in $\mathbb{P}^{\mathcal{X}}$ through them. Take the union of all such lines obtained for all pairs of points on $\mathrm{Seg}(r_1, \ldots, r_m)$. The *first secant variety* $\mathrm{Sec}(r_1, \ldots, r_m)$ is defined as the Zariski closure in $\mathbb{P}^{\mathcal{X}}$ of this union. This variety is parameterized by two copies of each of $\mathbb{P}^{\mathcal{X}_1}, \ldots, \mathbb{P}^{\mathcal{X}_m}$ that give us two points on $\mathrm{Seg}(r_1, \ldots, r_m)$, and by $\mathbb{P}^1$, with coordinates s_1, s_2, which tells us where on the line between these two points we are. The parameterization is

$$((s_1, s_2), (t_{1,0}^{(i)}, \ldots, t_{1,k_1}^{(i)})_{i=1}^2, \ldots, (t_{m,0}^{(i)}, \ldots, t_{m,k_m}^{(i)})_{i=1}^2) \mapsto$$

$$\mapsto (y_x = \textstyle\sum_{i=1}^2 s_i \prod_{j=1}^m t_{j,x_j}^{(i)})_{x \in \mathcal{X}},$$

where $(t_{j,0}^{(i)}, \ldots, t_{j,k_j}^{(i)}) \in \mathbb{P}^{\mathcal{X}_j}$.

Recall the definition of the flattening map in Definition 2.55. The following was conjectured in Garcia et al. [2005] and proved by Raicu [2012].

Theorem 2.59. *Let* $r_1, \ldots, r_m = 1$. *The ideal defining the secant variety* $\mathrm{Sec}(1, \ldots, 1)$ *is generated by all* 3×3 *minors of all the (matrix) flattenings.*

Example 2.60. If $m = 3$, then all flattenings are 2×4 matrices and hence there are no 3×3 minors so in this case $\mathrm{Sec}(1, 1, 1)$ is equal to $\mathbb{P}^{\mathcal{X}}$. If $m = 4$, then there are three non-trivial flattenings corresponding to splits $12|34$, $13|24$ and $14|23$.

The concept of the first secant variety can be generalized. The r-th secant variety $\mathrm{Sec}_r(k_1,\ldots,k_m) \subset \mathbb{P}^{\mathcal{X}} = \mathbb{P}^{(k_1+1)\cdots(k_m+1)-1}$ is parameterized by $\mathbb{P}^r$ and $r+1$ copies of $\mathbb{P}^{k_1},\ldots,\mathbb{P}^{k_m}$ via the map, which is a direct generalization of the $r = 1$ case above. Using tensor notation, this variety consists of points of the form

$$\sum_{i=1}^{r+1} s_i \left(t_1^{(i)} \otimes \cdots \otimes t_m^{(i)}\right). \tag{2.15}$$

In the applications, we are often interested in another related variety, when the parameterization in (2.15) is constrained to $\mathbb{RP}^r_{\geq}$ and $r+1$ copies of $\mathbb{RP}^{\mathcal{X}_1}_{\geq},\ldots,\mathbb{RP}^{\mathcal{X}_m}_{\geq}$. In this case, the variety is called the *mixture model*. The difference between the image of this parameterization and its Zariski closure is of a major interest in algebraic statistics. The description of the image of this parameterization is not very complicated if $r = 1$ but it requires developing some theory; see Section 7.3.4.

2.5   Birational geometry

In this section we present some classical constructions of algebraic geometry which are used in statistical learning theory.

2.5.1   *Rational functions on a variety*

The polynomial ring $k[x_1,\ldots,x_n]$ is included as a subring in the field of rational functions

$$k(x_1,\ldots,x_m) \;=\; \left\{\frac{f}{g} : f,g \in k[x_1,\ldots,x_n], g \neq 0\right\}.$$

The word "rational" refers to the fact that in the definition we consider quotients of polynomials.

Let $V \subset k^m$ and $W \subset k^n$ be irreducible affine varieties. A *rational mapping* from V to W is a function ϕ represented by

$$\phi(x_1,\ldots,x_n) \;=\; \left(\frac{f_1(x_1,\ldots,x_m)}{g_1(x_1,\ldots,x_m)},\ldots,\frac{f_n(x_1,\ldots,x_m)}{g_n(x_1,\ldots,x_m)}\right),$$

where

(i) ϕ is defined at some point of V.

(ii) For every $(a_1,\ldots,a_m) \in V$ where ϕ is defined, $\phi(a_1,\ldots,a_m) \in W$.

Note that a rational mapping ϕ from V to W may fail to be a function from V to W in the usual sense because ϕ may not be defined everywhere on V. For this reason we write a rational map with a dashed line $\phi : V \dashrightarrow W$.

Definition 2.61. Let $\phi,\psi : V \dashrightarrow W$ be rational mappings represented by

$$\phi = \left(\frac{f_1}{g_1},\ldots,\frac{f_n}{g_n}\right) \text{ and } \psi = \left(\frac{f_1'}{g_1'},\ldots,\frac{f_n'}{g_n'}\right).$$

Then we say that $\phi = \psi$, if for each i, $1 \leq i \leq n$,

$$f_i g_i' - f_i' g_i \in \mathcal{I}(V).$$

We say that two irreducible varieties $V \subset k^m$ and $W \subset k^n$ are *birationally equivalent* if there exist rational mappings $\phi : V \dashrightarrow W$ and $\psi : W \dashrightarrow V$ such that $\phi \circ \psi$ is defined (i.e., there exists a point p in W such that ψ is defined at p and ϕ is defined at $\psi(p)$) and equal to the identity map id_W (as in Definition 2.61), and similarly for $\psi \circ \phi$. Less formally, we say that V and W are birationally equivalent if they are isomorphic everywhere outside of the subsets where $\psi \circ \phi$ and $\phi \circ \psi$ are not defined.

2.5.2   Blow-up and resolution of singularities

An important example of birational equivalence is the *blow-up of a Euclidean space at a point p*. The idea is to leave k^n unaltered except at a point p, which is then replaced by the set of all lines through p and hence by $\mathbb{P}^{n-1}$. To make this precise, choose a suitable coordinate system for k^n so that p can be assumed to be the origin.

Definition 2.62. Let B be the set of all pairs (x, l), where $x \in k^n$ and $l \in \mathbb{P}^{n-1}$ is a line through the origin containing x, so

$$B = \{(x, l) \in k^n \times \mathbb{P}^{n-1} : x \in l\} \subset k^n \times \mathbb{P}^{n-1}.$$

The *blow-up* of k^n at p is by definition the projection to the affine factor

$$\begin{aligned} B &\xrightarrow{\pi} k^n, \\ (x, l) &\mapsto x. \end{aligned}$$

Let $x = (x_1, \dots, x_n) \in k^n$ and $l = (y_1 : \dots : y_n)$. Then x lies on l if and only if the matrix

$$\begin{bmatrix} x_1 & \cdots & x_n \\ y_1 & \cdots & y_n \end{bmatrix}$$

has rank less than or equal to 1. This is precisely when all the 2×2 minors vanish and hence

$$B \;=\; V_k(x_i y_j - x_j y_j : 0 \leq i < j \leq n) \subseteq k^n \times \mathbb{P}^{n-1}. \qquad (2.16)$$

Recall that $\mathbb{P}^n$ is covered by n copies of k^n. It is instructive to see how the blow-up looks on one of the copies of $k^n \times k^n$ covering $k^n \times \mathbb{P}^n$. For example, on U_1 (c.f. Section 2.1) where $y_1 \neq 0$, we can assume $y_1 = 1$ and hence from (2.16) we see that, in particular,

$$(x_1, \dots, x_n) \;=\; (x_1, x_1 y_2, \cdots, x_1 y_n). \qquad (2.17)$$

The operation of blowing-up is closely related to the resolution of singularities. Its importance is due to the fundamental theorem of Heisuke Hironaka on the resolution of singularities.

Theorem 2.63 (Hironaka's Theorem). *Let $f : \mathbb{R}^d \to \mathbb{R}$ be a real analytic function in the neighborhood of the origin such that $f(0) = 0$. Then there exists a neighborhood of the origin W and a proper real analytic map $\pi : U \to W$ where U is a d-dimensional real analytic manifold such that the following holds:*

1. *The map π is an isomorphism between $U \setminus U_0$ and $W \setminus W_0$, where $W_0 = \{x \in W : f(x) = 0\}$ and $U_0 = \{u \in U : f(\pi(u)) = 0\}$.*

2. *For an arbitrary point $P \in U_0$, there is a local coordinate system $u = (u_1, \ldots, u_d)$ of U in which P is the origin and*

$$f(\pi(u)) = a(u) u_1^{r_1} \cdots u_d^{r_d}, \qquad (2.18)$$

where $a(u)$ is a nowhere vanishing function on this local chart and $r_1, \ldots, r_d$ are nonnegative integers, and the Jacobian of $x = \pi(u)$ is

$$\pi'(u) = b(u) u_1^{h_1} \cdots u_d^{h_d},$$

where, again, $b(u) \neq 0$ and $h_1, \ldots, h_d$ are nonnegative integers.

Moreover, π can be always obtained as a composition of blow-ups along smooth centers.

Example 2.64. Let $f(x_1, \ldots, x_d) = x_1^2 + \cdots + x_c^2$ for some $c \leq d$. The blow-up of $\mathbb{R}^c$, with coordinates given by $x_1, \ldots, x_c$, at the origin satisfies all the properties of Theorem 2.63. Let W be an ϵ-ball around the origin in $\mathbb{R}^c$. We have $\pi : U \to W$, where $U \subset \mathbb{R}^c \times \mathbb{RP}^{c-1}$. Over $\mathbb{R}^c \times U_1$ by (2.17) we have

$$f(\pi(u)) = u_1^2 + u_1^2 u_2^2 + \cdots + u_1^2 u_c^2 = u_1^2(1 + u_2^2 + \cdots + u_c^2)$$

and $a_1(u) = 1 + u_2^2 + \cdots + u_c^2$ never vanishes if ϵ is sufficiently small. Similarly over each of the charts $\mathbb{R}^c \times U_i$ for $i = 1, \ldots, c$ the function $f(\pi(u)) = u_i^2 a_i(u)$, where $a_i(u)$ never vanishes. The Jacobian of π on each $\mathbb{R}^c \times U_i$ is equal to u_i^{c-1}.

Example 2.65. Let $f(x, y, z) = x^2 y^2 + x^2 z^2 + y^2 z^2$. Then $f(\pi(u)) = u_1^4(u_2^2 + u_3^2 + u_2^2 u_3^2)$ with the Jacobian of π equal to u_1^2. We make another blow-up along $\mathbb{R}^2$, which gives $f(\tilde{\pi}(v)) = u_1^4 v_2^2(1 + v_3^2 + v_2^2 v_3^2)$. Here again $1 + v_3^2 + v_2^2 v_3^2$ never vanishes in a small neighborhood of the origin and hence $f(\tilde{\pi}(v))$ is of the form in (2.18).

2.6   Bibliographical notes

The basic reference for algebraic geometry is Cox et al. [2007] and Smith et al. [2000]. The first book is written on the undergraduate level and it has an algebraic flavor. The second book elegantly introduces some main geometric concepts of algebraic geometry. Section 2.1 is almost entirely based on these two books. A good idea is also to read introductory chapters of other books on algebraic statistics such as those written by Pachter and Sturmfels [2005], Pistone et al. [2001], Watanabe [2009]. For a more detailed discussion on

Zariski topology; see for example Chapter 1 in Smith et al. [2000] or Chapter 1 in Shafarevich [1994]. We presented only a very informal discussion of real analytic manifolds. A more detailed treatment is given in [Krantz and Parks, 2002, Section 2.7] and Spivak [1965]. Our presentation of the basics of the real algebraic geometry is based on Basu et al. [2006], Bochnak et al. [1998]. The section on real toric varieties is based on Sottile [2003]. Standard references on toric varieties are Cox et al. [2011], Fulton [1993], Sturmfels [1996]. Toric cubes are introduced and discussed in Engström et al. [2012]; see also Basu et al. [2012], Basu et al. [2010]. There are several good references on tensors. For statisticians, McCullagh [1987] is a natural choice. A more algebraic introduction is given in [Roman, 2008, Chapter 14] and [Eisenbud, 1995, Appendix 2]. The most complete reference is Landsberg [2012]. Most of the classical examples in Section 2.4 can be looked up in any of the books cited above. A good introduction to secant varieties in the statistical context is given in Drton et al. [2009]. A good introduction to birational geometry is given, for example, in Shafarevich [1994]. Basic constructions are described in Smith et al. [2000]. Following Watanabe [2009], we formulated the resolution of singularities theorem of Hironaka in the analytic form given by Atiyah [1970].

In this introduction we completely ignored the concept of Gröbner bases, which forms an essential tool of computational algebraic geometry. Another algebro-geometric notion which was almost entirely ignored here was the theory of toric varieties. Together with Gröbner bases, toric geometry lies behind the great success of algebraic statistics over the last years, which started with the seminal paper by Diaconis and Sturmfels [1998]. The reason why we decided to skip these two important topics in this book is mainly because they are used here only implicitly. A well-written introduction to Gröbner bases is provided for example in Cox et al. [2007], Sturmfels [1996]. Also, Pistone et al. [2001] is a good place to start for a statistician. Standard references on toric varieties are Cox et al. [2011] and Fulton [1993]. However, Sturmfels [1996] may be more suitable for purposes of algebraic statistics.

Chapter 3

Algebraic statistical models

In this chapter we define the algebraic statistical model. Our focus is on models of conditional independence and in particular on graphical models. In this context we discuss some standard notions in statistics like maximum likelihood estimation and model identifiability.

3.1 Discrete measures

3.1.1 Discrete statistical models

Let $X = (X_1, \ldots, X_m)$ be a random vector such that each X_i takes $r_i + 1$ possible values, where each $r_i \geq 1$ is finite. Thus X takes values in a finite discrete set $\mathcal{X} = \prod_{i=1}^{m} \mathcal{X}_i \subseteq \mathbb{R}^m$ such that $|\mathcal{X}_i| = r_i + 1$ for $i = 1, \ldots, m$. Without loss of generality $\mathcal{X}_i = \{0, \ldots, r_i\}$ and thus

$$\mathcal{X} = \{0, \ldots, r_1\} \times \cdots \times \{0, \ldots, r_m\}. \tag{3.1}$$

Any probability distribution of X can be written as a point (or tensor) $P = [p(x)] \in \mathbb{R}^{\mathcal{X}}$ such that $p(x) \geq 0$ for all $x \in \mathcal{X}$ and $\sum_{x \in \mathcal{X}} p(x) = 1$. The set of all such points is called the *probability simplex* and it is denoted by $\Delta_{\mathcal{X}}$. By Lemma 2.37,

$$\Delta_{\mathcal{X}} \simeq \mathbb{RP}_{\geq}^{\mathcal{X}},$$

where $\mathbb{RP}_{\geq}^{\mathcal{X}}$ is the nonnegative part of the projective space $\mathbb{P}^{\mathcal{X}}$ as defined in Section 2.2.3.

Definition 3.1 (Statistical model). *A statistical model* of a random vector $X \in \mathcal{X}$ is any subset $\boldsymbol{M} \subseteq \Delta_{\mathcal{X}}$. We say that a discrete statistical model is *parametric* if there exists $\Theta \subseteq \mathbb{R}^d$ for some $d \geq 1$ called the *parameter space* and a map $p : \Theta \to \mathbb{R}^{\mathcal{X}}$, called the *parameterization*, such that $\boldsymbol{M} = p(\Theta)$. We indicate this by writing $\boldsymbol{M} = [p(x; \theta)]_{x \in \mathcal{X}}$.

In this book we are interested in a special type of parametric statistical model where the parameterization is given by polynomial maps.

Definition 3.2 (Algebraic statistical model). *An algebraic statistical model* is a statistical model $\boldsymbol{M}$ that forms a semialgebraic set. An algebraic statistical model is said to be *parametric* if Θ is a basic semialgebraic set and there exist polynomials $g_x \in \mathbb{R}[\theta]$ for $x \in \mathcal{X}$ such that the parameterization p satisfies:

$$p(x; \theta) = Z(\theta)^{-1} g_x(\theta) \qquad \text{for all } x \in \mathcal{X},$$

where $Z(\theta)$ is the normalizing constant, that is, $Z(\theta) = \sum_x g_x(\theta)$. The fact that a parametric algebraic statistical model also forms a semialgebraic set follows directly from Propositions 2.33 and 2.34.

Example 3.3 (Multinomial model). By $\mathcal{U} = \mathcal{U}(\mathcal{X}, n)$, denote the space of all $(r_1 + 1) \times \cdots \times (r_m + 1)$ tensors with nonnegative integer entries all summing to n. We have $u \in \mathcal{U}$ if $u \in \mathbb{N}_0^{\mathcal{X}}$ and $\sum_{x \in \mathcal{X}} u(x) = n$ (see also (2.14)). Define the *multinomial model* as the set of the probability distributions on $\mathcal{U}$ parameterized by $\Delta_{\mathcal{X}}$. Thus, for every $q \in \Delta_{\mathcal{X}}$ we have

$$p(u; q) = \frac{n!}{\prod_{x \in \mathcal{X}} u(x)!} \prod_{x \in \mathcal{X}} q(x)^{u(x)} \qquad \text{for every } u \in \mathcal{U}.$$

By Section 2.4 it follows that, modulo the constant terms, the multinomial model is the Veronese embedding of $\Delta_{\mathcal{X}}$ in $\Delta_{\mathcal{U}}$.

The algebraic constraints on statistical models may look restrictive. However, as we see later in this chapter, many known discrete statistical models are algebraic. In particular, discrete exponential families and their mixtures are algebraic.

Given an algebraic statistical model M that forms a basic semialgebraic set, its *semialgebraic description* is a set of polynomial equations and inequalities

$$f_1(p) = \cdots = f_r(p) = 0, \qquad g_1(p) \geq 0, \ldots, g_s(p) \geq 0$$

such that $p \in M$ if and only if p satisfies these constraints. Because the equation $\sum_{x \in \mathcal{X}} p(x) = 1$ and the inequalities $p(x) \geq 0$ for $x \in \mathcal{X}$ hold for all statistical models, they are called *trivial* constraints.

Remark 3.4. In many interesting cases a full description of algebraic statistical models is hard to obtain. In the algebraic statistics literature, a common approach is to replace the parameterization $g : \Theta \to \Delta_{\mathcal{X}}$ with $g : \mathbb{C}^d \to \mathbb{P}^{\mathcal{X}}$ and so assuming that parameters can take any complex values. The focus is then on the set $M_{\mathbb{C}} = g(\mathbb{C}^d) \cap \Delta_{\mathcal{X}}$ of all probability distributions in $g(\mathbb{C}^d)$. The analysis of this map follows the standard algebraic procedures. Although $M = g(\Theta) \subseteq M_{\mathbb{C}}$, typically the inclusion is strict; see Example 1.3.

Example 3.5. Let $m = 2$ and $r_1 = r_2 = 1$ so that $\mathcal{X} = \{0, 1\}^2$. Define $p_1(x_1) = p(x_1, 0) + p(x_1, 1)$ and $p_2(x_2) = p(0, x_2) + p(1, x_2)$. The model of independence is given by all probability distributions satisfying

$$p(x_1, x_2) = p_1(x_1)p_2(x_2)$$

for all $x = (x_1, x_2) \in \mathcal{X}$. This is a parametric model given by parameterization

$$\psi : [0, 1]^2 \to \Delta_{\mathcal{X}}, \qquad \theta = (\theta_1, \theta_2) \mapsto \Delta_{\mathcal{X}},$$

which is defined by $p(0, 0) = (1 - \theta_1)(1 - \theta_2)$, $p(0, 1) = (1 - \theta_1)\theta_2$, $p(1, 0) = \theta_1(1 - \theta_2)$ and $p(1, 1) = \theta_1\theta_2$. Thus, two free parameters θ_1, θ_2 correspond to $p_1(1)$ and $p_2(1)$, respectively. The model of independence is equal to the subset of all rank-one matrices in $\Delta_{\mathcal{X}}$, which is equal to the nonnegative part of the Segre variety $\mathrm{Seg}(1, 1)$. This set is depicted in Figure 3.1.

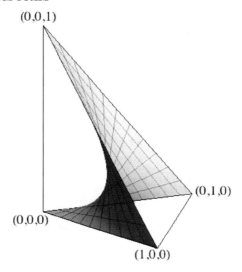

Figure 3.1: *The model of independence of two binary random variables.*

3.1.2 Independence and conditional independence

Let $A \subset [m]$, then the *marginal distribution* of $X_A = (X_i)_{i \in A} \in \mathcal{X}_A$ is a tensor $p_A \in \mathbb{R}^{\mathcal{X}_A}$ defined by

$$p_A(\tilde{x}) = \sum_{x \in \mathcal{X}:\, x_A = \tilde{x}} p(x) \qquad \text{for all } \tilde{x} \in \mathcal{X}_A. \tag{3.2}$$

Note in particular that $p_\emptyset \equiv 1$. Here, by writing that p_A is a tensor we mean only that it is an element of the vector space $\mathbb{R}^{\mathcal{X}_A}$. We exploit tensor algebra more in Chapter 4.

For any two disjoint subsets $A, B \subseteq [m]$, we say that X_A *and* X_B *are independent*, which we denote by $A \perp\!\!\!\perp B$ (or $X_A \perp\!\!\!\perp X_B$), if and only if

$$p_{A \cup B}(x_{A \cup B}) = p_A(x_A) p_B(x_B) \qquad \text{for all } x \in \mathcal{X}.$$

This can be generalized to $B_1 \perp\!\!\!\perp \cdots \perp\!\!\!\perp B_r$

In statistical modeling a more important concept is that of conditional independence. We often adopt another notational convention writing AB for $A \cup B$ and a for the singleton $\{a\}$. Let A, B be two disjoint subsets of $[m]$. We define the *conditional distribution* of X_A given $X_B = x_B$ as a tensor $p_{A|B}$ such that

$$p_{A|B}(x_A | x_B) = \frac{p_{AB}(x_A, x_B)}{p_B(x_B)}, \tag{3.3}$$

for every $x_A \in \mathcal{X}_A$ and $x_B \in \mathcal{X}_B$ such that $p_B(x_B) \neq 0$.

Definition 3.6. Given disjoint subsets $A, B, C \subset [m]$ we say that A, B are conditionally independent given C if

$$p_{AB|C}(x_A, x_B|x_C) = p_{A|C}(x_A|x_C)p_{B|C}(x_B|x_C)$$

for each $x_A \in \mathcal{X}_A$, $x_B \in \mathcal{X}_B$, $x_C \in \mathcal{X}_C$. $\qquad\square$

For many interesting statistical models the vector on which we condition is assumed to be hidden. In this case we are interested only in the marginal distribution over the observed variables. Let, for example, $A \perp\!\!\!\perp B|C$, then

$$p_{AB}(x_A, x_B) = \sum_{x_C \in \mathcal{X}_C} p_C(x_C)p_{A|C}(x_A|x_C)p_{B|C}(x_B|x_C). \qquad (3.4)$$

The following result is easy to prove.

Lemma 3.7. *If $A \perp\!\!\!\perp B$, then $p_{A|B}(x_A|x_B) = p_A(x_A)$ for every $x_A \in \mathcal{X}_A$, $x_B \in \mathcal{X}_B$.*

Definition 3.8. The *chain rule* states that the joint distribution of X can be written as a product of successive conditional distributions. Define $A_i := \{j < i\}$, then

$$p(x) = \prod_{i=1}^{m} p_{i|A_i}(x_i|x_{A_i}).$$

Any other total ordering on $[m]$ will lead to a different valid formula.

Example 3.9. Let $m = 3$ and $r_1 = r_2 = r_3 = 1$. The model of marginal independence $X_1 \perp\!\!\!\perp X_2$ is now given by all probability distributions satisfying $p_{12}(x_1, x_2) = p_1(x_1)p_2(x_2)$ for all $x_1 \in \mathcal{X}_1$, $x_2 \in \mathcal{X}_2$. There is a canonical parameterization of this model. Write the joint distribution of (X_1, X_2, X_3) using the chain rule

$$p(x_1, x_2, x_3) = p_1(x_1)p_{2|1}(x_2|x_1)p_{3|12}(x_3|x_1, x_2).$$

Since $X_1 \perp\!\!\!\perp X_2$, by Lemma 3.7, $p_{2|1}(x_2|x_1) = p_2(x_2)$, and the model is parameterized by $p : [0,1]^6 \to \Delta_\mathcal{X}$, where the six free parameters correspond to $p_1(1)$, $p_2(1)$ and $p_{3|12}(1|i,j)$ for $i,j = 0,1$.

Note that we do not develop the theory of the conditional independence (CI) in its full generality. For a complete introduction to the topic we suggest the monograph of Studený [2004]. For an exposition of the importance of this concept in the statistical theory; see Dawid [1979].

Define a *conditional independence model* (or *CI model*) as a statistical model defined by a collection of conditional independence statements. The following result shows that any conditional independence model is an algebraic statistical model.

Lemma 3.10. *The probability distribution $p = [p(x)]$ satisfies $A \perp\!\!\!\perp B|C$ for some disjoint $A, B, C \subset [m]$ if and only if*

$$p_{ABC}(x_A, x_B, x_C)p_{ABC}(x'_A, x'_B, x_C) -$$
$$p_{ABC}(x'_A, x_B, x_C)p_{ABC}(x_A, x'_B, x_C) = 0$$

holds for every $x_C \in \mathcal{X}_C$, $x_A, x'_A \in \mathcal{X}_A$ and $x_B, x'_B \in \mathcal{X}_B$.

Note that if $A \cup B \cup C = [m]$, then $A \perp\!\!\!\perp B | C$ is described by a quadratic equation

$$p(x_A, x_B, x_C)p(x'_A, x'_B, x_C) - p(x'_A, x_B, x_C)p(x_A, x'_B, x_C) = 0.$$

Therefore, every conditional independence model described only by conditional independencies of that form is toric, that is, defined by binomial equations; see Section 2.2.3. An important example is given in Section 3.4 by undirected graphical models.

3.1.3   Moments and moment aliasing

Let $\mathcal{X}$ be given as in (3.1) and let $\langle \cdot, \cdot \rangle$ be a scalar product on $\mathbb{R}^{\mathcal{X}}$. For any function $f : \mathcal{X} \to \mathbb{R}$ and any random variable $X \in \mathcal{X}$ with probability distribution p, the *expectation* of $f(X)$ is

$$\mathbb{E}[f(X)] \quad = \quad \langle p, f \rangle \quad = \quad \sum_{x \in \mathcal{X}} p(x)f(x).$$

In search of a convenient notation for dealing with moments, we first define a *multiset* as a set with multiplicities. More precisely, let A' be a set, called a *ground set*, and $f : A' \to \mathbb{N}_0$ be a function counting multiplicity of each element. Then the pair $A = (A', f)$ is a multiset. We typically write elements of a multiset as $a^{f(a)}$ ignoring the elements of multiplicity zero, so, for example, $\{1^3, 2^0, 3^1\} = \{1^3, 3^1\}$ is a multiset that we also sometimes write as $\{1, 1, 1, 3\}$. A *submultiset* of a multiset $A = (A', f)$ is a multiset $B = (B', g)$ such that B' is a subset of A' and $g(b) \leq f(b)$ for every $b \in B'$. So, for example, $\{1^2, 3^1\}$ is a submultiset of $\{1^3, 3^1\}$ but $\{1^2, 3^2\}$ is not.

Fix the ground set to be $[m]$. Then any multiset $A = \{i_1^{u_1}, \dots, i_k^{u_k}\}$ can be identified with a vector $(u_1, \dots, u_m)$ of multiplicities. Given A we define the corresponding *moment*

$$\mu_A \quad = \quad \mathbb{E}[\prod_{i=1}^{m} X_i^{u_i}]$$

and the *central moment*

$$\mu'_A \quad = \quad \mathbb{E}[\prod_{i=1}^{m} (X_i - \mu_i)^{u_i}].$$

In particular, $\mu_i = \mathbb{E}(X_i)$ is the mean of X_i, $\mu'_{ii} = \text{var}(X_i)$ is the variance of X_i and $\mu'_{ij} = \text{cov}(X_i, X_j)$ is the covariance between X_i and X_j. We always have $\mu'_i = 0$.

Example 3.11. Let X_1, X_2 be two binary random variables with the joint distribution $p = [p(i, j)]$. Then the covariance satisfies

$$\text{cov}(X_1, X_2) = p(1, 1) - p_1(1)p_2(1) = \det p.$$

In our geometric approach we can switch between probabilities and moments. To make it more formal, for each $u = (u_1, \ldots, u_m) \in \mathcal{X}$, denote the corresponding multiset as $\mathcal{A}(u)$, that is,

$$\mathcal{A}(u) = \{1^{u_1}, \ldots, m^{u_m}\} = \{\underbrace{1, \ldots, 1}_{u_1 \text{ times}}, \ldots, \underbrace{m, \ldots, m}_{u_m \text{ times}}\} \qquad (3.5)$$

and let

$$\mathcal{A}(\mathcal{X}) = \{\mathcal{A}(u) : u \in \mathcal{X}\}.$$

In particular, if $r_1 = \cdots = r_m = 1$, then $\mathcal{A}(\mathcal{X})$ is equal to the set of all subsets of $[m]$.

Let $u \in \mathbb{N}_0^m$ and $A = \mathcal{A}(u)$. A traditional way to compute the moment μ_A is using the *moment generating function*

$$M_X(t) = \mathbb{E}(\exp(\langle t, X \rangle)) = \sum_{x \in \mathcal{X}} p(x) \exp(\langle t, x \rangle),$$

where $t = (t_1, \ldots, t_m)$ and $\langle t, x \rangle = \sum_{i=1}^{m} t_i x_i$. We first compute the u-derivatives of M_X

$$D^u M_X(t) := \frac{\partial^{|u|}}{\partial t_1^{u_1} \cdots \partial t_n^{u_n}} M_X(t) = \mathbb{E}[X^u \exp(\langle t, X \rangle)].$$

If the moment generating function is analytic in the neighborhood of 0, then

$$\mathbb{E} X^u = \mu_{\mathcal{A}(u)} = D^u M_X(t) \Big|_{t=0}.$$

In this case we have

$$M_X(t) = \sum_{u \in \mathbb{N}_0^m} \frac{1}{u!} D^u M_X(t) \Big|_{t=0} t^u = \sum_{u \in \mathbb{N}_0^m} \frac{1}{u!} \mu_{\mathcal{A}(u)} t^u,$$

where $u! = u_1! \cdots u_n!$.

In the setting of Example 3.11, the moment generating function is

$$M_X(t) = p(0,0) + p(0,1)e^{t_2} + p(1,0)e^{t_1} + p(1,1)e^{t_1+t_2}.$$

We have

$$\frac{\partial}{\partial t_1} M_X(t) = p(1,0)e^{t_1} + p(1,1)e^{t_1+t_2}, \qquad \frac{\partial^2}{\partial t_1 \partial t_2} M_X(t) = p(1,1)e^{t_1+t_2}$$

and hence $\mu_1 = p(1,0) + p(1,1)$ and $\mu_{12} = p(1,1)$.

The cardinality of the set $\mathcal{A}(\mathcal{X})$ is $(r_1 + 1) \cdots (r_n + 1)$, which is equal to the cardinality of $\mathcal{X}$ and hence $\mathbb{R}^{\mathcal{A}(\mathcal{X})} \simeq \mathbb{R}^{\mathcal{X}}$. Consider two tensors in $\mathbb{R}^{\mathcal{A}(\mathcal{X})}$ given by moments $\mu = [\mu_{\mathcal{A}(x)}]_{x \in \mathcal{X}}$ and by central moments $\mu' = [\mu'_{\mathcal{A}(x)}]_{x \in \mathcal{X}}$.

Proposition 3.12 (The moment aliasing principle). *There exists a polynomial isomorphism between $p = [p(x)]_{x \in \mathcal{X}}$ and μ. The equation $\sum_{x \in \mathcal{X}} p(x) = 1$ becomes $\mu_\emptyset = 1$.*

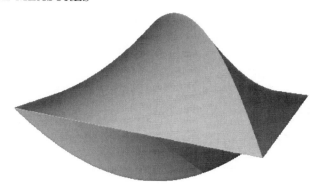

Figure 3.2 *The image of the probability simplex in the space of μ_1, μ_2 and the central moment μ'_{12}.*

In particular, every model $M \subseteq \Delta_{\mathcal{X}}$, after a change of coordinates, can be equivalently expressed in terms of μ. As we will see later in Section 4.2.1, the list of other helpful coordinate systems includes cumulants and their generalizations.

Example 3.13. Let $m = 2$, $r_1 = r_2 = 1$ and hence $\mathcal{X}_i = \{0, 1\}$ for $i = 1, 2$ and $\mathcal{A}(\mathcal{X}) = \{\emptyset, \{1\}, \{2\}, \{1, 2\}\}$. Consider the model of independence in Example 3.9. We leave it as an exercise to check that the determinant of the joint distribution matrix $p = [p(x)]_{x \in \mathcal{X}}$ is equal to the covariance μ'_{12}. This determinant vanishes if and only if $X_1 \perp\!\!\!\perp X_2$. Hence the independence model is linearized by the transformation to the central moments because it is described by a single equation $\mu'_{12} = 0$; see Figure 3.2.

The moment aliasing principle (see Pistone et al. [2001], Pistone and Wynn [2006]) implies that every moment μ_A, where A is any multiset of elements of $[m]$, can be written as a function of μ. This is particularly easy to see in the binary case when $r_1 = \cdots = r_m = 1$. In this case, $X_i^k = X_i$ for every i and for every $k \geq 1$ because $X_i \in \{0, 1\}$. Thus every monomial X^α is equal to $X^{\text{ind}(\alpha)}$, where the *indicator function* $\beta = \text{ind}(\alpha)$ such that $\beta_i = 1$ if $\alpha_i \neq 0$ and is zero otherwise. In particular, $\mu_A = \mu_{I(A)}$, where $I(A)$ is a subset of $[m]$ with the same elements as A (counted with multiplicity one).

We formulate the following alternative definition of independence; see Feller [1971], page 136.

Lemma 3.14. *Let X and $\mathcal{X}$ be as usual. For any two disjoint subsets $A, B \subseteq [m]$ we have $A \perp\!\!\!\perp B$, if and only if for every two functions $f : \mathbb{R}^{\mathcal{X}_A} \to \mathbb{R}$, $g : \mathbb{R}^{\mathcal{X}_B} \to \mathbb{R}$ we have*

$$\mathbb{E}[f(X_A)g(X_B)] = \mathbb{E}[f(X_A)]\mathbb{E}[g(X_B)].$$

The following formulation of independence in terms of moments will be helpful.

Lemma 3.15. *We have $A \perp\!\!\!\perp B$ for some disjoint sets $A, B \subseteq [m]$ if and only if*

$$\mu_{I \cup J} = \mu_I \mu_J \qquad \text{for all } I \in \mathcal{A}(\mathcal{X}_A), J \in \mathcal{A}(\mathcal{X}_B).$$

Proof. The "if" direction of the lemma is immediate by Lemma 3.14. The "only if" direction uses the fact that the set of values of X is discrete and finite. In this case any function on $\mathcal{X}$ is a polynomial function (can be represented as a polynomial in the entries of X), where the terms of these polynomials are $\prod_{i \in I} X_i$ for all $I \in \mathcal{A}(\mathcal{X})$. Thus, to check if $A \perp\!\!\!\perp B$, it remains to check if

$$\mathbb{E}[f(X_A)g(X_B)] = \mathbb{E}[f(X_A)]\mathbb{E}[g(X_B)]$$

for all polynomials f, g such that each f has only terms $\prod_{i \in I} X_i$ for all $I \in \mathcal{A}(\mathcal{X}_A)$ and g has only terms $\prod_{i \in J} X_i$ for all $J \in \mathcal{A}(\mathcal{X}_B)$. By expanding the terms of f and g and using linearity of the expectation, it suffices to check that this property holds for each monomial, which is true by assumption. □

Example 3.16. Suppose $\mathcal{X}_1 = \{0, 1\}$ and $\mathcal{X}_2 = \{0, 1, 2\}$. The distribution of $X = (X_1, X_2)$ is uniquely identified by the value of $\mu_1 = \mathbb{E}[X_1]$, $\mu_2 = \mathbb{E}[X_2]$, $\mu_{22} = \mathbb{E}[X_2^2]$, $\mu_{12} = \mathbb{E}[X_1 X_2]$ and $\mu_{122} = \mathbb{E}[X_1 X_2^2]$. We have $1 \perp\!\!\!\perp 2$ if and only if $\mu_{12} - \mu_1 \mu_2 = 0$ and $\mu_{122} - \mu_1 \mu_{22} = 0$.

Lemma 3.15 easily generalizes.

Lemma 3.17. *We have $B_1 \perp\!\!\!\perp B_2 \perp\!\!\!\perp \cdots \perp\!\!\!\perp B_r$ for some disjoint sets $B_1, \ldots, B_r \subseteq [m]$ if and only if*

$$\mu_{A_1 \cup \cdots \cup A_r} = \prod_{i=1}^{r} \mu_{A_i}, \qquad \text{for all } A_i \in \mathcal{A}(\mathcal{X}_{B_i}), i = 1, \ldots, r. \qquad (3.6)$$

One of the central concept of this book will be the *conditional expectation*. From the probabilistic point of view the conditional expectation should be thought of as the expectation with respect to the conditional distribution. Let X_1 and X_2 be two discrete random variables and $f : \mathcal{X}_1 \to R$. Then the *conditional expectation of $f(X_1)$ given $X_2 = x_2$* is defined as

$$\mathbb{E}[f(X_1)|X_2 = x_2] = \sum_{x_1 \in \mathcal{X}_1} f(x_1) \, p_{1|2}(x_1|x_2). \qquad (3.7)$$

For each $x_2 \in \mathcal{X}_2$ such that $p_2(x_2) > 0$ the right-hand side is a well-defined number. We should think of the conditional expectation as a function on $\mathcal{X}_2$, which we denote by $\mathbb{E}[X_1|X_2]$. Since X_2 is a random variable, $\mathbb{E}[X_1|X_2]$ is also a random variable with its own distribution. There are some immediate consequences of (3.7).

Proposition 3.18. *For any two functions $f : \mathcal{X}_1 \to \mathbb{R}$, $g : \mathcal{X}_2 \to \mathbb{R}$ and $a \in \mathbb{R}$,*

1. $\mathbb{E}[\mathbb{E}[f(X_1)|X_2]] = \mathbb{E}[f(X_1)]$,
2. $\mathbb{E}[f(X_1) - a|X_2] = \mathbb{E}[f(X_1)|X_2] - a$, *and*

3. $\mathbb{E}[f(X_1)g(X_2)|X_2] = g(X_2)\mathbb{E}[f(X_1)|X_2].$
Moreover, if $X_1 \perp\!\!\!\perp X_2$*, then*

$$\mathbb{E}[f(X_1)|X_2] = \mathbb{E}[f(X_1)].$$

Proof. Left as an exercise.   □

Since $\mathbb{E}[X_1|X_2]$ is a function on $\mathcal{X}_2$, it has a polynomial representation and this polynomial has degree r_2. It is instructive to write the conditional expectation in the binary case, which we then generalize in Chapter 4. Because X_2 is binary, the conditional expectation $\mathbb{E}(X_1|X_2)$ is a linear function of X_2, which without loss of generality can be written for some $a, b \in \mathbb{R}$ as

$$\mathbb{E}[X_1|X_2] \quad = \quad a + b(X_2 - \mathbb{E}[X_2]). \tag{3.8}$$

Using the first formulas in Proposition 3.18 we find that $a = \mathbb{E}[X_1]$. This can be further rewritten as

$$\mathbb{E}[X_1 - \mathbb{E}[X_1]|X_2] \quad = \quad b(X_2 - \mathbb{E}[X_2]).$$

Now multiply both sides by $X_2 - \mathbb{E}[X_2]$ and use the third formula in Proposition 3.18 to obtain

$$\mathbb{E}[(X_1 - \mathbb{E}[X_1])(X_2 - \mathbb{E}[X_2])|X_2] \quad = \quad b(X_2 - \mathbb{E}[X_2])^2.$$

Taking expectations yields

$$\mathbb{E}[(X_1 - \mathbb{E}[X_1])(X_2 - \mathbb{E}[X_2])] \quad = \quad b\mathbb{E}(X_2 - \mathbb{E}[X_2])^2,$$

which gives

$$b = \frac{\text{cov}(X_1, X_2)}{\text{var}(X_2)} \tag{3.9}$$

whenever X_2 is not a *degenerate random variable*, that is, if $\text{var}(X_2) > 0$. In this way we have obtained the formula for the conditional expectation when X_2 is a binary random variable:

$$\mathbb{E}[X_1|X_2] = \mathbb{E}[X_1] + \text{cov}(X_1, X_2)\text{var}(X_2)^{-1}(X_2 - \mathbb{E}[X_2]). \tag{3.10}$$

A general discrete random variable is non-degenerate if its distribution has no zeros. Suppose that X_i is degenerate. This means that X_i can never take at least one of its $r_i + 1$ possible values. In that case we can consider a copy of X_i with a reduced state space. Hence, without loss of generality, we always assume that X is a random vector such that all X_i are non-degenerate and in particular $\text{var}(X_i) > 0$. Now we can define $\bar{X} = (\bar{X}_1, \ldots, \bar{X}_m)$, where $\bar{X}_i = \frac{X_i - \mathbb{E}(X_i)}{\text{var}(X_i)}$. Let $\rho_{ij} = \text{corr}(X_i, X_j) = \mathbb{E}(\bar{X}_i\bar{X}_j)$ denote the correlation between X_i and X_j. In the binary case (3.10) can be rewritten as

$$\mathbb{E}[\bar{X}_1|\bar{X}_2] \quad = \quad \rho_{12}\bar{X}_2. \tag{3.11}$$

Let X_1, X_2, X_3 be binary random variables. If they are non-degenerate, then $X_1 \perp\!\!\!\perp X_3 | X_2$ implies

$$\rho_{13} = \rho_{12}\rho_{23}.$$

This can be further generalized. For a multiset A of elements of $\{1, \ldots, m\}$, define the *standardized moment* ρ_A by

$$\rho_A \quad = \quad \mathbb{E}[\prod_{i \in A} \bar{X}_i].$$

For example, $\rho_{ii} = 1$ and $\rho_{iii} = \mathbb{E}\bar{X}_i^3$ is the *skewness* of the distribution of X_i.

Lemma 3.19. *Let $X_1, \ldots, X_m$ be binary random variables. If they are non-degenerate, then $X_1 \perp\!\!\!\perp \cdots \perp\!\!\!\perp X_{m-1} | X_m$ implies*

$$\rho_A = \rho_{m \cdots m} \prod_{i \in A} \rho_{im} \quad \text{for every } A \subseteq \{1, \ldots, m\},$$

where $m \cdots m$ is a sequence of length $|A|$.

3.1.4 Identifiability of statistical models

Let $p(\theta) = (p(x; \theta))_{x \in \mathcal{X}}$ and let $\boldsymbol{M} = \{p(\theta) : \theta \in \Theta\}$ be a parametric statistical model. We say that $\boldsymbol{M}$ is *identifiable* if and only if $p(\theta_0) = p(\theta_1)$ implies that $\theta_0 = \theta_1$.

Example 3.20. Let $\mathcal{X} = \{0, \ldots, n\}$ and consider the binomial model $\text{Bin}(n, \theta)$ parameterized by

$$p(x; \theta) \quad = \quad \binom{n}{x}(1-\theta)^{n-x}\theta^x \qquad \text{for } x \in \mathcal{X}, \ \theta \in [0, 1].$$

The fact that this model is identifiable follows from the fact that $p(n; \theta) = \theta^n$ and hence $p(n; \theta_0) = p(n; \theta_1)$ already implies that $\theta_0 = \theta_1$.

Remark 3.21. Establishing identifiability is important since it informs, for example, how the maximum likelihood method or Bayes estimation might behave; see e.g., Rothenberg [1971]. For instance, in Bayesian statistics identifiability affects the asymptotics for the posterior distribution of parameters; see e.g., Gustafson [2009], Kadane [1974], Poirier [1998], Sahu and Gelfand [1999]. Moreover, if the model is not identified, the posterior analysis may strongly depend on the prior specification even for small sample sizes.

In our algebraic setting, identifiability can be formulated as follows. Let $\boldsymbol{M}$ be an algebraic discrete parametric model given by the map $p : \Theta \to \Delta_{\mathcal{X}}$. Then $\boldsymbol{M}$ is *identifiable* if and only if the map p is one-to-one onto its image. In other words, if $\theta_1 \neq \theta_2$, then $p(x; \theta_1) \neq p(x; \theta_2)$ for some $x \in \mathcal{X}$. In practice, there is another useful concept related to identifiability.

Definition 3.22. A model $\boldsymbol{M}$ is *locally identifiable* if for each θ_0 there exists a neighborhood W of θ_0 in Θ such that for every $\theta \in W$ we have $P_\theta = P_{\theta_0}$ if and only if $\theta = \theta_0$.

A necessary condition for local identifiability is that

$$p^{-1}(q) \quad = \quad \{\theta \in \Theta : p(\theta) = q\}$$

is a finite set for each $q \in M$. This motivates the following definition.

Definition 3.23. Let $q \in M = p(\Theta)$. The preimage $\Theta_q := p^{-1}(q) \subseteq \Theta$, that is, the set of values of the parameters which are consistent with the known probability model q, is called the *q-fiber*.

Although for many parametric statistical models used in practice identifiability does not hold, the following milder version may still hold.

Definition 3.24. We say that a model is *generically identifiable* if the set of points q whose q-fiber is not finite forms a lower dimensional subvariety of the model.

The q-fiber is always a semialgebraic set by Proposition 2.33 and the fact that $p : \Theta \to \Delta_{\mathcal{X}}$ is a polynomial map. The geometry of q-fibers affects statistical inference. In particular, we are often interested when q-fibers are finite, when they are smooth subsets of Θ, and when they are singular. These ideas link to the singular learning theory. We refer to Watanabe [2009] for details and to Drton and Plummer [2013] for more recent ideas.

Example 3.25. Consider again the binomial model in Example 3.20. Because the model is identifiable, the preimage of every $[p(x; \theta)]$ in the model has exactly one point. Consider now the mixture of this model like in Example 1.2. This model is not (locally, generically) identifiable. Geometrically, it can be seen in Figure 1.3 that every point in the interior of the shaded area can be written as a convex combination of many pairs of points in the binomial model.

An example of a model that is generically identifiable but not identifiable is given by latent tree models, which we analyze in the second part of the book. Establishing generic identifiability for various models with latent data became an important part of tensor decomposition methods; see for example Allman et al. [2009].

3.2 Exponential families and their mixtures

Discrete exponential families form a very important family of discrete statistical models. Discrete exponential families are usually described by a parameterization of the form

$$p(x; \theta) \quad = \quad q(x) \exp(\sum_{i=1}^{d} \theta_i a_i(x) - \log Z(\theta)), \quad \theta \in \mathbb{R}^d, \qquad (3.12)$$

where $x \in \mathcal{X}$, $Z(\theta)$ is a normalizing constant, q is a *base measure*, and $A = [a_i(x)]$ is called the *sufficient statistic*. For given A, the corresponding discrete exponential family is denoted by $\mathcal{E}_A$. The discrete exponential family $\mathcal{E}_A$ is unchanged under the row operations on A. Also, because of the normalization, we can without loss assume that the vector $(1, \ldots, 1)$ is one of the rows of A.

The name and representation in (3.12) suggests that these discrete models are not algebraic. However, it turns out that the model can be reformulated so that its description is given by polynomials. Let $A = [a_i(x)]$ be a matrix with $|\mathcal{X}|$ columns such that all $a_i(x)$ are integers and $a(x) = [a_i(x)]$ is the column of A related to $x \in \mathcal{X}$. Define the set of probability distributions given as the image of the following map

$$\phi_A : \quad \mathbb{R}^d_> \to \Delta_{\mathcal{X}}, \quad \mathbf{t} = (t_1, \ldots, t_d) \mapsto Z(\mathbf{t})^{-1} \left(q(x)\mathbf{t}^{a(x)} \right)_{x \in \mathcal{X}}, \qquad (3.13)$$

where $\mathbf{t}^{a(x)} = \prod_{i=1}^d t_i^{a_i(x)}$. To see that classes of models in (3.12) and (3.13) are identical, apply the transformation $t_i = \exp(\theta_i)$.

Remark 3.26. The representation of closures of discrete exponential families is induced from (3.13) by considering $\phi_A : \mathbb{R}^d_{\geq 0} \to \Delta_{\mathcal{X}}$. In this case we adopt the convention that $\mathbf{t}^0 = 1$ for $\mathbf{t} \geq 0$. By Proposition 2.39 these families of probability distributions correspond to the nonnegative parts of toric varieties. This enables us to investigate the boundary of the closure of exponential families using the moment map introduced in Section 2.2; see [Kähle, 2010, Theorem 1.2.11].

By Remark 2.40, the semialgebraic description of discrete exponential families is obtained from the algebraic description of their Zariski closure by adding the nonnegativity constraints defining the probability simplex.

Proposition 3.27. *Let M be a discrete exponential family defined by ϕ_A as in (3.13) and let $\overline{M}$ be the Zariski closure of M. Then*

$$M = \overline{M} \cap \Delta^0_{\mathcal{X}}.$$

In particular, the only inequalities describing M as a semialgebraic set are trivial inequalities.

Proof. Since $M \subseteq \overline{M}$ and $M \subseteq \Delta^0_{\mathcal{X}}$, one inclusion is immediate. For the opposite inclusion, note that $\overline{M} \subseteq \mathbb{P}^{\mathcal{X}}$ is an image of a toric variety Y_A obtained by scaling each coordinate by $q(x) > 0$. This toric variety is parameterized by $\mathbf{t} \mapsto (\mathbf{t}^{a(x)})_{x \in \mathcal{X}}$. By Proposition 2.39, the nonnegative part $(Y_A)_{\geq}$ of this variety is the Euclidean closure of the image of $\tilde{\phi} : \mathbb{R}^d_> \to \mathbb{RP}^{\mathcal{X}}_{\geq 0}$, or equivalently it is equal to $Y_A \cap \mathbb{RP}^{\mathcal{X}}_{\geq 0}$. Each of the points in $(Y_A)_{\geq} \setminus \tilde{\phi}(\mathbb{R}^d_>)$ has at least one coordinate zero. Therefore, all points in $Y_A \cap \mathbb{RP}^{\mathcal{X}}_{>0}$ lie necessarily in $\tilde{\phi}(\mathbb{R}^d_>)$. Scaling again by $q(x)$ gives us that $\overline{M} \cap \mathbb{RP}^{\mathcal{X}}_{>0}$ is given by $\phi_A(\mathbb{R}^d_>)$. Finally, by Lemma 2.37, $\mathbb{RP}^{\mathcal{X}}_{>0} \simeq \Delta^0_{\mathcal{X}}$. $\qquad\square$

In this context it is easy to explain the name of the moment map in (2.6). The sufficient statistics A is a random variable with values $a(x)$ for $j \in \mathcal{X}$ obtained with probability $p(x; \mathbf{t}) = Z^{-1}(\mathbf{t})\mathbf{t}^{a(x)}$. Its mean $\mathbb{E}A$ is

$$\mathbb{E}A \quad := \quad \sum_{x \in \mathcal{X}} p(x; \mathbf{t})a(x) \quad = \quad \sum_{a \in \mathcal{A}} p(a; \mathbf{t})a$$

where $\mathcal{A} = \{a(x) : x \in \mathcal{X}\}$ and $p(a(x); \mathbf{t}) := p(x; \mathbf{t})$. This expectation map is precisely the algebraic moment map defined in (2.6). In the theory of exponential families, the parameterization in terms of the moment of A is called the *mean parameterization*. From the general properties of the exponential families it follows that for the *cumulant function* $K(\theta) = \log Z(\theta)$ we have

$$\mathbb{E}A = \nabla_\theta K(\theta) = Z^{-1}(\theta)\nabla_\theta Z(\theta).$$

By Proposition 2.19, to obtain the implicit description of a discrete exponential family, it suffices to identify the defining ideal $\ker(\phi_A^*)$. In the statistical context, when the input data is given by the A matrix, this can be done efficiently using the computer software 4ti2; see Hemmecke et al. [2005].

Example 3.28. Let $X = (X_1, X_2, X_3, X_4)$ be a vector of four binary random variables, so that $\mathcal{X}_i = \{0, 1\}$ for all i. Consider an exponential family with the parameterization $\phi_A : \mathbb{R}_{>}^{12} \to \Delta_\mathcal{X}$ defined by

$$A = \begin{bmatrix}
1 & 0 & 1 & 0 & 1 & 0 & 1 & 0 & 0 & 0 & 0 & 0 & 0 & 0 & 0 & 0 \\
0 & 1 & 0 & 1 & 0 & 1 & 0 & 1 & 0 & 0 & 0 & 0 & 0 & 0 & 0 & 0 \\
0 & 0 & 0 & 0 & 0 & 0 & 0 & 0 & 1 & 0 & 1 & 0 & 1 & 0 & 1 & 0 \\
0 & 0 & 0 & 0 & 0 & 0 & 0 & 0 & 0 & 1 & 0 & 1 & 0 & 1 & 0 & 1 \\
1 & 0 & 1 & 0 & 0 & 0 & 0 & 0 & 1 & 0 & 1 & 0 & 0 & 0 & 0 & 0 \\
0 & 1 & 0 & 1 & 0 & 0 & 0 & 0 & 0 & 1 & 0 & 1 & 0 & 0 & 0 & 0 \\
0 & 0 & 0 & 0 & 1 & 0 & 1 & 0 & 0 & 0 & 0 & 0 & 1 & 0 & 1 & 0 \\
0 & 0 & 0 & 0 & 0 & 1 & 0 & 1 & 0 & 0 & 0 & 0 & 0 & 1 & 0 & 1 \\
1 & 0 & 0 & 0 & 1 & 0 & 0 & 0 & 1 & 0 & 0 & 0 & 1 & 0 & 0 & 0 \\
0 & 1 & 0 & 0 & 0 & 1 & 0 & 0 & 0 & 1 & 0 & 0 & 0 & 1 & 0 & 0 \\
0 & 0 & 1 & 0 & 0 & 0 & 1 & 0 & 0 & 0 & 1 & 0 & 0 & 0 & 1 & 0 \\
0 & 0 & 0 & 1 & 0 & 0 & 0 & 1 & 0 & 0 & 0 & 1 & 0 & 0 & 0 & 1
\end{bmatrix}.$$

This model is a special graphical model that plays an important role in this book; see also Example 5.40. Here the columns correspond to elements of $\mathcal{X}$ ordered lexicographically so that $(0, 0, 0, 0)$ corresponds to the first column, $(0, 0, 0, 1)$ to the second, $(0, 0, 1, 0)$ to the third, and finally $(1, 1, 1, 1)$ to the last column. The first block of rows corresponds to parameters s_{00}, s_{01}, s_{10}, and s_{11}, respectively. The second block of rows corresponds to parameters t_{00}, t_{01}, t_{10}, and t_{11}. The last block corresponds to parameters u_{00}, u_{01}, u_{10}, and u_{11}. The first column is the vector of exponents of the parameter vector in the expression for $p((0, 0, 0, 0); \mathbf{t})$. The other columns are obtained in the same way. The Zariski closure of the model can be described by quadratic equations of the form

$$p(0, 0, 1, 1)p(1, 1, 0, 1) - p(0, 1, 0, 1)p(1, 0, 1, 1) = 0.$$

The complete list can be easily obtained using 4ti2.

Mixtures of exponential families A powerful approach to probabilistic modeling involves supplementing a set of observed variables with additional hidden variables. By defining a joint distribution over visible and latent variables, the corresponding distribution of the observed variables is then obtained by marginalization. This allows relatively complex distributions to be expressed in terms of more tractable joint distributions over the expanded variable space. One well-known example of a hidden variable model is the mixture distribution in which the hidden variable is the discrete component label.

Note that mixtures of discrete exponential families are also algebraic models. Suppose that the system under consideration consists of both the observed variables X and the hidden variables H. Let $(X, H) \in \mathcal{X} \times \mathcal{H}$, where

$$\mathcal{H} = \{0, \ldots, s_1\} \times \cdots \times \{0, \ldots, s_n\}.$$

Suppose that the model $p(x, h; \theta)$ for (X, H) is a discrete exponential family. We call it a *fully observed model*. In practice we never observe H and therefore we are interested in the family $p(x; \theta)$ of possible marginal probability distributions of X, where

$$p(x; \theta) \quad = \quad \sum_{h \in \mathcal{H}} p(x, h; \theta) \quad \text{for all } x \in \mathcal{X}.$$

Geometrically, the model for X is a projection of a toric variety defining the model for (X, H). It is significantly harder to describe than a discrete exponential family, and in particular, there is no equivalent of Proposition 3.27. A simple example is given in Example 1.2. A more general example is given by latent tree models, which are studied in the second part of this book.

Example 3.29 ($X_1 \perp\!\!\!\perp X_2 | H$). Consider a simple model with $\mathcal{X} = \{0, 1\}^2$, $\mathcal{H} = \{0, 1\}$ given by

$$p(x_1, x_2, h) = p_H(h) p_{1|H}(x_1|h) p_{2|H}(x_2|h),$$

where p_H denotes the marginal distribution of H and $p_{i|H}$ for $i = 1, 2$ denotes the conditional distribution of X_i given H. Since H is hidden, we are interested in the marginal distribution of (X_1, X_2). Thus the parameterization is given by a polynomial map, $p : [0, 1]^5 \to \Delta_{\mathcal{X}}$ given for each $x_1, x_2 \in \{0, 1\}$ by

$$p_{12}(x_1, x_2) = p_H(0) p_{1|H}(x_1|0) p_{2|H}(x_2|0) + p_H(1) p_{1|H}(x_1|1) p_{2|H}(x_2|1),$$

where the five free parameters are $p_H(1)$, $p_{1|H}(1|h)$ and $p_{2|H}(1|h)$ for $h \in \{0, 1\}$. As it is shown by Gilula [1979], every possible probability distribution for (X_1, X_2) can be obtained in this way for some parameters or in other words $p([0, 1]^5) = \Delta_{\mathcal{X}}$. Using a more statistical language we say that the model is *saturated*.

3.3 Maximum likelihood of algebraic models

3.3.1 The likelihood function

Let X be a finite discrete random variable with values in $\mathcal{X}$ and probability distribution $q \in M \subseteq \Delta_{\mathcal{X}}$, where M is a parametric algebraic statistical model with parameterization $p(x; \theta)$. In particular, there exists $\theta^* \in \Theta$ such that

$$q(x) \quad = \quad p(x; \theta^*)$$

for every $x \in \mathcal{X}$. Suppose that we observe a *random sample* $X^{(1)} = x^{(1)}, \ldots, X^{(n)} = x^{(n)}$ of n independent copies of X from $q \in \boldsymbol{M}$. The aim of statistical inference is to obtain valid estimates of the "true" distribution q or equivalently a "true" parameter θ^* based on the observed data. In this section we focus on the maximum likelihood inference.

Because observations $X^{(1)} = x^{(1)}, \ldots, X^{(n)} = x^{(n)}$ are independent and identically distributed, the probability of observing this sample can be rewritten as

$$\prod_{i=1}^{n} p(x^{(i)}; \theta) \quad = \quad \prod_{x \in \mathcal{X}} p(x; \theta)^{u(x)}, \tag{3.14}$$

where

$$u(x) \quad = \quad \#\{i : x^{(i)} = x\}$$

are entries of the *contingency table* $u = [u(x)]$. By construction $u(x) \geq 0$ and $\sum_{x \in \mathcal{X}} u(x) = n$ and hence $u \in \mathcal{U} = \mathcal{U}(\mathcal{X}, n)$ as defined in (2.14).

Remark 3.30. The assumption that our data are collected in this way is just a convention called the *multinomial sampling* assumption. It is, however, often done in statistics and we are not going to consider other *sampling schemes*.

The most popular method to make inference about q or θ^* is the *maximum likelihood method*. The idea is to estimate θ^* by the parameter maximizing the probability of observing random sample $x^{(1)}, \ldots, x^{(n)}$. By (3.14) this means maximizing the likelihood function $L : \Theta \to \mathbb{R}$ given by

$$L(\theta; u) \quad = \quad \prod_{x \in \mathcal{X}} p(x; \theta)^{u(x)}. \tag{3.15}$$

The global maximum of $L(\theta; u)$, if it exists, is called the *maximum likelihood estimator* (or the *MLE*) and is denoted by $\hat{\theta}$.

To maximize $L(\theta; u)$ it is convenient to take the logarithm of (3.15), which does not change the optima. The resulting *log-likelihood function* is

$$\ell(\theta; u) \quad = \quad \sum_{x \in \mathcal{X}} u(x) \log p(x; \theta) \qquad \text{for all } \theta \in \Theta \subseteq \mathbb{R}^d,$$

and hence it is a linear function in $\lambda(x; \theta) := \log p(x; \theta)$. Define the *sample proportions* $\hat{p}$ by

$$\hat{p}(x) \quad = \quad \frac{u(x)}{n} \qquad \text{for all } x \in \mathcal{X}.$$

For a fixed sample size n, there is a one-to-one correspondence between $\hat{p}$ and u. We then can write

$$\ell(\theta; \hat{p}) \quad = \quad n \langle \lambda(\theta), \hat{p} \rangle \qquad \text{for all } \theta \in \Theta \subseteq \mathbb{R}^d,$$

where $\langle \lambda, p \rangle = \sum_{x \in \mathcal{X}} \lambda(x) p(x)$ denotes the usual scalar product in $\mathbb{R}^{\mathcal{X}}$. To maximize, we compute the derivatives

$$\frac{\partial \ell(\theta; \hat{p})}{\partial \theta_i} = n \sum_{x \in \mathcal{X}} \frac{\hat{p}(x)}{p(x; \theta)} \frac{\partial p(x; \theta)}{\partial \theta_i} \quad \text{for } i = 1, \ldots, d, \qquad (3.16)$$

and if the global maximum lies in the interior of Θ, then it is given by the solution of

$$\frac{\partial \ell(\theta; \hat{p})}{\partial \theta_i} = 0 \qquad \text{for all } i.$$

The maximum likelihood estimate can be alternatively obtained by minimizing the *Kullback–Leibler divergence*. For two discrete probability distributions over $\mathcal{X}$ that are strictly positive we define the Kullback–Leibler divergence $K(p||q)$ as

$$K(p||q) \quad := \quad \sum_{x \in \mathcal{X}} \log\left(\frac{p(x)}{q(x)}\right) p(x).$$

It is well known that $K(p||q) \geq 0$ and is zero precisely when $p(x) = q(x)$ for all $x \in \mathcal{X}$. The following proposition links the Kullback–Leibler divergence to maximum likelihood estimation.

Proposition 3.31. *Let $M = \{p(\theta) : \theta \in \Theta\}$ be a discrete model over $\mathcal{X}$ parameterized by the set Θ. Let $\hat{p}$ be sample proportions obtained from a random sample from some true distribution. Then θ^* maximizes $\ell(\theta; \hat{p})$ if and only if it minimizes $K(\hat{p}||p(\theta))$.*

Proof. Directly from the definition

$$K(\hat{p}||p(\theta)) = \sum_{x \in \mathcal{X}} \log(\hat{p}(x))\hat{p}(x) - \sum_{x \in \mathcal{X}} \log(q(x; \theta))\hat{p}(x).$$

The first term does not depend on θ and the second term is $-\frac{1}{n}\ell(\theta; \hat{p})$. $\square$

A special case of (3.15) is when M is equal to the interior of $\Delta_{\mathcal{X}}$, which we denote by $\Delta_{\mathcal{X}}^0$. Then $\Theta = \Delta_{\mathcal{X}}^0$ and we chose the parameters to be $\bar{\lambda} = [\bar{\lambda}(x)]$, where $\bar{\lambda}(x) = \lambda(x) - \lambda(\mathbf{0})$. In this case the log-likelihood function, called the *multinomial log-likelihood*, is given by

$$\ell(\bar{\lambda}; \hat{p}) \quad = \quad n\langle \lambda, \hat{p} \rangle \quad = \quad n(\langle \bar{\lambda}, \hat{p} \rangle - K(\bar{\lambda})), \qquad (3.17)$$

where $K(\bar{\lambda}) = -\lambda(\mathbf{0}) = \log[1 + \langle \mathbf{1}, \exp(\bar{\lambda}) \rangle]$ is the cumulant function and hence it is convex.

Proposition 3.32. *Given $\hat{p}(x) > 0$ for all $x \in \mathcal{X}$, the multinomial log-likelihood is a concave function with the unique maximum p^* given by the sample proportions, that is, $p^* = \hat{p}$.*

Proof. Since $K(\bar{\lambda})$ is convex, $\ell(\bar{\lambda}; \hat{p})$ is concave in $\bar{\lambda}$. Therefore it has at most one maximum. The form of the maximum can be obtained by differentiation. $\square$

3.3.2 MLE for exponential families

An important feature of discrete exponential families is that they admit a unique MLE. In this case the likelihood can be written as

$$L(\mathbf{t}; u) \quad = \quad Z(\mathbf{t})^{-1} \mathbf{t}^{A \cdot u},$$

where $u : \mathcal{X} \to \mathbb{N}_0^{\mathcal{X}}$ is the tensor of sample counts and $t_i = e^{\theta_i}$. A concept closely related to the maximum likelihood estimation for exponential families is that of the *maximum entropy principle*. For every probability distribution p, define its *Shannon entropy*, as

$$H(p) \quad := \quad -\sum_{x \in \mathcal{X}} p(x) \log p(x).$$

The following result is the central result of the theory of exponential families.

Theorem 3.33 (Maximum entropy principle). *Let $\hat{p}$ be the sample proportions such that $\hat{p}(x) > 0$ for all $x \in \mathcal{X}$. If $\mathbf{M}$ is a discrete exponential family, then the likelihood function always has a unique maximum. It is given by the unique point p^* in the model satisfying $A\hat{p} = Ap^*$.*

Proof. See, for example, Wainwright and Jordan [2008], Theorem 3.4. □

Note that in Theorem 3.33 we do not require p^* to lie in $\mathbf{M}$. The fact that indeed $p^* \in \mathbf{M}$ is part of the theorem. This also shows that we obtain two equivalent ways to describe the MLE for discrete exponential families. *Remark* 3.34. Theorem 3.33 is true also if we extend $\mathbf{M}$ by adding its limiting distributions.

In many interesting situations there exist closed form formulas for the unique MLE in Theorem 3.33. In the case when such a formula is not available a popular way to find the maximum likelihood estimate is to use the *Iterative Scaling* algorithm; see Darroch and Ratcliff [1972].

3.3.3 Constrained multinomial likelihood

There are two alternative ways of thinking about the likelihood function of a parametric algebraic statistical model $\mathbf{M}$. The first is to define it as a function on Θ like in (3.15). The second uses that fact that the model $\mathbf{M}$ is parameterized by $\Theta \to \mathbf{M} \subseteq \Delta_{\mathcal{X}}$. Write the multinomial likelihood

$$L_m(p; u) \quad = \quad \prod_{x \in \mathcal{X}} p(x)^{u(x)}.$$

Comparing this with (3.15) shows that alternatively the maximum likelihood estimate θ^* can be obtained by finding a constrained maximum p^* of the multinomial likelihood constrained to the image of Θ in the probability simplex and then mapping p^* back to Θ. In this case we call the likelihood function the *constrained multinomial likelihood*.

These two ways of thinking about the multinomial likelihood can be useful when we toggle between the parametric and implicit representation of the model. It gives a good framework for understanding the model selection problems. Thinking about the likelihood function as a constrained multinomial likelihood function also gives intuition about how intrinsic multimodality arises in certain type of models. For example, even if the constrained multinomial likelihood is unimodal but the parameterization of the model is not one-to-one, the likelihood function of the model will not be unimodal; see Example 3.39 below. In that case the best solution is to reparameterize the model to get rid of this non-uniqueness.

The following proposition follows immediately from the observation that M is a subset of $\Delta_\mathcal{X}$.

Proposition 3.35. *Let M be given by $p : \Theta \to \Delta_\mathcal{X}$ be an algebraic parametric model. If $\hat{p} \in M$ such that all entries of $\hat{p}$ are strictly positive, then any point in the $\hat{p}$-fiber $\Theta_{\hat{p}}$ (c.f. Definition 3.23) is the maximum likelihood estimator for the unknown true parameter θ^*.*

Typically, M has a much smaller dimension than the ambient probability simplex $\Delta_\mathcal{X}$. In this case, $\hat{p} \notin M$ almost surely and consequently we cannot use the above proposition. However, this insight will be useful anyway for our understanding of the likelihood geometry. An example is the following basic result.

Proposition 3.36. *Let M be a statistical model whose intersection with $\Delta_\mathcal{X}^0$ is non-empty. If $\hat{p} > 0$, then the maximum of the multinomial likelihood constrained to M lies in the interior of the probability simplex.*

Proof. By Proposition 3.32 the multinomial likelihood is concave and has a unique maximum at $\hat{p}$. Its value goes to $-\infty$ at the boundary of the probability simplex. Therefore, the constrained maximum must be attained in $\Delta_\mathcal{X}^0$. $\square$

We now discuss a number of examples in which we give a brief overview of possible obstacles in the maximum likelihood estimation.

Example 3.37. Consider the $\text{Bin}(2, \theta)$ model from Example 1.1. Suppose that the experiment was repeated n times out of which u_0 times no success was observed, u_1 times exactly one success was observed, and u_2 times two successes were observed. The multinomial likelihood in this case is given by

$$L_m(p; u) = p_0^{u_0} p_1^{u_1} p_2^{u_2}.$$

The model is given in $\Delta_\mathcal{X}$ by the single equation $p_1^2 - 4p_0p_1 = 0$. Project everything on $\mathbb{R}^2$ by using the fact that $p_0 = 1 - p_1 - p_2$. The probability simplex $\Delta_\mathcal{X}$ becomes a triangle $p_1, p_2 \geq 0$, $p_1 + p_2 \leq 1$. Consider three situations: $u = [8, 6, 5]$, $u = [6, 3, 5]$, and $u = [6, 10, 6]$. The multinomial likelihood and the corresponding constrained multinomial likelihood are depicted in Figure 3.3. In the last case, the MLE under the constrained model lies very close to the multinomial MLE. In the second case, both estimators lie far apart.

Observe also that both unconstrained and constrained likelihood functions have unique maxima; see Theorem 3.33.

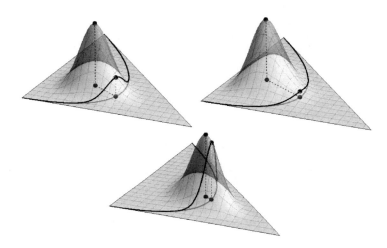

Figure 3.3 *The multinomial likelihood and the constrained multinomial likelihood on the* $\mathrm{Bin}(2,\theta)$ *model (given by the solid black).*

Remark 3.38. The ratio between the multinomial (unconstrained) likelihood at the maximum $\hat{p}$ and the constrained likelihood at its maximum $\hat{\theta}$ gives some evidence of how likely our model is to hold. A formal treatment of this idea is given by the *likelihood ratio* statistics. For a geometric insight into this notion; see Fan et al. [2000].

For more general models with non-trivial inequality constraints, the situation may of course get more complicated even in this low-dimensional case.

Example 3.39. Consider the mixture model from Example 1.2. The model is given in Δ_3 by a single inequality $p_1^2 - 4p_0p_2 \leq 0$. Consider three situations: $u = [8, 6, 5]$, $u = [3, 9, 5]$, and $u = [4, 20, 8]$. The multinomial likelihood and the corresponding constrained multinomial likelihood are depicted in Figure 3.4. In the first situation, the multinomial likelihood and the constraint multinomial likelihood share the maximum. In the second situation, the global and the constraint maxima are different but close to each other. In the last situation, the multinomial likelihood has its unique maximum far outside the constrained model space. In the last two cases, the constrained maximum lies on the boundary of the model space. Thus, we have two possible situations; either the global and the constrained optima of the multinomial likelihood function coincide or not. In the second situation, the constrained optimum lies always on the boundary of the constrained model space, which is given by $p_1^2 - 4p_0p_1 = 0$ and hence corresponds to the binomial model. The maximum of the constrained multinomial likelihood function is always uniquely defined, however, the likelihood function of the model is more complicated.

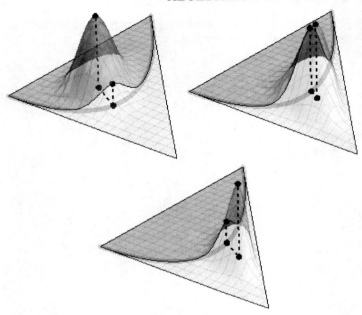

Figure 3.4 *The multinomial likelihood and the constrained multinomial likelihood on the mixture model of the binomial model* $\mathrm{Bin}(2, \theta)$.

In Example 3.39 the situation is still relatively simple. The constrained model forms a convex set. In this case the maximum of the constrained multinomial likelihood is always unique by the following result.

Proposition 3.40. *Let L be the multinomial likelihood $L_m : \Delta_\mathcal{X} \to \mathbb{R}$ constrained to a convex set $\boldsymbol{M} \subseteq \Delta_\mathcal{X}$. If $\hat{p} > 0$, then L has a unique maximum. If $\boldsymbol{M}$ is full dimensional, then this maximum lies in the interior of $\boldsymbol{M}$ if L_m has a global maximum in $\boldsymbol{M}$ or on the boundary of $\boldsymbol{M}$ otherwise.*

Proof. Since L_m is strictly concave over $\Delta_\mathcal{X}^0$, it remains strictly concave over any convex subset. For details; see for example [Davis–Stober, 2009, Theorem 1] or [Boyd and Vandenberghe, 2004, Section 4.2.2].   □

In this book we are interested in models for which finding the maximum likelihood estimator is far more challenging than in the above examples. The two main reasons are that these models are non-convex subsets of $\Delta_\mathcal{X}$ and their dimension is typically much smaller than the dimension of the ambient space. The problem is not only that the likelihood function is multimodal but also that we can have many distant maxima that give similar value of the likelihood function. A similar problem will arise in Bayesian statistics when studying the maximum aposteriori estimator.

3.3.4 Hidden variables and the EM algorithm

Let $(X, H) \in \mathcal{X} \times \mathcal{H}$ be a random vector, where X is its observed part and H is hidden. For simplicity, assume that the model $p(x, h; \theta)$ for (X, H) is a discrete exponential family. Our task is to maximize the likelihood function $L(\theta; u)$ in (3.15) for the marginal model of X. In the presence of hidden data, even in simple cases the numerical maximization can be hard, as explained in the previous section. On the other hand, maximizing the fully observed likelihood

$$L(\theta; x, h) = \prod_{(x,h) \in \mathcal{X} \times \mathcal{H}} p(x, h; \theta)^{u(x,h)}, \qquad (3.18)$$

is relatively simple because it always has a unique solution by Theorem 3.33. The maximum can be found using a closed form formula if such a formula is available, or numerically by the Iterative Scaling algorithm.

Of course, in practice, h is never known. The idea of the EM algorithm is as follows. First, make an initial guess on the parameter vector θ. For this fixed θ we compute the expectation of $u(x, H = h)$ given the observed data. This is called the expectation step (E-step). Now we maximize the fully observed likelihood this estimation of the complete data table $u(x, h)$. This step is called the *maximization step* (or M-step). Let θ^* be the optimal solution found at the M-step. We now replace this value for θ and repeat the E-step. We repeat this procedure until we reach some optimality criterion, typically, when changes in the estimated value of θ in subsequent M-steps are very small.

Algorithm 3.41 (EM Algorithm).
Input: A tensor $p(x, h; \theta)$ of polynomials in θ for $(x, h) \in \mathcal{X} \times \mathcal{H}$ representing the fully observed model and the observed data $u(x)$ for $x \in \mathcal{X}$.
Output: A local maximum $\hat{\theta} \in \Theta$ of the likelihood function $L(\theta; u)$.
Step 0:   Select a threshold $\epsilon > 0$ and select starting parameters $\theta_0 \in \Theta$ satisfying $p(x, h; \theta_0) > 0$ for all $(x, h) \in \mathcal{X} \times \mathcal{H}$.
E-step: Define the *expected hidden data* by

$$u(x, h) \quad := \quad u(x)p(h|x; \theta) \quad = \quad \frac{u(x)}{p(x; \theta_0)} p(x, h; \theta_0)$$

for all $(x, h) \in \mathcal{X} \times \mathcal{H}$.
M-step: Find the unique maximum $\theta^* \in \Theta$ of the fully observed likelihood function in (3.18).
Step 3:   If $L(\theta^*; x) - L(\theta; u) > 0$, then set $\theta_0 := \theta^*$ and go back to the E-step.
Step 4:   Output the parameter $\hat{\theta} := \theta^*$ and the corresponding probability distribution $p(x; \hat{\theta})$ for $x \in \mathcal{X}$.

What is less obvious is that in every iteration, the likelihood increases.

Theorem 3.42 (Dempster et al. [1977]). *The value of the likelihood function $L(\theta; u)$ strictly increases with each iteration of the EM-algorithm until a local maximum is attained.*

This result has good and bad implications. On one hand, we are guaranteed to always find a local optimum. On the other hand, we often get stuck in a local minimum, and to find the global maximum we need to repeat the procedure many times from different starting points. For this reason it is important to study fixed points of the EM algorithm for interesting model classes and develop procedures to obtain good starting points.

3.4   Graphical models

Graphical models form a popular family of statistical models, which enable us to efficiently encode large sets of conditional independence statements in terms of graphs. The aim of this chapter is to introduce only very basic models and elementary related concepts. We focus on two model classes: *undirected graphical models*, also known as *Markov random fields*, and *directed graphical models*, also known as *Bayesian networks*. To study the latter we also briefly discuss the *chain graph models*. Most of this section is based on Lauritzen [1996]. We refer there for the proofs and a more detailed discussion.

3.4.1   Basic graphical notions

Let $G = (V, E)$ be a graph with set of vertices V and set of edges $E \subseteq V \times V$. We always assume there are no loops and no multiple edges. A directed edge from i to j is denoted by $i \rightarrow j$ and an undirected edge between i and j is denoted by $i - j$. We write $i \cdots j$, and say that i and j are *linked*, whenever we mean that either $i \rightarrow j$, or $i \leftarrow j$, or $i - j$. A *hybrid graph* is a graph with both directed and undirected edges; see for example Figure 3.5. We exclude also a situation when two vertices are connected by an undirected and a directed edge.

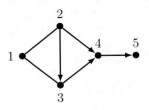

Figure 3.5: *An example of a hybrid graph with both directed and undirected edges.*

A *path* between i and j in a hybrid graph G is any sequence $k_1, \ldots, k_n$ of vertices such that $k_1 = i$, $k_n = j$ and $k_i \cdots k_{i+1}$ in G for every $i = 1, \ldots, n-1$. We say that a path is *undirected* if $k_i - k_{i+1}$ in G for every $i = 1, \ldots, n-1$. It is said to be *directed* if $k_i \rightarrow k_{i+1}$ in G for every $i = 1, \ldots, n-1$. A *semi-directed path* between i and j is any sequence $k_1, \ldots, k_n$ of vertices such that $k_1 = i$, $k_n = j$ and either $k_i - k_{i+1}$ or $k_i \rightarrow k_{i+1}$ in G for every $i = 1, \ldots, n-1$ and $k_i \rightarrow k_{i+1}$ for at least one i. A *semi-directed cycle* in a hybrid graph G

is a sequence $k_1, \ldots, k_{n+1} = k_1$, $n \geq 3$ of vertices in G such that $k_1, \ldots, k_n$ are distinct, and this sequence forms a semi-directed path. In a similar way we define a *undirected cycle* and directed cycle.

Definition 3.43. We say that G is undirected if it contains only undirected edges. A graph is *directed* if it contains only directed edges. A *chain graph* is a hybrid graph without semi-directed cycles. A *directed acyclic graph* (or *DAG*) is a directed graph without directed cycles.

For example, the hybrid graph in Figure 3.5 is not a chain graph because $2 \to 3 - 1 - 2$ forms a semi-directed cycle. An example of a DAG and a chain graph is given in Figure 3.6. Of course, every DAG and every undirected graph are also chain graphs.

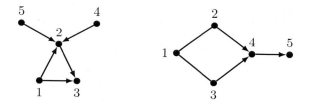

Figure 3.6: *An example of a DAG and a chain graph.*

We say that two vertices i, j are *neighbors* if $i - j$. In this case we say that the edge (i, j) is *incident with* i and j. If G is a directed graph and $i \to j$ in G (and hence when $(i, j) \in E$), then we call i a *parent* of j and j a *child* of i. Let $i \in V$, the *degree* of i is denoted by $\deg(i)$, and is the number of vertices j such that $i \cdots j$ in G.

Let $A \subset V$, then G_A denotes the *induced subgraph* with vertices A and edges $E \cap (A \times A)$. A *skeleton* of a directed graph G is the undirected graph with all arrows of G replaced with undirected edges. By a *moral graph G^m* of G we mean an undirected graph obtained from G by first joining by an undirected edge any two vertices sharing a child, and then replacing all arrows with undirected edges.

3.4.2 Undirected models and decomposable graphs

Consider a random vector $X = (X_i)_{i \in V}$ together with a graph $G = (V, E)$ whose vertices index the components of X. Lack of an edge between X_i and X_j indicates some form of conditional independence between X_i and X_j. These constraints are known as the *Markov properties* of the graph G. The graphical model associated with G is a family of multivariate probability distributions for which these Markov properties hold.

Undirected graphical models. Let G be an undirected graph and let $X \in \mathcal{X} = (\mathcal{X}_v)_{v \in V}$ be a discrete random vector. The *undirected pairwise Markov property* associates the following conditional independence constraints with the non-edges of G:

$$X_i \perp\!\!\!\perp X_j | X_{V \setminus \{i,j\}}, \quad \text{for all } (i,j) \notin E. \tag{3.19}$$

A more general concept is that of the *undirected global Markov property*. For any three disjoint subsets A, B, C of V we define a relation $\perp_G$ and we write $A \perp_G B | C$ if and only C *separates A and B in G*, i.e., every path from a vertex in A to a vertex in B necessarily crosses C. The undirected global Markov properties (UG) are given by

$$X_A \perp\!\!\!\perp X_B | X_C, \quad \text{for all } A \perp_G B | C. \tag{3.20}$$

Obviously (3.20) implies (3.19) but the opposite implication in general does not hold.

Definition 3.44. Let $X = (X_v)_{v \in V}$ be a random vector and let $G = (V, E)$ be an undirected graph. The *undirected graphical model of G* is the CI model satisfying all global Markov properties in (3.20).

By Lemma 3.10, every discrete graphical model is an algebraic statistical model. Equations in Lemma 3.10 together with the trivial constraint defining $\Delta_{\mathcal{X}}$ give the implicit description of the model. Undirected graphical models also have their parametric version.

Definition 3.45. Let G be an undirected graph. A *clique* $C \subseteq V$ in an undirected graph is a collection of vertices such that $(i, j) \in E$ for every pair $i, j \in C$. The set of maximal cliques of G is denoted by $\mathcal{C}(G)$.

The probability distribution p on $\mathcal{X}$ *factorizes according to G*, which we denote by (UF), if there exist a parameter vector $[\theta_C(x_C)]_{x_C \in \mathcal{X}_C} \in \mathbb{R}_{\geq 0}^{\mathcal{X}_C}$ for all $C \in \mathcal{C}(G)$ such that

$$p(x) = \frac{1}{Z(\theta)} \prod_{C \in \mathcal{C}(G)} \theta_C(x_C) \quad \text{for all } x \in \mathcal{X}, \tag{3.21}$$

where $Z(\theta)$ is the normalizing constant. This class of models, when restricted to positive distributions, lies in the class of discrete exponential families. However, it is important to note that in general the family of distributions induced by this parameterization need not exactly coincide with the undirected graphical model. One inclusion follows from the following proposition.

Proposition 3.46. *For any undirected graph G and any probability distribution on $\mathcal{X}$ it holds that*

$$(UF) \Longrightarrow (UG) \Longrightarrow (UP).$$

These three families of probability distributions coincide on the subset $\Delta_{\mathcal{X}}^0$ of positive probability distributions by the following important result.

Theorem 3.47 (Hammersley–Clifford Theorem). *Let $p \in \Delta_{\mathcal{X}}^0$, then p satisfies (UP) with respect to an undirected graph G if and only if it satisfies (UF).*

This shows that the difference between the undirected graphical model (as given in Definition 3.44) and the models defined by (UF) and (UP) can occur only on the boundary of the probability simplex $\Delta_{\mathcal{X}}$. For distributions with zeros, this characterization in general is not clear and the reason is precisely that $X_A \perp\!\!\!\perp X_B | X_C$ is not well defined for $z \in \mathcal{X}_C$ such that $p_C(z) = 0$.

Decomposable models *Decomposable models* form an important family of undirected graphical models.

Definition 3.48. An undirected graph is said to be *decomposable* (or *triangulated*) if it contains no induced cycle of length greater than 3.

Note that if G is decomposable and $A \subseteq V$, then G_A is decomposable. For decomposable graphs we mention the following important result.

Proposition 3.49. *Let G be decomposable. Then*

$$(UF) \Longleftrightarrow (UG).$$

Decomposable graphs admit a useful decomposition of its set of vertices. Let $B_1, \ldots, B_k$ be a sequence of subsets of the vertex set V of an undirected graph G. Let

$$H_j = B_1 \cup \cdots \cup B_j, \quad R_j = B_j \setminus H_{j-1}, \quad S_j = H_{j-1} \cap B_j.$$

The sequence is said to be *perfect* if the following conditions are fulfilled:

(i) for all $i > 1$ there is a $j < i$ such that $S_i \subset B_j$,

(ii) the sets S_i are complete for all i.

Call the sets H_j *histories*, R_j the *residuals*, and S_j the *separators* of the sequence.

Lemma 3.50. *Let $B_1, \ldots, B_k$ be a perfect sequence of sets which contains all cliques of an undirected graph G. Then for every j, S_j separates $H_{j-1} \setminus S_j$ from R_j in G_{H_j}.*

The aim of this section is to show why decomposable graphs are important especially from the inferential point of view. We have the following proposition.

Proposition 3.51. *Assume A, B, S are three disjoint subsets of V such that S separates A from B and S forms a clique. Then a probability distribution P factorizes with respect to G if and only if both marginal distributions $P_{A \cup S}$ and $P_{B \cup S}$ factorize with respect to $G_{A \cup S}$ and $G_{B \cup S}$, respectively, and the probability mass functions satisfy*

$$p(x)p_S(x_S) = p_{AS}(x_{AS})p_{BS}(x_{BS}) \tag{3.22}$$

for every $x \in \mathcal{X}$.

In Proposition 3.51 we described how the factorization (and hence the global Markov property) behaves across decompositions of the graph. Using the factorization, an especially simple expression for the vector of means can be obtained. Consider a graphical model with a decomposable graph. By [Lauritzen, 1996, Proposition 2.17] the cliques of G can be numbered to form a perfect sequence exactly when G is decomposable. The repeated use of (3.22) gives the formula

$$p(x) = \frac{\prod_{j=1}^{k} p_{C_j}(x_{C_j})}{\prod_{j=2}^{k} p_{S_j}(x_{S_j})},$$

where $S_j = H_{j-1} \cap C_j$ is the sequence of separators. Alternatively, we can collect the terms in the denominator in groups and obtain the formula

$$p(x) = \frac{\prod_{C \in \mathcal{C}} p_C(x_C)}{\prod_{S \in \mathcal{S}} p_S(x_S)^{\nu(S)}}, \qquad (3.23)$$

where $\mathcal{C}$ is the set of maximal cliques of G, $\mathcal{S}$ is the set of separators, and $\nu(S)$ is an index that counts the number of times a given separator S occurs in a perfect sequence. The main application of this formula is the following proposition.

Proposition 3.52. *Let u be the data tensor. In a decomposable graphical model with graph G, the maximum likelihood estimate of the mean vector is given as*

$$\hat{u}(x) = \frac{\prod_{C \in \mathcal{C}} u(x_C)}{\prod_{S \in \mathcal{S}} u(x_S)^{\nu(S)}},$$

where $u(x_A)$ is the number of times the event $\{X_A = x_A\}$ was observed.

Suppose that $X = (X_v) \in \mathcal{X}$ has distribution in $M(G)$, the graphical model of G. Consider the multinomial likelihood $L_m(q; u)$ on $\Delta_{\mathcal{X}}$ as in (3.17). The likelihood function $L(\theta; u)$ of $M(G)$ can be treated as a constrained version of L_m on $\Delta_{\mathcal{X}}$ as explained in Section 3.3.3. Then L_m is maximized in sample frequencies and its constrained version is maximized in $\frac{1}{n}\hat{u}(x)$, where $\hat{u}$ is given in Proposition 3.52.

Corollary 3.53. *Let G be a decomposable graph. Suppose that some of the vertices represent observed variables and some of them are hidden. Then the M step in Algorithm 3.41 can be effectively implemented using Proposition 3.52.*

3.4.3 *DAGs, chain graphs, and DAG equivalence*

Bayesian networks Let $G = (V, E)$ be a directed acyclic graph (DAG). For each $v \in V$, the set $\mathrm{pa}(v)$ denotes the set $\{u \in V : (u, v) \in E\}$ of parents of v. The set $\mathrm{de}(v)$ of *descendants* is the set of vertices w such that there is a directed path from v to w in G. The *non-descendants* of v are $\mathrm{nd}(v) = V \setminus (v \cup \mathrm{de}(v))$. For a subset $C \subset V$, we define $\mathrm{an}(C)$ to be the set of vertices w that are *ancestors* of some vertex $v \in C$. Here, w is an ancestor of v if there is a directed path from w to v.

The conditional independence statements defining a Bayesian network are derived from the graph in a different way than in the undirected case. The *directed local Markov property* (DL) associates the conditional independence constraints

$$X_v \perp\!\!\!\perp X_{\mathrm{nd}(v)\backslash\mathrm{pa}(v)} | X_{\mathrm{pa}(v)}, \quad v \in V \tag{3.24}$$

with the directed acyclic graph G. For every triple A, B, C of pairwise disjoint subsets of V, consider the set $\mathrm{an}(A \cup B \cup C)$ and the induced moral graph $G^m_{\mathrm{an}(A\cup B\cup C)}$. The *directed global Markov properties* (DG) are given by

$$X_A \perp\!\!\!\perp X_B | X_C \quad \text{for all } A \perp\!\!\!\perp_H B|C, \ H = G^m_{\mathrm{an}(A\cup B\cup C)} \tag{3.25}$$

and hence by all A, B, C such that C separates A and B in the moral graph $G^m_{\mathrm{an}(A\cup B\cup C)}$.

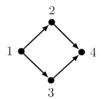

Figure 3.7: *This DAG encodes:* $1 \perp\!\!\!\perp 4 | \{2, 3\}$ *and* $2 \perp\!\!\!\perp 3 | 1$.

Remark 3.54. To check if $A \perp\!\!\!\perp B|C$ in a given DAG model we study the moral graph $G^m_{\mathrm{an}(A\cup B\cup C)}$. In particular, it is not enough to check if C separates A and B in G^m. As an example, consider the graph in Figure 3.7. We have $2 \perp\!\!\!\perp 3 | 1$ even though $2 - 3$ in the moral graph G^m.

We say that a probability distribution satisfies the *recursive factorization property* (DF) if there exists a parameter vector $[\theta_{j|\mathrm{pa}(j)}(x_j|x_{\mathrm{pa}(j)})]_{x_j \in \mathcal{X}_j} \in \mathbb{R}^{\mathcal{X}_j}_{\geq 0}$ for all $x_{\mathrm{pa}(j)} \in \mathcal{X}_{\mathrm{pa}(j)}$ and $j \in V$ such that $\sum_{x_j} \theta_{j|\mathrm{pa}(j)}(x_j|x_{\mathrm{pa}(j)}) = 1$ and

$$p(x) = \prod_{j=1}^{m} \theta_{j|\mathrm{pa}(j)}(x_j|x_{\mathrm{pa}(j)}), \qquad \text{for all } x \in \mathcal{X}. \tag{3.26}$$

For example, factorization in Example 3.9 corresponds to the parameterization of a simple Bayesian network of the form $\overset{1}{\bullet} \to \overset{3}{\bullet} \leftarrow \overset{2}{\bullet}$.

In the case of directed acyclic graphs, all three characterizations of Bayesian networks are equivalent.

Theorem 3.55. *Let G be a directed acyclic graph. For a probability distribution p on $\mathcal{X}$ the following conditions are equivalent:*

(DF) *p admits a recursive factorization according to G,*

(DG) *p obeys the directed global Markov property relative to G,*

(DL) *p obeys the directed local Markov property relative to G.*

DAG equivalence   Theorem 3.55 shows that at least in the case of various
Markov properties the DAG model seems to be easier to work with than
models of undirected graphs. However, this model class has a drawback not
present in the undirected case, namely, two different DAGs may define the
same statistical model. For example, the three DAGs in Figure 3.8 all lead
to the same statistical model defined by a single conditional independence
statement $X_1 \perp\!\!\!\perp X_3 | X_2$. This is equivalent to the undirected graphical model
$\overset{1}{\bullet} - \overset{2}{\bullet} - \overset{3}{\bullet}$.

Figure 3.8: *Three equivalent DAGs.*

Let G be a DAG. Write $[G]$ for the equivalence class of all DAGs defining
the same statistical model as G. It turns out that this equivalence class
has a very simple description. Define an *immorality* as an induced subgraph
$\overset{i}{\bullet} \to \overset{j}{\bullet} \leftarrow \overset{k}{\bullet}$ (there is no arrow between i and k in G). We have the result,
which has been discovered independently by Frydenberg [1990] and Verma
and Pearl [1991].

Theorem 3.56. *Two DAGs with the same set of vertices are equivalent if
and only if they have the same skeleton and the same immoralities.*

From the point of view of model selection, this non-uniqueness of DAGs
was worrying. For that reason, researchers often work with a special chain
graph that represents the equivalence class $[G]$. This graph is called the *essential graph* and is denoted by G^*.

To understand equivalence classes of DAGs, we need to introduce *chain
graph models*. Recall that a chain graph is a hybrid graph without semi-
directed cycles. Chain graph models form a family of graphical models which
generalize both the undirected models and Bayesian networks. In this section
we only very briefly introduce this important model class.

Two non-equivalent definitions of chain graph models can be found in
the literature and they are referred to as LWF or AMP chain graph models
in Andersson et al. [2001], which refers to: Lauritzen-Wermuth-Frydenberg
Frydenberg [1990], Lauritzen and Wermuth [1989] and Andersson-Madigan-
Perlman Andersson et al. [2001]. These two definitions differ in how exactly
a graph encodes the defining set of conditional independence statements.
Define a *flag* as an induced subgraph of the form $i \to j - k$. In this book we
consider only *chain graphs without flags* (NF-CG). This subfamily is sufficient
for our purposes of discussing the equivalence of DAGs. Moreover, it has the
important property that for NF-CGs both the LWF and AMP chain graph
models coincide; see [Andersson et al., 2001, Theorem 1, Theorem 4].

For every $A \subset V$, denote by $\mathrm{An}(A)$ the set of vertices containing A
together with all vertices i in V such that there exists either a directed,
undirected, or semi-directed path from i to some $a \in A$. The set of global

Markov properties (CG) induced by a chain graph G is given by $A \perp\!\!\!\perp B | C$ whenever C separates A from B in the moral graph $G^m_{\text{An}(A \cup B \cup C)}$.

Theorem 3.57. *Let G be a DAG. The essential graph G^* is the unique chain graph with no flags, with the same skeleton and immoralities as G and with the maximal number of undirected edges. Moreover, the statistical models defined by the DG properties on G and CG properties on G^* are equal.*

By definition, G^* has the same skeleton as G, and $i \to j$ in G^* if and only if $i \to j$ in every $H \in [G]$, all other edges are undirected. For example, the essential graph for any of the graphs in Figure 3.8 is the undirected graph $\overset{1}{\bullet} - \overset{2}{\bullet} - \overset{3}{\bullet}$, whereas the essential graph of $G = \overset{1}{\bullet} \to \overset{2}{\bullet} \leftarrow \overset{3}{\bullet}$ is G itself. Another example is given in Figure 3.9. By Theorem 3.56, every arrow that participates in an immorality in G is essential, but G^* may contain other arrows. For example, in the DAG in Figure 3.10 is equal to its essential graph G^* even though not all arrows participate in immoralities.

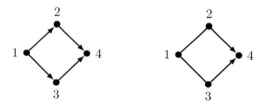

Figure 3.9: *A DAG and its essential graph.*

Figure 3.10 *A NF-CG whose edges are all essential but not all part of immoralities.*

Note that if a chain graph is an undirected graph, then the global undirected Markov properties coincide with chain graph Markov properties.

Corollary 3.58. *Let G be a DAG such that the essential graph G^* is an undirected graph. Then*

$$(DF) \quad \Longleftrightarrow \quad (UG).$$

3.5 Bibliographical notes

All basic definitions like marginal distribution and conditional are completely classical. Good textbooks on basic probability and statistics are, for example, Casella and Berger [1990], Grimmett and Stirzaker [2001], Young and Smith

[2005]. Discrete statistical models are discussed in detail, for example, in Agresti [1996, 2002]. The notion of an algebraic statistical model was defined in Drton et al. [2009], Pachter and Sturmfels [2005]. Its most refined version can be found in Drton and Sullivant [2007]. The moment structures in the algebraic setting are discussed for example in Pistone et al. [2001]. The moment aliasing principle was formulated in [Pistone and Wynn, 2006, Lemma 3]. The notion of identifiability and its importance in statistics was discussed for example by Rothenberg [1971]. It turns out that identifiability is also a problem in the Bayesian setting, which was raised by Gustafson [2009], Kadane [1974], Poirier [1998], Sahu and Gelfand [1999]. Recently a more mathematical, general treatment of identifiability for models with hidden data has been given by Allman et al. [2009]. The best reference for conditional independence models is Dawid [1979], Studený [2004]. A general reference on exponential families is Barndorff–Nielsen [1978], Brown [1986], Efron [1978]. Discrete exponential families are linked to toric geometry in Geiger et al. [2006], Diaconis and Sturmfels [1998]. More general links between exponential families and algebraic geometry has been recently developed by Michałek et al. [2014]. The moment map can then be used to analyze possible supports of a model Kähle [2010], Rauh et al. [2009]. Mixture models are treated in detail for example in Bartholomew et al. [2011], Lindsay [1995]. The algebraic geometry of the mixture model is discussed in Garcia et al. [2005], Drton et al. [2009], Lindsay [1995]. Also here the analysis of possible supports is possible; see Montúfar [2013]. The maximum likelihood method is described in any textbook on statistics like Casella and Berger [1990], Young and Smith [2005]. A more algebraic treatment is given in Catanese et al. [2006], which led to unexpected links in algebraic geometry by Huh [2012], Huh and Sturmfels [2014]. The constrained multinomial likelihood theory is summarized in Davis–Stober [2009]. The EM algorithm was proposed by Dempster et al. [1977]. We use its special version, explicitly written in the most convenient form, which can be also found in Pachter and Sturmfels [2005]. The use of graphs to represent statistical models dates back to Wright [1921]. The books Whittaker [1990], Pearl [2000], Neapolitan [1990], Lauritzen [1996], Koller and Friedman [2009] summarize these developments. Recently the variational approach to graphical models has become popular; see Wainwright and Jordan [2008]. Most of the results of Section 3.4 are based on Lauritzen [1996]; see also references therein. The analysis of the equivalence classes of DAGs and the corresponding link to chain graph models is given in Andersson et al. [1997], Frydenberg [1990], Roverato [2005], Studený [2004], Verma and Pearl [1991].

Chapter 4

Tensors, moments, and combinatorics

[]

In this chapter we develop techniques to work efficiently with arbitrary discrete distributions. We create deep links between statistics and geometry, which provide a set of tools for symbolic computations useful in the analysis of various algebraic varieties. The final aim, however, is to construct a coordinate system suitable for the study of hidden tree models in the second part of this book.

4.1 Posets and Möbius functions

4.1.1 Basic concepts

A *partially ordered set* $\mathcal{P}$ (or *poset*) is a set, together with a binary relation denoted $\leq$, satisfying the following three axioms:

1. For all $x \in \mathcal{P}$, $x \leq x$ (reflexivity).
2. If $x \leq y$ and $y \leq x$, then $x = y$ (antisymmetry).
3. If $x \leq y$ and $y \leq z$, then $x \leq z$ (transitivity).

We use the obvious notation $x \geq y$ to mean $y \leq x$, $x < y$ to mean $x \leq y$, and $x \neq y$. We consider only finite posets.

Example 4.1. The following two examples are very important.

a. Let $m \in \mathbb{N}_0$. We can make the set of all subsets of $[m]$ into a poset $\mathbf{B}_m$. For any two subsets $A, B \subseteq [m]$ we say that $A \leq B$ in $\mathbf{B}_m$ if $A \subseteq B$.

b. Let $m \in \mathbb{N}$. We say that $\pi = B_1 | \cdots | B_r$ is a *set partition* (or *partition*) of $[m]$, if the *blocks* $B_i \neq \emptyset$ are disjoint sets whose union is $[m]$. The set $\mathbf{\Pi}_m$ of all set partitions of $[m]$ becomes a poset by defining $\pi \leq \nu$ in $\mathbf{\Pi}_m$ whenever every block of π is contained in a block of ν. For instance, if $n = 9$, $\pi = 137|2|46|58|9$ and $\nu = 13467|2589$, then $\pi \leq \nu$. It follows that $[m]$ (the one-block partition) is always the maximal element of $\mathbf{\Pi}_m$ and the partition $1|2|\cdots|m$ into singletons is the minimal element of $\mathbf{\Pi}_m$.

Two posets $\mathcal{P}_1, \mathcal{P}_2$ are *isomorphic*, which we denote by $\mathcal{P}_1 \simeq \mathcal{P}_2$, if there exists a *bijection* $\phi : \mathcal{P}_1 \to \mathcal{P}_2$ such that

$$x \leq y \text{ in } \mathcal{P}_1 \iff \phi(x) \leq \phi(y) \text{ in } \mathcal{P}_2.$$

By a *subposet* of $\mathcal{P}$, we mean a subset $\mathcal{P}'$ of $\mathcal{P}$ together with a partial ordering of $\mathcal{P}'$ such that for $x, y \in \mathcal{P}'$ we have $x \leq y$ in $\mathcal{P}'$ if and only if $x \leq y$ in $\mathcal{P}$. A special type of a subposet of $\mathcal{P}$ is the (closed) *interval* $[x, y] = \{z \in \mathcal{P} : x \le z \le y\}$, defined whenever $x \le y$.

If $x, y \in \mathcal{P}$, then we say that y *covers* x if and only if $x < y$ and $[x, y] = \{x, y\}$. The *Hasse diagram* of a finite poset $\mathcal{P}$ is the graph whose vertices are the elements of $\mathcal{P}$, whose edges the cover relations, and such that if $x < y$, then y is drawn above x.

Example 4.2. Consider the poset $\mathbf{\Pi}_3$ of all set partitions of $\{1, 2, 3\}$. The poset has five elements $1|2|3$, $1|23$, $2|13$, $12|3$, and 123. The poset $\mathbf{\Pi}_4$ has 15 elements. The Hasse diagrams of both posets are given in Figure 4.1.

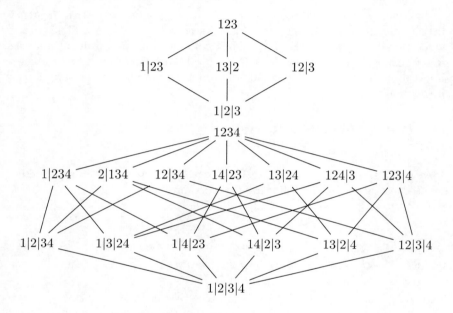

Figure 4.1: *The Hasse diagrams of $\mathbf{\Pi}_3$ and $\mathbf{\Pi}_4$.*

4.1.2   The Möbius function

Let $\mathcal{P}$ be a poset. We say that a function $f : \mathcal{P} \times \mathcal{P} \to \mathbb{R}$ is *triangular* if $f(x, y) \ne 0$ only if $(x, y) \in \mathcal{P} \times \mathcal{P}$ is such that $x \le y$. For any two triangular functions $f, g : \mathcal{P} \times \mathcal{P} \to \mathbb{R}$ we define their *product* as another triangular function given by

$$fg(x, y) = \sum_{x \le z \le y} f(x, z)g(z, y) \qquad \text{for all } x, y \in \mathcal{P}.$$

The identity of this product is a function $\delta(x, y)$ defined by $\delta(x, x) = 1$ and $\delta(x, y) = 0$ for all $x \ne 0$. For every triangular function f we define its *inverse* as the function satisfying $ff^{-1} = f^{-1}f = \delta$.

We say that a total order $\prec$ of elements of $\mathcal{P}$ is *consistent* with the ordering of $\mathcal{P}$ if for every $x, y \in \mathcal{P}$ if $x < y$ in $\mathcal{P}$, then $x \prec y$ in this total ordering. For example, for $\mathbf{\Pi}_3$ the total ordering $1|2|3 \prec 1|23 \prec 2|13 \prec 3|12 \prec 123$ is consistent with the ordering of $\mathbf{\Pi}_3$. In the theory of partially ordered sets, every such a total ordering is called a *linear extension*.

Given any linear extension of $\mathcal{P}$ every triangular function f can be represented uniquely by an upper triangular matrix $F = [F_{xy}] \in \mathbb{R}^{\mathcal{P} \times \mathcal{P}}$ such that $F_{xy} = f(x, y)$ and the rows and columns of F are ordered according to the given linear extension. For example, for the linear extension of $\mathbf{\Pi}_3$ given above the possible triangular matrices are of the form

$$\begin{bmatrix} * & * & * & * & * \\ 0 & * & 0 & 0 & * \\ 0 & 0 & * & 0 & * \\ 0 & 0 & 0 & * & * \\ 0 & 0 & 0 & 0 & * \end{bmatrix},$$

where stars indicate entries that can take nonzero values.

For any two triangular functions f, g with corresponding matrices F, G, the matrix of their product fg is given by the matrix product FG. Moreover, the matrix of the inverse f^{-1} is given by the inverse matrix F^{-1}. From now on we are going to use the same notation for a triangular function and the corresponding upper triangular matrix.

The *zeta function* ζ is a triangular function defined by

$$\zeta(x, y) = \begin{cases} 1 & \text{if } x \leq y \\ 0 & \text{otherwise.} \end{cases} \tag{4.1}$$

The matrix representing the zeta function ζ of a finite poset $\mathcal{P}$ is invertible; its inverse ζ^{-1} is called the *Möbius function of $\mathcal{P}$* and denoted by $\mathfrak{m}$ (or $\mathfrak{m}_{\mathcal{P}}$ if there is a possible ambiguity). We can define $\mathfrak{m}$ inductively. The matrix relation $\mathfrak{m} = \zeta^{-1}$ is equivalent to

$$\begin{aligned} \mathfrak{m}(x, x) &= 1, &&\text{for all } x \in \mathcal{P} \\ \mathfrak{m}(x, y) &= -\textstyle\sum_{x \leq z < y} \mathfrak{m}(x, z) &&\text{for all } x < y \text{ in } \mathcal{P}. \end{aligned} \tag{4.2}$$

Directly from (4.2), or equivalently from the fact that $\mathfrak{m}\zeta = \delta$, it follows that

$$(\mathfrak{m}\zeta)(x, y) = \sum_{x \leq z \leq y} \mathfrak{m}(x, z) = \begin{cases} 0 & \text{if } x < y \\ 1 & \text{if } x = y. \end{cases} \tag{4.3}$$

Example 4.3. A *chain poset* is a totally ordered set $C_m = \{a_1, \ldots, a_m\}$ such that a_i covers a_{i-1} for every $i = 2, \ldots, m$. The Möbius function on C_m satisfies $\mathfrak{m}(a_i, a_i) = 1$, $\mathfrak{m}(a_i, a_{i+1}) = -1$ and is zero otherwise.

The importance of the Möbius function is due to the following basic result.

Proposition 4.4 (Möbius inversion formula). *Let $\mathcal{P}$ be a finite poset. Let $f, g : \mathcal{P} \to \mathbb{R}$. Then*

$$g(x) \;=\; \sum_{y \leq x} f(y), \qquad \text{for all } x \in \mathcal{P},$$

if and only if

$$f(x) \;=\; \sum_{y \leq x} g(y)\mathfrak{m}(y, x) \qquad \text{for all } x \in \mathcal{P}.$$

Proof. The Möbius inversion formula is nothing but the following statement for upper triangular matrices

$$\zeta f = g \quad \Longleftrightarrow \quad f = \zeta^{-1}g = \mathfrak{m}g.$$

$\square$

Example 4.5. For the posets in Example 4.1, the Möbius functions are well known.

(i) The Möbius function on $\mathbf{B}_m$ is defined by

$$\mathfrak{m}(A, B) \;=\; (-1)^{|B \setminus A|} \qquad \text{for all } A \subseteq B.$$

(ii) Let $\pi \in \mathbf{\Pi}_m$ be a set partition of $[m]$. Then

$$\mathfrak{m}(\pi) \;:=\; \mathfrak{m}(\pi, [m]) \;=\; (-1)^{|\pi|-1}(|\pi| - 1)!,$$

where $|\pi|$ denotes the number of blocks of π.

Example 4.6. Consider the poset $\mathbf{\Pi}_3$ in Example 4.2. We order the elements of $\mathbf{\Pi}_3$ by $1|2|3 \prec 1|23 \prec 2|13 \prec 3|12 \prec 123$. The zeta function is represented by the following matrix

$$\zeta = \begin{bmatrix} 1 & 1 & 1 & 1 & 1 \\ 0 & 1 & 0 & 0 & 1 \\ 0 & 0 & 1 & 0 & 1 \\ 0 & 0 & 0 & 1 & 1 \\ 0 & 0 & 0 & 0 & 1 \end{bmatrix}.$$

Its inverse is representing the Möbius function

$$\mathfrak{m} = \begin{bmatrix} 1 & -1 & -1 & -1 & 2 \\ 0 & 1 & 0 & 0 & -1 \\ 0 & 0 & 1 & 0 & -1 \\ 0 & 0 & 0 & 1 & -1 \\ 0 & 0 & 0 & 0 & 1 \end{bmatrix}. \tag{4.4}$$

We easily verify that this complies with the recursive formula in (4.2).

For any two posets $\mathcal{P}_1, \mathcal{P}_2$ we define the poset $\mathcal{P}_1 \times \mathcal{P}_2$ as a set with the ordering $(x, y) \leq (x', y')$ if $x \leq x'$ and $y \leq y'$.

Proposition 4.7. *Let $\mathcal{P}_1 \times \mathcal{P}_2$ be a product of two finite posets $\mathcal{P}_1, \mathcal{P}_2$. If $(x, y) \leq (x', y')$ in $\mathcal{P}_1 \times \mathcal{P}_2$, then*

$$\mathfrak{m}_{\mathcal{P}_1 \times \mathcal{P}_2}((x, y), (x', y')) = \mathfrak{m}_{\mathcal{P}_1}(x, x')\mathfrak{m}_{\mathcal{P}_2}(y, y').$$

Proof. See [Stanley, 2002, Proposition 3.1.2]. □

This proposition gives a very efficient way of computing the Möbius function for certain posets.

Lemma 4.8. *The Boolean lattice $\boldsymbol{B}_m$ is isomorphic to the product $(C_2)^m$, where C_2 is the chain poset of length 2; see Example 4.3.*

Recall that the Möbius function on C_2 satisfies $\mathfrak{m}(a_1, a_1) = \mathfrak{m}(a_2, a_2) = 1$ and $\mathfrak{m}(a_1, a_2) = -1$. Now Lemma 4.8 and Proposition 4.7 give another way to confirm the form of the Möbius function on $\boldsymbol{B}_m$ given in Example 4.5.

Remark 4.9. If $\prec_1, \prec_2$ are total orderings consistent with the ordering of $\mathcal{P}_1$ and $\mathcal{P}_2$, then there is a natural (lexicographic) total ordering of $\mathcal{P}_1 \times \mathcal{P}_2$ given by $(x, y) \prec (x', y')$ if either $x \prec_1 x'$ or $x = x'$ and $y \prec_2 y'$. Given any such ordering, Proposition 4.7 phrased in terms of matrix representations of triangular functions states that the matrix of $\mathfrak{m}_{\mathcal{P}_1 \times \mathcal{P}_2}$ is the Kronecker product of matrices $\mathfrak{m}_{\mathcal{P}_1}$ and $\mathfrak{m}_{\mathcal{P}_2}$.

Example 4.10. To illustrate Remark 4.9, let $\mathcal{P}_1 = \boldsymbol{\Pi}_3$ and let $\mathcal{P}_2$ be $\boldsymbol{\Pi}(\{4, 5\}) \simeq \boldsymbol{\Pi}_2$. The product $\mathcal{P}_1 \times \mathcal{P}_2$ is isomorphic to the interval $[1|2|3|4|5, 123|45]$ in $\boldsymbol{\Pi}_5$. The matrix $\mathfrak{m}_{\mathcal{P}_1}$ is given in (4.4). The Kronecker product (we use notation of Section 2.3.3) of matrices $\mathfrak{m}_{\mathcal{P}_1}$ and $\mathfrak{m}_{\mathcal{P}_2}$ is

$$\text{mat} \left(\begin{bmatrix} 1 & -1 & -1 & -1 & 2 \\ 0 & 1 & 0 & 0 & -1 \\ 0 & 0 & 1 & 0 & -1 \\ 0 & 0 & 0 & 1 & -1 \\ 0 & 0 & 0 & 0 & 1 \end{bmatrix} \otimes \begin{bmatrix} 1 & -1 \\ 0 & 1 \end{bmatrix} \right).$$

In particular,

$$\mathfrak{m}_{\mathcal{P}_1}(1|2|3, 123) \cdot \mathfrak{m}_{\mathcal{P}_2}(4|5, 4|5) = 2 \cdot 1 = 2,$$

which is equal to $\mathfrak{m}_{\boldsymbol{\Pi}_5}(1|2|3|4|5, 123|4|5)$. Similarly,

$$\mathfrak{m}_{\mathcal{P}_1}(1|2|3, 123) \cdot \mathfrak{m}_{\mathcal{P}_2}(4|5, 45) = 2 \cdot (-1) = -2,$$

which is equal to $\mathfrak{m}_{\boldsymbol{\Pi}_5}(1|2|3|4|5, 123|45)$.

In the previous example we use the fact that the Möbius function on an interval in $\mathcal{P}$ is naturally induced from the Möbius function on $\mathcal{P}$; see for example [Rota, 1964, Proposition 4].

Proposition 4.11. *The Möbius function of any interval $[x, y]$ of a poset $\mathcal{P}$ equals the restriction to $[x, y]$ of the Möbius function on $\mathcal{P}$.*

4.1.3 Lattices and partition lattices

We say that poset $\mathcal{P}$ has a $\hat{0}$ if there exists an element $\hat{0} \in \mathcal{P}$ such that $x \geq \hat{0}$ for all $x \in \mathcal{P}$. Similarly, $\mathcal{P}$ has a $\hat{1}$ if there exists $\hat{1} \in \mathcal{P}$ such that $x \leq \hat{1}$ for all $x \in \mathcal{P}$. If x and y belong to a poset $\mathcal{P}$, then a *least upper bound* of x and y is an element z such that $z \geq x$, $z \geq y$ and $z \leq w$ for any other w such that $w \geq x$, $w \geq y$. If a least upper bound of x and y exists, then it is clearly unique and it is denoted by $x \vee y$ and called the *join* of x and y. Dually, one can define the *greatest lower bound* $x \wedge y$, which is called the *meet* of x and y.

A *lattice* is a poset L such that for every $x, y \in L$ both $x \vee y$ and $x \wedge y$ exist. A *sublattice* of a lattice L is a nonempty subset of L which is a lattice with *the same* meet and join operations as L. Clearly, all finite lattices have a $\hat{0}$ and $\hat{1}$. We say that S is a *meet-semilattice* if S is a poset such that $x \wedge y$ exists for all $x, y \in S$. The following result will be useful.

Lemma 4.12. *If S is a finite meet-semilattice with $\hat{1}$, then S is a lattice.*

Proof. Let $x, y \in S$. A natural definition for $x \vee y$ is as a meet of all the elements of $A := \{z \in S : x, y \leq z\}$. Since $x, y \leq \hat{1}$, A is always nonempty. By constriction, $w = \bigwedge_{z \in A} z$ is the smallest element satisfying $w \geq x, y$. $\square$

Recall that $\pi = B_1 | \ldots | B_k$ is called a set partition of $[m]$, if the *blocks* $B_i \neq \emptyset$ are disjoint sets, whose union is $[m]$. Equivalently, a partition of $[m]$ corresponds to an equivalence relation $\sim_\pi$ on $[m]$ where $i \sim_\pi j$ if i and j lie in the same block. Now let A be a multiset $A = \{i_1, \ldots, i_d\}$ of elements of $[m]$ such that $1 \leq i_1 \leq \cdots \leq i_d \leq n$. We can define a partition of A using a partition of $[d]$ by $i_j \sim_\pi i_k$ if $j \sim_\pi k$ in $\mathbf{\Pi}_d$. The set of all partitions of A is denoted by $\mathbf{\Pi}(A)$. For example, if $A = \{1, 1, 1\} = \{1^3\}$, then $\mathbf{\Pi}(A)$ contains three different partitions of the form $1|11$.

The poset $\mathbf{\Pi}_m$ has the $\hat{0}$ given by the m-block partition $1|2|\cdots|m$ and the $\hat{1}$ given by the one-block partition $[m]$. If $\pi = B_1|\cdots|B_r$ and $\nu = B_1'|\cdots|B_s'$, then $\pi \wedge \nu$ is a partition with blocks given by all the non-empty sets of the form $B_i \cap B_j'$ for $i = 1, \ldots, r$, $j = 1, \ldots, s$. This shows that $\mathbf{\Pi}_m$ is a meet-semilattice and by Lemma 4.12 it forms a lattice.

Any set partition with two blocks is called a *split*. Let $\mathcal{S}$ be a set of splits. We say that $\pi \in \mathbf{\Pi}_m$ is *generated* by $\mathcal{S}$ if $\pi = \pi_1 \wedge \cdots \wedge \pi_d$, where $\pi_i \in \mathcal{S}$. By convention, the one-block partition $[m]$ is generated by $\mathcal{S}$ as it can be written as the empty meet of elements from $\mathcal{S}$. The set of all partitions generated by $\mathcal{S}$ is denoted by $\langle \mathcal{S} \rangle$. By construction, $\langle \mathcal{S} \rangle$ forms a finite meet-semilattice and hence a lattice by Lemma 4.12. If $\mathcal{S}$ is the set of *all* splits in $\mathbf{\Pi}_m$, then $\langle \mathcal{S} \rangle = \mathbf{\Pi}_m$. For special subsets $\mathcal{S}$ we obtain a useful generalization of the lattice of all set partitions.

Definition 4.13. *A partition lattice is the lattice $\langle \mathcal{S} \rangle$, where $\mathcal{S}$ is any set of splits in $\mathbf{\Pi}_m$ such that $\hat{0} \in \langle \mathcal{S} \rangle$.*

Note that a partition lattice need not be a sublattice of $\mathbf{\Pi}_m$ in the sense that the join operators need not coincide.

Definition 4.14. We say that a split $A|B$ is *trivial* if $\min\{|A|,|B|\} = 1$. The set of trivial splits is denoted by $\mathcal{S}_0$.

Definition 4.15. The following is a list of interesting set partition lattices:

(1) An *interval partition* of $[m]$ is a partition π such that if $i \sim_\pi j$ for $i < j$, then $i \sim_\pi k$ for every $i < k < j$. This poset of all interval partitions is denoted by $\mathcal{I}_m$. It is a partition lattice generated by splits of the form $1 \cdots k|(k+1) \cdots m$ for $k = 1, \ldots, m-1$. It is an easy exercise to show that $\mathcal{I}_m$ is a sublattice of $\mathbf{\Pi}_m$, isomorphic to the Boolean lattice of $[m-1]$.

(2) A partition $\pi \in \mathbf{\Pi}_m$ is called a *one-cluster partition* if it contains at most one block of size greater than one. In particular, the one-block partition $[m]$ and the minimal partition $1|2|\cdots|n$ are one-cluster partitions. The poset of all one-cluster partitions forms a lattice $\mathcal{C}_m$, which is not a sublattice of $\mathbf{\Pi}_m$. It is isomorphic to the poset of all subsets of $[m]$ excluding singletons. We have $\mathcal{C}_m = \langle \mathcal{S}_0 \rangle$.

(3) A partition $\pi \in \mathbf{\Pi}_m$ is *non-crossing* if there is no quadruple of elements $i < j < k < l$ such that $i \sim_\pi k$, $j \sim_\pi l$, and $i \not\sim_\pi j$. The noncrossing partitions of $[m]$ form a lattice which we denote by $\mathbf{NC}_m$. This lattice is not a sublattice of $\mathbf{\Pi}_m$. We have $\mathcal{S}_0 \subseteq \mathbf{NC}_m$.

For every lattice, denote $\mathfrak{m}(\pi) := \mathfrak{m}(\pi,\hat{1})$. Later we will see that it is important to identify values of $\mathfrak{m}(\pi)$ for various partition lattices. By Example 4.5, for $\mathbf{\Pi}_m$ we have

$$\mathfrak{m}(\pi) = (-1)^{|\pi|-1}(|\pi|-1)!$$

The lattice of interval partitions $\mathcal{I}_m$ is isomorphic to the Boolean lattice of all subsets of $[m-1]$ and hence $\mathfrak{m}(\pi) = (-1)^{|\pi|-1}$. For the lattice of one-cluster partitions we have

$$\mathfrak{m}(\pi) = \begin{cases} (-1)^{n-1}(n-1) & \text{if } \pi = 1|2|\cdots|n, \text{ and} \\ (-1)^{|\pi|-1} & \text{otherwise.} \end{cases} \tag{4.5}$$

For any two $\pi \leq \nu \in \mathbf{\Pi}_m$, the interval $[\pi,\nu]$ can be written as the following product

$$[\pi,\nu] \simeq \prod_{B \in \nu} [\pi(B), B],$$

where $\pi(B)$ is the restriction of π to elements in B. For an arbitrary partition lattice a weaker result holds.

Lemma 4.16. *Let L be a partition lattice; see Definition 4.13. Then for any $\pi \in L$*

$$[\hat{0},\pi] \simeq \prod_{B \in \pi} L(B).$$

In particular, by Proposition 4.7

$$\mathfrak{m}(\hat{0},\pi) = \prod_{B \in \pi} \mathfrak{m}_B(\hat{0}, B) \qquad \text{for any } \pi \in L,$$

where $\mathfrak{m}_B$ is the Möbius function on $L(B)$.

Proof. Let $\delta \in [\hat{0}, \pi]$. Since $\delta(B) \in L(B)$ for all $B \in \pi$, there is a canonical order-preserving map from $[\hat{0}, \pi]$ to $\prod_{B \in \pi} L(B)$. We now construct the inverse of this map. Let $\mathcal{S}$ be the set of splits generating L. For every $B \in \pi$, define $\mathcal{S}(B) = \{\nu \in \mathcal{S} : \nu(B) \neq B\}$ and $\mathcal{S}_B = \{\nu(B) : \nu \in \mathcal{S}(B)\}$. Then $L(B)$ is a partition lattice generated by splits in $\mathcal{S}_B$ and there is an obvious bijection between $\mathcal{S}_B$ and $\mathcal{S}(B)$. In addition, define $\mathcal{S}_\pi = \{\nu \in \mathcal{S} : \nu \geq \pi\}$ and note that $\mathcal{S}_\pi = \mathcal{S} \backslash \bigcup_{B \in \pi} \mathcal{S}(B)$. Suppose we are given a collection of partitions $\delta_B \in L(B)$ for $B \in \pi$, which is an element of $\prod_{B \in \pi} L(B)$. Each δ_B is generated by some partitions in $\mathcal{S}(B)$. Now, for each $B \in \pi$, we are using the isomorphism between $\mathcal{S}_B$ and $\mathcal{S}(B)$ to get the corresponding splits in $\mathcal{S}$. The meet of all these splits together with the splits in $\mathcal{S}_\pi$ defines some partition $\delta \in [\hat{0}, \pi]$. So the defined map is the inverse of the first map and is also order preserving. $\square$

4.1.4 *Lattices and their Möbius rings*

The aim of this section is to introduce the main technical lemma of this chapter given by Lemma 4.19. A standard way of proving this result can be derived from the proof of Theorem 4.27 in Aigner [1997]. Here, we present an alternative proof extending ideas in Section 3.9 of Stanley [2002]. Let L be a lattice and let $\mathbb{R}[L]$ be the polynomial ring over $\mathbb{R}$ with indeterminates $x \in L$. By $A(L)$ we denote the quotient (see Definition 2.12) $\mathbb{R}[L]/I$, where I is the ideal generated by $x \cdot y - x \wedge y$ for $x, y \in L$. We call $A(L)$ the *Möbius ring* of L.

For $f \in \mathbb{R}[L]$ we denote by $[f]$ the equivalence class of f in $A(L)$. As in (2.3) the operations in the Möbius ring are defined so that we have that for all $a, b \in \mathbb{R}$ and $f, g \in \mathbb{R}[L]$

$$[af + bg] = a[f] + b[g]$$
$$[f \cdot g] = [f] \cdot [g].$$

Also, by construction, for every $x, y \in L$ we have $[x \cdot y] = [x \wedge y]$.

Example 4.17. Let L be the Boolean lattice $\boldsymbol{B}_3$. Consider two linear polynomials in $\mathbb{R}[\boldsymbol{B}_3]$:

$$\rho_{12} = \{1,2\} - \{1\} - \{2\} + \emptyset \qquad \text{and} \qquad \rho_1 = \{1\} - \emptyset.$$

Then

$$\begin{aligned}[\rho_1] \cdot [\rho_{12}] &= [(\{1\} - \emptyset)] \cdot [(\{1,2\} - \{1\} - \{2\} + \emptyset)] = \\ &= [\{1\} \cdot \{1,2\}] - [\{1\} \cdot \{1\}] - [\{1\} \cdot \{2\}] + [\{1\} \cdot \emptyset] - \\ &\quad - [\emptyset \cdot \{1,2\}] + [\emptyset \cdot \{1\}] + [\emptyset \cdot \{2\}] - [\emptyset \cdot \emptyset] = \boldsymbol{0},\end{aligned}$$

where $\boldsymbol{0}$ is the zero of the ring $A(L)$.

We generalize the above example in the following way. Define for $x \in L$ the element $\rho_x \in \mathbb{R}[L]$ by

$$\rho_x = \sum_{y \leq x} \mathfrak{m}(y, x) y. \tag{4.6}$$

By the Möbius inversion formula,

$$x = \sum_{y \leq x} \rho_y. \tag{4.7}$$

Proposition 4.18. *For every $x \in L$, let ρ_x be given by (4.6). Then $[\rho_x] \cdot [\rho_x] = [\rho_x]$ and for every $x \neq y$*

$$[\rho_x] \cdot [\rho_y] = \mathbf{0}.$$

Proof. From (4.7) and the definition of $A(L)$ it follows that for every $x, y \in L$

$$[x] \cdot [y] \quad = \quad [x \wedge y] \quad = \quad \sum_{z \leq x \wedge y} [\rho_z]. \tag{4.8}$$

On the other hand, we can expand separately x and y in terms of the ρ's, which implies that for every $x, y \in L$

$$[x] \cdot [y] \quad = \quad \sum_{u \leq x,\, w \leq y} [\rho_u] \cdot [\rho_w]. \tag{4.9}$$

First note that $[\rho_{\hat{0}}] \cdot [\rho_{\hat{0}}] = [\rho_{\hat{0}}]$. Taking $y = \hat{0}$ and comparing (4.8) and (4.9) gives that

$$\sum_{u \leq x} [\rho_u] \cdot [\rho_{\hat{0}}] = [\rho_{\hat{0}}] \cdot [\rho_{\hat{0}}].$$

If x covers $\hat{0}$ in L, then we obtain $[\rho_x] \cdot [\rho_{\hat{0}}] = [\rho_{\hat{0}}] \cdot [\rho_x] = \mathbf{0}$. Now by taking $y = x$ in (4.9) we obtain

$$[x] = [\rho_{\hat{0}}] \cdot [\rho_{\hat{0}}] + [\rho_x] \cdot [\rho_x].$$

Because $[\rho_{\hat{0}}] \cdot [\rho_{\hat{0}}] = [\rho_{\hat{0}}] = [\hat{0}]$ and $[\rho_x] = [x] - [\hat{0}]$ we obtain $[\rho_x] \cdot [\rho_x] = [\rho_x]$.

Let now x, y be arbitrary. By induction we may assume that $[\rho_u] \cdot [\rho_w] = \mathbf{0}$ for all $u < x$ and $w \leq y$ or for all $u \leq x$ and $w < y$. We can also assume that $[\rho_z] \cdot [\rho_z] = [\rho_z]$ for all $z < x \wedge y$. If $x = y$, then equating (4.8) with (4.9) gives

$$\sum_{z \leq x} [\rho_z] \quad = \quad \sum_{z < x} [\rho_z] + [\rho_x] \cdot [\rho_x],$$

which implies that $[\rho_x] = [\rho_x] \cdot [\rho_x]$. If $x \neq y$, then again equating (4.8) with (4.9) gives

$$\sum_{z \leq x \wedge y} [\rho_z] \quad = \quad \sum_{z < x \wedge y} [\rho_z] + [\rho_{x \wedge y}] \cdot [\rho_{x \wedge y}] + [\rho_x] \cdot [\rho_y].$$

Since, by induction, $[\rho_{x \wedge y}] \cdot [\rho_{x \wedge y}] = [\rho_{x \wedge y}]$ (both if $x \wedge y < x, y$ and if $x \wedge y \in \{x, y\}$) we conclude that $[\rho_x] \cdot [\rho_y] = \mathbf{0}$. $\square$

As an immediate corollary we get the following result.

Lemma 4.19. *Let L be a finite lattice with at least two elements, and let $\hat{1} \neq a \in L$. Then for any $y \neq \hat{1}$*

$$\sum_{x:x \wedge a=y} \mathfrak{m}(x, \hat{1}) = 0.$$

Proof. By (4.7) and by Proposition 4.18 we have

$$[a] \cdot [\rho_{\hat{1}}] = \left(\sum_{x \leq a} [\rho_x] \right) \cdot [\rho_{\hat{1}}] = \mathbf{0}, \text{ since } a \neq \hat{1}. \qquad (4.10)$$

On the other hand,

$$[a] \cdot [\rho_{\hat{1}}] = [a] \cdot \sum_{x \in L} \mathfrak{m}(x, \hat{1})[x] = \sum_{x \in L} \mathfrak{m}(x, \hat{1})[a \wedge x]. \qquad (4.11)$$

Since $a \cdot \rho_{\hat{1}}$ is an element in $\mathbb{R}[L]$, it is of the form $\sum_{y \in L} c_y y$ and hence $[a \cdot \rho_{\hat{1}}] = \sum_{y \in L} c_y [y]$. From (4.10) we conclude that $c_y = 0$ for all $y \in L$ and from (4.11) that $c_y = \sum_{x:x \wedge a=y} \mathfrak{m}(x, \hat{1})$. $\qquad \square$

4.2   Cumulants and binary L-cumulants

In this section, we discuss the links between cumulants and combinatorics. This will enable us to propose a generalization of cumulants, which in particular will give us a better understanding of tree models in the second part of the book.

4.2.1   Cumulants

Cumulants, like moments, give an alternative way of storing information about probability distributions. Their importance comes from the observation that many properties of random variables can be better represented by cumulants than by moments. The key features of cumulants are:

 (i) For independent random variables, the cumulant of a sum becomes the sum of cumulants.

 (ii) For random vectors with independent components, the joint cumulant is zero.

(iii) Where approximate normality is involved, high-order cumulants can be neglected.

Let $X = (X_1, \ldots, X_m)$ be a finite discrete random vector with values in $\mathcal{X} = \mathcal{X}_1 \times \cdots \times \mathcal{X}_m$, where $\mathcal{X}_i = \{0, \ldots, r_i\}$. Let

$$M_X(t) = \mathbb{E}(\exp(\langle t, X \rangle)),$$

where $t = (t_1, \ldots, t_m) \in \mathbb{R}^m$, and $\langle t, X \rangle = \sum_{i=1}^m t_i X_i$ be the moment generating function of X as defined in Section 3.1.3. The *cumulant generating function* is defined as

$$K_X(t) = \log M_X(t).$$

The moment generating function M_X and the cumulant generating function K_X are convex and analytic. The *cumulants of X* are defined in a similar way as the moments by derivatives of the cumulant generating function. Thus, for every $u \in \mathbb{N}_0^m$, denote by $\mathcal{A}(u)$ the multiset defined in (3.5). Then

$$k_{\mathcal{A}(u)} \quad := \quad D^u K_X(t)\Big|_{t=0} \quad = \quad \frac{\partial^{|u|}}{\partial t_1^{u_1} \cdots \partial t_n^{u_n}} K_X(t)\Big|_{t=0} \qquad (4.12)$$

is the corresponding cumulant of the vector $X_{\mathcal{A}(u)} = (X_i)_{i \in \mathcal{A}(u)}$. Hence the cumulant generating function can be written as

$$K_X(t) \quad = \quad \sum_{u \in \mathbb{N}_0^m} \frac{1}{u!} k_{\mathcal{A}(u)} t^u.$$

Remark 4.20. It follows from Marcinkiewicz [1939], Lukacs [1958] that the normal distribution is the only probability distribution such that all cumulants of order greater than a certain number vanish. In a more algebraic language, the normal distribution is the only distribution such that its cumulant generating function $K_X(t)$ is a polynomial function.

The relationship between cumulants and moments can be found by taking the Taylor expansion of the both sides of the identity $K_X(t) = \log(M_X(t))$. For example, in the univariate case it is easily checked that

$$k_1 = \mu_1, \quad k_{11} = \mu_{11} - \mu_1^2, \quad k_{111} = \mu_{111} - 3\mu_{11}\mu_1 + 2\mu_1^3. \qquad (4.13)$$

We now explore this relationship in a more combinatorial context; see for example [McCullagh, 1987, Section 2.3].

Theorem 4.21. *The cumulant of the vector X is defined as*

$$k_{[m]} \quad = \quad \sum_{\pi \in \mathbf{\Pi}_m} (-1)^{|\pi|-1}(|\pi|-1)! \prod_{B \in \pi} \mu_B, \qquad (4.14)$$

where the sum is over all set partitions of $[m]$, the product is over all blocks of a partition, and $|\pi|$ denotes the number of blocks of π.

For example, if $m = 3$, then there are five partitions in $\mathbf{\Pi}_3$: 123, 1|23, 2|13, 12|3, and 1|2|3 and (4.14) gives

$$k_{123} \quad = \quad \mu_{123} - \mu_1\mu_{23} - \mu_2\mu_{13} - \mu_{12}\mu_3 + 2\mu_1\mu_2\mu_3. \qquad (4.15)$$

The formula for $k_{[m]}$ in Theorem 4.21 can be easily extended to any k_A for $A \subseteq [m]$ just by replacing $\mathbf{\Pi}_m$ with $\mathbf{\Pi}(A)$. What is less clear is that the same procedure can be used to generalize (4.14) to any multiset $A = \{1^{u_1}, \ldots, m^{u_m}\}$. Let $d = \sum_i u_i$ and write A as $\{i_1, \ldots, i_d\}$. We use the bijection between A and $[d]$ to write

$$k_A \quad = \quad \sum_{\pi \in \mathbf{\Pi}([d])} (-1)^{|\pi|-1}(|\pi|-1)! \prod_{B \in \pi} \mu_{i_B}, \qquad (4.16)$$

where $i_B = \{i_j : j \in B\}$. Hence, for instance, the formula for k_{111} in (4.13) can be derived from the formula for k_{123} in (4.15) by replacing $\{1, 2, 3\}$ by $\{1, 1, 1\}$. Similarly,

$$k_{112} \quad = \quad \mu_{112} - 2\mu_1\mu_{12} - \mu_{11}\mu_2 + 2\mu_1^2\mu_2. \qquad (4.17)$$

Formula (4.16) also shows that k_A depends only on μ_B for which B is a submultiset of A. Thus, in our computation, instead of working with the generating function $M(t)$, we can take the truncation of its power series up to order $u \in \mathbb{N}_0^m$, where u is such that $A = \mathcal{A}(u)$. So, for example, to obtain formulas in (4.13) it is enough to consider the polynomial

$$1 + \mu_1 t + \frac{1}{2}\mu_{11}t^2 + \frac{1}{6}\mu_{111}t^3.$$

We obtain k_1, k_{11}, and k_{111} by computing derivatives and evaluating them at zero. Equation (4.13) can be verified using the following Mathematica code

```
M[t_]:=1+Sum[m[i]*t^i/Factorial[i],{i,1,3}]
List[D[Log[M[t]],{t,1}],D[Log[M[t]],{t,2}],
                 D[Log[M[t]],{t,3}]] /. t-> 0
```

Similarly, to confirm (4.17) we run

```
M[s_,t_]:=Sum[m[i,j]*s^i*t^j/(Factorial[i]*
              Factorial[j]),{i,0,3},{j,0,3}] /. m[0,0]->1
D[Log[M[s,t]],{s,2},{t,1}] /. {s-> 0,t-> 0}
```

In certain situations we may want to generalize the well-known formula for the covariance in terms of conditional covariance

$$k_{12} = \mathrm{cov}(X_1, X_2) = \mathrm{cov}(\mathbb{E}[X_1|Y], \mathbb{E}[X_2|Y]) + \mathbb{E}[\mathrm{cov}(X_1, X_2|Y)].$$

It is not hard to provide a formula for general conditional cumulants; see Brillinger [1969]. For any $A \subseteq [m]$ denote the corresponding conditional cumulant by $k_{A|Y}$. Each of these is a random variable itself and we have

$$k_A \quad = \quad \sum_{\pi \in \mathbf{\Pi}(A)} k((k_{B|Y})_{B\in\pi}),$$

where $k((k_{B|Y})_{B\in\pi})$ denotes the order $|\pi|$ cumulant of random variables $k_{B|Y}$ for $B \in \pi$. In particular, if all X_i are conditionally independent given Y, then $k_{B|Y} = 0$ unless $|B| = 1$, which simplifies the above sum to one term.

4.2.2   Binary L-cumulants

We assume in this section that $\mathcal{X} = \{0, 1\}^m$, in which case $\mathcal{A}(\mathcal{X})$, defined by (3.5), becomes the set of all subsets of $[m]$. Let $L \subseteq \mathbf{\Pi}_m$ be a partition lattice, c.f. Definition 4.13. For every $A \subseteq [m]$, consider $L(A)$ as the subposet of $\mathbf{\Pi}(A)$ obtained from L by constraining each partition to the subset A.

The Möbius function on $L(A)$ is also denoted by $\mathfrak{m}$ unless it may lead to ambiguity, in which case we write explicitly $\mathfrak{m}_A$.

A *multiplicative function* on a partition lattice L is any function such that there exists a collection $f_B \in \mathbb{R}$ for $B \subseteq [m]$, and for every $\pi \in L$

$$f(\pi) = \prod_{B \in \pi} f_B.$$

For every $\nu \in \mathbf{\Pi}_m$, define

$$k(\nu) := \sum_{\pi \leq \nu} \mathfrak{m}(\pi, \nu)\mu(\pi), \tag{4.18}$$

where $\mu(\pi) = \prod_{B \in \pi} \mu_B$ is a multiplicative function and the sum is taken over all π in the interval $[\hat{0}, \nu]$ in $\mathbf{\Pi}_m$. The Möbius function on $\mathbf{\Pi}_m$ satisfies $\mathfrak{m}(\pi) := \mathfrak{m}(\pi, [m]) = (-1)^{|\pi|-1}(|\pi| - 1)!$ for all $\pi \in \mathbf{\Pi}_m$. It follows, by (4.16), that the cumulant $k_{[m]}$ is equal to $k([m])$ defined by (4.18) with $\nu = [m]$. The formula for k_A, $A \subseteq [m]$, is obtained in the same way by replacing $\mathbf{\Pi}_m$ with $\mathbf{\Pi}(A)$.

To get the inverse formula for moments in terms of cumulants we need the following result.

Lemma 4.22. *Let $k(\nu)$ be given by (4.18). For every $\nu \in \mathbf{\Pi}(A)$, we have $k(\nu) = \prod_{B \in \nu} k_B$.*

Proof. By Lemma 4.16, every interval $[\hat{0}, \nu] \subseteq \mathbf{\Pi}(A)$ is isomorphic to the product $\prod_{B \in \nu} \mathbf{\Pi}(B)$. By Proposition 4.7, the Möbius function on a product of posets is equal to the product of Möbius functions for each individual factor. Hence, (4.18) can be rewritten as

$$k(\nu) = \prod_{B \in \nu} \left(\sum_{\delta \in \mathbf{\Pi}(B)} \mathfrak{m}_B(\delta)\mu(\delta) \right) = \prod_{B \in \nu} k_B,$$

which finishes the proof. $\qquad\square$

Now the inverse formula for moments in terms of cumulants follows directly by the Möbius inversion formula in Proposition 4.4 and Lemma 4.22. For every $A \subseteq [m]$ we have

$$\mu_A = \sum_{\pi \in \mathbf{\Pi}(A)} k(\pi) = \sum_{\pi \in \mathbf{\Pi}(A)} \prod_{B \in \pi} k_B, \tag{4.19}$$

where the second equation follows by Lemma 4.22. By Proposition 3.12, the probability distribution of a binary vector X is uniquely identified by the set of moments μ_A for all $A \subseteq [m]$. We call this set of moments *binary moments*. Similarly, the set of cumulants k_A for $A \subseteq [m]$ is called *binary cumulants*.

We now generalize the combinatorial definition of binary cumulants to *binary L-cumulants* ℓ_A for $A \subseteq [m]$.

Definition 4.23 (Binary L-cumulants). Let $X = (X_1, \ldots, X_m)$ be a binary random vector and let L be a partition lattice of $[m]$. For any $A \subseteq [m]$ and $\nu \in L(A)$ define

$$\ell(\nu) \quad = \quad \sum_{\pi \leq \nu} \mathfrak{m}_A(\pi, \nu)\mu(\pi), \qquad (4.20)$$

where $\mu(\pi) = \prod_{B \in \pi} \mu_B$. Then $\ell_A := \ell(A)$ is the L-*cumulant of* X_A . In other words,

$$\ell_A \quad = \quad \sum_{\pi \in L(A)} \mathfrak{m}_A(\pi) \prod_{B \in \pi} \mu_B \qquad \text{for every } A \subseteq [m]. \qquad (4.21)$$

The map in (4.20) is invertible with the inverse given by the Möbius inversion formula. In particular, for every $A \subseteq [m]$,

$$\mu_A \quad = \quad \sum_{\pi \in L(A)} \ell(\pi) \quad = \quad \sum_{\pi \in L(A)} \prod_{B \in \pi} \ell_B. \qquad (4.22)$$

The fact that $\ell(\pi)$ is equal to the product $\prod_{B \in \pi} \ell_B$ is proved in a similar way as in Lemma 4.22.

Suppose that we are given a statistical model M for a binary vector $X \in \mathcal{X} = \{0, 1\}^m$. Binary L-cumulants offer a polynomial change of coordinates from the raw probabilities. The map from $p(x)$ for $x \in \mathcal{X}$ to moments μ_B for $B \subseteq [m]$ is a simple linear map of the form

$$\mu_B \quad = \quad \sum_{x:\, \mathcal{A}(x) \supseteq B} p(x). \qquad (4.23)$$

By definition, for every $A \subseteq [m]$, the maximal and minimal element of the lattice $L(A)$ coincide with the minimal and maximal element of $\mathbf{\Pi}(A)$. In particular, for every L we have $\ell_i = \mu_i$ for $i = 1, \ldots, n$; and $\ell_{ij} = \mu_{ij} - \mu_i \mu_j$ for all $1 \leq i < j \leq n$. However, already when $m = 3$ not all L-cumulants coincide with cumulants.

Example 4.24. Let $m = 3$ and consider L-cumulants induced by the lattice of interval partitions. The lattice $\mathcal{I}_3$ has four elements: 123, $1|23$, $12|3$, and $1|2|3$. The Möbius function satisfies $\mathfrak{m}(\pi) = (-1)^{|\pi|-1}$ and we have

$$\ell_{123} \quad = \quad \mu_{123} - \mu_1 \mu_{23} - \mu_{12} \mu_3 + \mu_1 \mu_2 \mu_3.$$

Compare this with the formula for k_{123} in (4.15) to note that not only term $\mu_2 \mu_{13}$ is missing now in the formula for ℓ_{123}, but also the coefficient of $\mu_1 \mu_2 \mu_3$ is 1 not 2.

Unlike in the case of cumulants, for general L-cumulants no generating function is known. It may be then useful to realize that L-cumulants can be expressed in terms of classical cumulants in a rather simple manner.

Proposition 4.25. *Let $L \subseteq \mathbf{\Pi}_m$ be a partition lattice and let $\mathbf{\Pi}^*$ denote the set of elements $\pi \in \mathbf{\Pi}_m$ such that $[\pi, [m]] \cap L = \{[m]\}$, where the interval $[\pi, [m]]$ is taken in $\mathbf{\Pi}_m$. We have*

$$\ell_{[m]} = \sum_{\pi \in \mathbf{\Pi}^*} k(\pi) = \sum_{\pi \in \mathbf{\Pi}^*} \prod_{B \in \pi} k_B.$$

The formula for ℓ_A for $A \subseteq [m]$ is obtained in the same way.

Proof. In this proof, $\delta \leq_\Pi \pi$ means that $\delta \leq \pi$ and $\delta \in \mathbf{\Pi}_m$. Similarly, $\pi \geq_L \delta$ denotes $\pi \geq \delta$ and $\pi \in L$. Expressing the L-cumulant in terms of moments and then the moments in terms of classical cumulants gives

$$\ell_{[m]} = \sum_{\pi \in L} \mathfrak{m}(\pi) \prod_{B \in \pi} \left(\sum_{\delta_B \in \mathbf{\Pi}(B)} \prod_{C \in \delta_B} k_C \right) =$$
$$= \sum_{\pi \in L} \mathfrak{m}(\pi) \sum_{\delta \leq_\Pi \pi} \prod_{B \in \delta} k_B.$$

For every $\delta \in \mathbf{\Pi}_m$, let $\bar{\delta}$ denote the smallest element of L such that $\delta \leq_\Pi \bar{\delta}$. Then, by changing the order of summation, the above equation can be rewritten as

$$\ell_{[m]} = \sum_{\delta \in \mathbf{\Pi}_m} \prod_{B \in \delta} k_B \left(\sum_{\pi \geq_L \bar{\delta}} \mathfrak{m}(\pi) \right).$$

By (4.3), the sum in brackets vanishes whenever $\bar{\delta} \neq [m]$ and is equal to 1 if $\bar{\delta} = [m]$. Therefore the whole expression is equal to $\sum_{\delta \in \mathbf{\Pi}^*} \prod_{B \in \delta} k_B$. □

Example 4.26. Consider again L-cumulants defined by interval partitions in Example 4.24. If $m = 3$, we have $\mathbf{\Pi}^* = \{123, 2|13\}$ and hence, by Proposition 4.25, $\ell_{123} = k_{123} + k_2 k_{13}$, which we easily verify directly.

4.2.3   Basic properties of binary L-cumulants

Lemma 3.17 shows how, under marginal independences, joint moments factorize. In the binary case this gives the following result.

Lemma 4.27. *Let $X = (X_1, \ldots, X_m)$ be a binary random vector. If $\pi_0 = B_1 | \cdots | B_r$ is a partition of $[m]$, then $B_1 \perp\!\!\!\perp \cdots \perp\!\!\!\perp B_r$ if and only if*

$$\mu_A = \mu(\pi_0(A)) = \mu_{B_1 \cap A} \cdots \mu_{B_r \cap A} \qquad \text{for all } A \subseteq [m],$$

where $\mu(\pi) = \prod_{B \in \pi} \mu_B$ and $\pi(A)$ is the partition π constrained to elements in A.

The following result is well known for binary cumulants.

Proposition 4.28. *Let X be a binary random vector. There exists a partition $\pi_0 = B_1 | \cdots | B_r$ of $[m]$ such that $B_1 \perp\!\!\!\perp \cdots \perp\!\!\!\perp B_r$ if and only if $k_A = 0$ for all $A \subseteq [m]$ unless $A \subseteq B_i$ for some i.*

Proof. The proof will follow from Theorem 4.31. □

As an example consider the cumulant k_{123} given in (4.15). If $1 \perp\!\!\!\perp \{2,3\}$, then, by Lemma 4.27, $\mu_{123} = \mu_1\mu_{23}$, $\mu_{12} = \mu_1\mu_2$, and $\mu_{13} = \mu_1\mu_3$. It follows that (4.15) becomes

$$k_{123} = \mu_1\mu_{23} - \mu_1\mu_{23} - \mu_1\mu_2\mu_3 - \mu_1\mu_2\mu_3 + 2\mu_1\mu_2\mu_3 = 0.$$

It turns out that Proposition 4.28 is a special instance of a more general result linking vanishing L-cumulants and the independence statements.

Definition 4.29. We say that a partition lattice L of $[m]$ is *saturated* if the set $\mathcal{S}_0$ of all trivial splits (see Definition 4.15) is contained in L.

Among the partition lattices in Definition 4.15, only the lattice of interval partitions is *not* saturated. Note also that for every $m \leq 3$ there exists precisely one saturated partition lattice.

Lemma 4.30. *If L is saturated, then all one-cluster partitions lie in L.*

Proof. By definition, every partition lattice shares the meet operator with $\mathbf{\Pi}_m$. Hence, $\mathcal{S}_0 \subset L$ implies that $\langle \mathcal{S}_0 \rangle \subseteq L$. The result follows because $\mathcal{C}_m = \langle \mathcal{S}_0 \rangle$. □

Theorem 4.31. *Let $L \subseteq \mathbf{\Pi}_m$ be a partition lattice and consider the L-cumulant of $X = (X_1, \ldots, X_m)$ as in Definition 4.23. Let $\pi_0 = B_1 | \cdots | B_r$ be a fixed partition in L. The following are equivalent:*

(i) $B_1 \perp\!\!\!\perp \cdots \perp\!\!\!\perp B_r$.

(ii) $\mu_A = \mu(\pi_0(A))$ for all $A \subseteq [m]$.

(iii) $\ell_A = 0$ unless A is contained in a block of π_0.

Moreover, if L is saturated, then the above three conditions are equivalent to any of the following:

(iv) $\mu(\pi) = \mu(\pi \wedge \pi_0)$ for every $\pi \in L$.

(v) $\ell(\pi) = 0$ for all $\pi \not\leq \pi_0$.

Proof. The equivalence of (i) and (ii) is given by Lemma 4.27. To show that (ii)⇒(iii), note that for every $A \subseteq [m]$

$$\ell_A = \sum_{\pi \in L(A)} \mathfrak{m}_A(\pi) \prod_{B \in \pi} \mu_B \overset{(ii)}{=} \sum_{\pi \in L(A)} \mathfrak{m}_A(\pi) \prod_{B \in \pi} \mu(\pi_0(B)) =$$

$$= \sum_{\pi \in L(A)} \mathfrak{m}_A(\pi)\mu(\pi \wedge \pi_0(A)) \overset{a:=\pi_0(A)}{=} \sum_{\delta \leq a} \left(\sum_{\pi \in L(A), \pi \wedge a = \delta} \mathfrak{m}_A(\pi) \right) \mu(\delta),$$

where, by Lemma 4.19, each summand in the last formula is zero whenever

$a = \pi_0(A) \neq A$ or equivalently A is not contained in a block of π_0, which proves (iii). To show that (iii) implies (ii), use (4.22) to write

$$\mu_A = \sum_{\pi \in L(A)} \prod_{B \in \pi} \ell_B \overset{(iii)}{=} \sum_{\pi \leq \pi_0(A)} \prod_{B \in \pi} \ell_B = \prod_{B \in \pi_0(A)} \left(\sum_{\pi \in L(B)} \prod_{C \in \pi} \ell_C \right),$$

which implies (ii).

Now suppose that L is saturated. If (ii) holds, then

$$\mu(\pi) \overset{\text{def.}}{=} \prod_{B \in \pi} \mu_B \overset{(ii)}{=} \prod_{B \in \pi} \mu(\pi_0(B)) = \mu(\pi \wedge \pi_0)$$

and hence (iv) holds. To show (iv)$\Rightarrow$(ii) we use the fact that L is saturated. In this case, by Lemma 4.30, the one-cluster partition with the cluster given by A lies in L. Fix π to be that partition. By (iv), $\mu(\pi) = \mu(\pi \wedge \pi_0)$. Dividing both sides of this equation by $\prod_{i \in [m] \setminus A} \mu_i$ yields (ii). Here, we could divide by $\prod_{i \in [m] \setminus A} \mu_i$ because X is non-degenerate and hence $\mu_i \neq 0, 1$. We now prove (iv)$\Rightarrow$(v). Using Definition 4.23 we obtain

$$\ell(\nu) \overset{\text{def.}}{=} \sum_{\pi \leq \nu} \mathfrak{m}(\pi, \nu) \mu(\pi) \overset{(iii)}{=} \sum_{\pi \leq \nu} \mathfrak{m}(\pi, \nu) \mu(\pi \wedge \pi_0) =$$

$$= \sum_{\delta \leq \nu} \left(\sum_{\pi \wedge \pi_0 = \delta} \mathfrak{m}(\pi, \nu) \right) \mu(\delta),$$

where the inner sum in the last expression is over all $\pi \in L$ such that $\pi \leq \nu$ and $\pi \wedge \pi_0 = \delta$ (or $\pi \wedge (\pi_0 \wedge \nu) = \delta$). If $\nu \not\leq \pi_0$, then

- $\pi_0 \wedge \nu \neq \nu$,
- $L' = [\hat{0}, \nu]$ is a lattice with at least two elements, and
- $\pi \wedge \pi_0 = (\pi \wedge (\pi_0 \wedge \nu))$ for every $\pi \in L'$ ($\pi \leq \nu$).

Denoting $\tilde{\pi}_0 = \pi_0 \wedge \nu$ we use Lemma 4.19 for L' to obtain that for every $\delta \neq \nu$ ($\delta \in L'$) the sum $\sum_{\pi \wedge \tilde{\pi}_0 = \delta} \mathfrak{m}(\pi, \nu)$ is zero. Since $\pi \wedge \pi_0 = \pi \wedge \tilde{\pi}_0$, $\sum_{\pi \wedge \pi_0 = \delta} \mathfrak{m}(\pi, \nu)$ vanishes, which shows that $\ell(\nu) = 0$ unless $\nu \leq \pi_0$ and hence (v). To show (v)$\Rightarrow$(iv), note that if $\ell(\delta) = 0$ for all $\delta \not\leq \pi_0$, then for every $\pi \in L$

$$\mu(\pi) = \sum_{\delta \leq \pi} \ell(\delta) = \sum_{\delta \leq \pi \wedge \pi_0} \ell(\delta) = \mu(\pi \wedge \pi_0).$$

$\square$

Remark 4.32. We want to stress that the equivalence in Theorem 4.31 breaks down if $\pi_0 \notin L$. More precisely, if $\pi_0(A) \notin L(A)$, then ℓ_A is generically nonzero even if (i) and (ii) in Theorem 4.31 hold.

Example 4.33. Consider the situation of Example 4.24, where $m = 3$ and L-cumulants are defined by the lattice of interval partitions. If $X_1 \perp\!\!\!\perp (X_2, X_3)$, then $\mu_{123} = \mu_1 \mu_{23}$, $\mu_{12} = \mu_1 \mu_2$, and $\mu_{13} = \mu_1 \mu_3$. It follows that $\ell_{12} = \ell_{13} = \ell_{123} = 0$. On the other hand, the condition $X_2 \perp\!\!\!\perp (X_1, X_3)$ does not imply that $\ell_{123} = 0$ because in this case

$$\ell_{123} = \mu_2 \mu_{13} - \mu_1 \mu_2 \mu_3,$$

which is zero only when in addition $\mu_{13} = \mu_1 \mu_3$ and hence when $X_1 \perp\!\!\!\perp X_3$. Here there is no contradiction with Theorem 4.31 because $2|13 \notin \mathcal{I}_3$.

Theorem 4.31 shows one of the important applications of L-cumulants. Because Π_m contains all set partitions, all marginal independences imply that $k_{[m]} = 0$. By Remark 4.32, in the case of L-cumulants, only some of the independences imply vanishing. Hence, this new coordinate system can be designed to better fit the model under consideration. This concept will be explained in more detail for tree cumulants in the second part of the book.

An important property of central moments and cumulants is that they are invariant with respect to translations of the random vector X. It turns out that L-cumulants have the same feature whenever L is saturated.

Proposition 4.34. *Suppose that L is a saturated partition lattice. For a binary random vector X, let $\widetilde{X} = X + a$, where $a \in \mathbb{R}^n$ and, for every $A \subseteq [m]$, by $\widetilde{\ell}_A$ denote the corresponding L-cumulant of the subvector $\widetilde{X}_A$. Then $\widetilde{\ell}_i = \ell_i + a_i$ for all $i = 1, \ldots, n$ and $\widetilde{\ell}_A = \ell_A$ for any $A \subseteq [m]$ such that $|A| \geq 2$.*

Proof. Because $\ell_i = \mu_i = \mathbb{E}[X_i]$, it is clear that $\widetilde{\ell}_i = \mathbb{E}[X_i + a_i] = \ell_i + a_i$. Suppose that $|A| \geq 2$ and without loss of generality assume $A = [m]$. Since $a = \sum a_i e_i$, where the e_i's are the unit vectors in $\mathbb{R}^n$, it suffices to prove this result only in the case when a is such that a_1 is the only non-zero entry. In this case write $\widetilde{X}_1 = X_1 + a_1$ as $X_1 - \mu_1 + (a_1 + \mu_1)$, where $\mu_1 = \mathbb{E}X_1$ and $a_1 + \mu_1 = \mathbb{E}\widetilde{X}_1$. Let $\pi_0 = 1|\{2, \ldots m\} \in L$, then for every $\pi \in L$,

$$\widetilde{\mu}(\pi) = \mu(\pi) - \mu(\pi \wedge \pi_0) + \widetilde{\mu}(\pi \wedge \pi_0).$$

It follows that

$$\widetilde{\ell}_{[m]} = \sum_{\pi \in L} \mathsf{m}(\pi)\mu(\pi) - \\ - \sum_{\pi \in L} \mathsf{m}(\pi)\mu(\pi \wedge \pi_0) + \sum_{\pi \in L} \mathsf{m}(\pi)\widetilde{\mu}(\pi \wedge \pi_0). \quad (4.24)$$

Since L is a lattice and $\pi_0 \neq [m]$, by Lemma 4.19 we have that $\sum_{\pi \wedge \pi_0 = \nu} \mathsf{m}(\pi) = 0$ for each $\nu \in L$ and hence the second and third summands in (4.24) are zero. The proof is completed because the first summand is exactly $\ell_{[m]}$. $\square$

4.2.4   Central cumulants

Define *central binary L-cumulants* by replacing moments μ_B in (4.21) by central moments μ'_B. For every $A \subseteq [m]$ the corresponding central binary L-cumulant is denoted by ℓ'_A.

Lemma 4.35. *If L is saturated, then $\ell'_A = \ell_A$ for every $A \subseteq [m]$ such that $|A| \geq 2$.*

Proof. Central binary L-cumulants of X can be alternatively defined as binary L-cumulants of $\widetilde{X}$, where $\widetilde{X}_i = X_i - \mathbb{E}X_i$. The lemma follows from Proposition 4.34. $\qquad\square$

We now show that central moments are L-cumulants induced by the lattice of one-cluster partitions $\mathcal{C}_m = \langle \mathcal{S}_0 \rangle$.

Proposition 4.36. *Let X be a random vector with values in $\mathcal{X} = \{0,1\}^m$. Then the central moment μ'_A for $|A| \geq 2$ is equal to the corresponding L-cumulants induced by $\mathcal{C}(A)$.*

Proof. Let $A \in \mathcal{A}(\mathcal{X})$ be such that $|A| \geq 2$. Denote the L-cumulants induced by $\mathcal{C}(A)$ by $\mathfrak{c}_A$. Since $\mathcal{C}(A)$ is saturated, by Lemma 4.35, we can write $\mathfrak{c}_A$ in terms of the central moments

$$\mathfrak{c}_A \;=\; \sum_{\pi \in \mathcal{C}(A)} \mathfrak{m}(\pi) \prod_{B \in \pi} \mu'_B \qquad \text{for all } |A| \geq 2.$$

However, $\mu'_i = 0$ for every $i \in [m]$ and hence the only non-zero term of the above sum is where $\pi = A$, which proves that $\mathfrak{c}_A = \mu'_A$. $\qquad\square$

The correspondence between the lattice of one-cluster partitions and central moments gives also the following explicit formula for central moments in terms of moments.

Proposition 4.37. *Let X be a random vector with values in $\mathcal{X}$. For every $A \in \mathcal{A}(\mathcal{X})$ such that $|A| \geq 2$ we have:*

$$\mu'_A \;=\; \sum_{B \subseteq A} (-1)^{|A \setminus B|} \mu_B \prod_{i \in A \setminus B} \mu_i = \tag{4.25}$$

$$=\; \sum_{B \subseteq A, |B| \geq 2} (-1)^{|A \setminus B|} \mu_B \prod_{i \in A \setminus B} \mu_i + (-1)^{|A|-1} (|A|-1) \prod_{i=1}^{m} \mu_i.$$

If L is saturated we sometimes use the following strategy to compute the corresponding L-cumulants.

Proposition 4.38. *Let $L = \langle \mathcal{S} \rangle$, where $\mathcal{S}$ is a set of splits of $[m]$ containing the set $\mathcal{S}_0$ of all trivial splits. If $K = \langle \mathcal{S}' \rangle$ is such that $\mathcal{S}' \cup \mathcal{S}_0 = \mathcal{S}$, then*

$$\sum_{\pi \in L} \mathfrak{m}_L(\pi) \prod_{B \in \pi} \mu_B \;=\; \sum_{\pi \in K} \mathfrak{m}_K(\pi) \prod_{B \in \pi} \mu'_B.$$

In particular, the formula for L-cumulants in terms of moments can be written as a composition of the change from moments to central moments and then from central moments to K-cumulants.

Proof. Since $\mathcal{S}_0 \subseteq \mathcal{S}$, then by Lemma 4.35

$$\sum_{\pi \in L} \mathfrak{m}_L(\pi) \prod_{B \in \pi} \mu_B \;=\; \sum_{\pi \in L} \mathfrak{m}_L(\pi) \prod_{B \in \pi} \mu'_B.$$

Since $\mu'_i = 0$ for every $i \in [m]$, we can replace "$\pi \in L$" with "$\pi \in K$" in the sum on the right. It now suffices to show that $\mathfrak{m}_L(\pi) = \mathfrak{m}_K(\pi)$ for all $\pi \in K \subseteq L$ such that π does not contain singleton blocks. It is enough to show that the intervals $[\pi, \hat{1}]$ in K and L are equal (as sets). Let $\delta \in L$ be such that $\pi < \delta < \hat{1}$. Since π has no singleton blocks, δ also has no singleton blocks and hence δ is a *meet* of non-trivial splits in $\mathcal{S}$. However, by assumption, the set of non-trivial splits of $\mathcal{S}$ is contained in $\mathcal{S}'$. It follows that $\delta \in K$. By Proposition 4.11 the Möbius functions $\mathfrak{m}_L$ and $\mathfrak{m}_K$ are equal when constrained to the interval $[\pi, \hat{1}]$. $\qquad\square$

Proposition 4.38 is helpful in a situation, in which computing the Möbius function for L is complicated and it is simple for K. We will see such an example in Section 6.3.1.

4.3   Tensors and discrete measures

In this section we present how discrete probability distributions can be efficiently represented by tensors. We use the basic theory of tensors introduced in Section 2.3. The results presented below can be used in computations.

4.3.1   *Tensor notation in statistics*

Let $\mathcal{X} = \mathcal{X}_1 \times \cdots \times \mathcal{X}_m$, where $\mathcal{X}_i = \{0, \ldots, r_i\}$ and let $p \in \mathbb{R}^{\mathcal{X}}$ be the probability distribution tensor of a random variable $X \in \mathcal{X}$. For $A \subseteq [m]$, let V_A denote the tensor space $\mathbb{R}^{\mathcal{X}_A}$. The marginal distribution p_A is a tensor in V_A defined in terms of coordinates by (3.2). The marginalization in (3.2) can be defined in terms of multilinear transformations. We illustrate this first with an example.

Example 4.39. Suppose that $m = 3$ and $r_1 = r_2 = r_3 = 1$ so that $X \in \{0, 1\}^3$. The marginal distribution p_{12} is a 2×2 tensor obtained from the joint distribution tensor p by $p_{12} = (I_2, I_2, \mathbf{1}_2^T) \cdot p$, where I_2 is the 2×2 identity matrix and $\mathbf{1}_2$ is a vector of ones of length two. Using the vec-mat notation defined in Section 2.3.3 we have $\text{vec}(p_{12}) = \text{mat}(I_2 \otimes I_2 \otimes \mathbf{1}_2^T) \cdot \text{vec}(p)$, where $\text{mat}(I_2 \otimes I_2 \otimes \mathbf{1}_2^T)$ is the Kronecker product of I_2, I_2 and $\mathbf{1}_2^T$:

$$
\begin{bmatrix} p_{12}(0,0) \\ p_{12}(0,1) \\ p_{12}(1,0) \\ p_{12}(1,1) \end{bmatrix}
=
\begin{bmatrix}
1 & 1 & 0 & 0 & 1 & 1 & 0 & 0 \\
0 & 0 & 1 & 1 & 0 & 0 & 1 & 1 \\
0 & 0 & 0 & 0 & 1 & 1 & 0 & 0 \\
0 & 0 & 0 & 0 & 0 & 0 & 1 & 1
\end{bmatrix}
\begin{bmatrix} p(0,0,0) \\ p(0,0,1) \\ p(0,1,0) \\ p(0,1,1) \\ p(1,0,0) \\ p(1,0,1) \\ p(1,1,0) \\ p(1,1,1) \end{bmatrix}.
$$

More generally, p_A is computed from p by a multilinear transformation $(A_1, \ldots, A_m)$, where $A_i = I_{r_i+1}$ if $i \in A$ and $A_i = \mathbf{1}_{r_i+1}^T$ otherwise.

For $p_B \in V_B$ we defined $p_{BB} = \mathrm{diag}_2(p_B) \in V_B \otimes V_B$ by

$$p_{BB}(x_B, y_B) = \begin{cases} p_B(x_B) & \text{if } x_B = y_B, \\ 0 & \text{otherwise.} \end{cases}$$

The inverse of p_{BB} is $p_{BB}^{-1} \in V_B^* \otimes V_B^*$, where V_B^* is the space dual to V_B. The conditional distribution defined by (3.3) is a tensor $p_{A|B} \in V_A \otimes V_B^*$, which in the tensor notation can be written as a contraction

$$(V_A \otimes V_B) \otimes (V_B^* \otimes V_B^*) \quad \to \quad V_A \otimes V_B^*,$$
$$p_{AB} \otimes p_{BB}^{-1} \quad \mapsto \quad p_{A|B}.$$

Indeed, by definition $p_{BB}(x_B', x_B) = \delta_{x_B x_B'} p_B(x_B)$ and hence writing this contraction explicitly in coordinates gives

$$p_{A|B} = \sum_{x_B'} p_{AB}(x_A, x_B') p_{BB}(x_B', x_B)^{-1} = p_{AB}(x_A, x_B) p_B(x_B)^{-1}.$$

Recall from Section 2.3.3 the definition of a flattening $p_{A;B}$ of a tensor p. In the probabilistic setting, $p_{A;B}$ is just the matrix representing the joint distribution of vectors X_A and X_B, where their possible values are ordered lexicographically. Using the vec-mat notation and flattenings we can rewrite this as

$$\mathrm{mat}(p_{A|B}) = p_{A;B} \cdot p_{B;B}^{-1},$$

where $p_{A;B}$ is defined by (2.12) and $p_{B;B}$ is the flattening of the tensor p_{BB}.

Recall from Section 3.1.2 that for any two disjoint subsets $A, B \subseteq [m]$ we have $A \perp\!\!\!\perp B$ (or $X_A \perp\!\!\!\perp X_B$), if and only if

$$p_{AB}(x_{AB}) = p_A(x_A) p_B(x_B) \qquad \text{for all } x \in \mathcal{X}.$$

In tensor notation this means that $p_{AB} \in V_{AB}$ can be written as

$$p_{AB} = p_A \otimes p_B \in V_A \otimes V_B.$$

Here $V_A \otimes V_B$ is isomorphic to V_{AB} and we identify both spaces. In the vec-mat notation we have

$$p_{A;B} = \mathrm{vec}(p_A) \cdot \mathrm{vec}(p_B)^T = (p_A \otimes p_B)_{A;B}.$$

The notion of independence can be generalized to the joint independence of any set partition of $[m]$. Given a partition $B_1 | \cdots | B_r$ of $B \subseteq [m]$ we have

$$B_1 \perp\!\!\!\perp \cdots \perp\!\!\!\perp B_r \qquad \text{if and only if} \qquad p_B = p_{B_1} \otimes \cdots \otimes p_{B_r}. \qquad (4.26)$$

The *full independence model* is given by $1 \perp\!\!\!\perp 2 \perp\!\!\!\perp \cdots \perp\!\!\!\perp m$. In geometry the full independence model corresponds to the Segre variety; see Section 2.4.

Proposition 4.40. *The nonnegative part of* $\mathrm{Seg}(\mathbb{P}^{r_1} \times \cdots \times \mathbb{P}^{r_n})$ *is isomorphic to the full independence model.*

This follows directly from the parameterization in (2.13).

Definition 4.41. Given disjoint subsets $B, C \subset [m]$ and a partition $B_1 | \cdots | B_r$ of B we say that $B_1, \ldots, B_r$ are conditionally independent given C if

$$p_{B|C}(\,\cdot\,|x_C) = p_{B_1|C}(\,\cdot\,|x_C) \otimes \cdots \otimes p_{B_r|C}(\,\cdot\,|x_C)$$

for each $x_C \in \mathcal{X}_C$. We write $B_1 \perp\!\!\!\perp \cdots \perp\!\!\!\perp B_r | C$. ☐

Suppose that $A \perp\!\!\!\perp B | C$ and X_C is not observed. We are then interested in the marginal distribution of (X_A, X_B). In tensor notation (3.4) can be written as a multilinear transformation

$$p_{AB} = (p_{A|C}, p_{B|C}) \cdot p_{CC}, \qquad (4.27)$$

where $p_{CC} = \mathrm{diag}_2(p_C) \in V_C \otimes V_C$ as defined in (2.8). Using (2.11) we can equivalently rewrite it using the vec-mat notation as

$$\mathrm{vec}(p_{AB}) = \mathrm{mat}(p_{A|C} \otimes p_{B|C}) \, \mathrm{vec}(p_{CC}).$$

This last representation admits a generalization to the case when $B_1 \perp\!\!\!\perp \cdots \perp\!\!\!\perp B_r | C$. We have

$$\mathrm{vec}(p_{B_1 \cdots B_r}) = \mathrm{mat}(p_{B_1|C} \otimes \cdots \otimes p_{B_r|C}) \, \mathrm{vec}(p_{C \cdots C}). \qquad (4.28)$$

Note that $\mathrm{vec}(p_{C \cdots C})$ is a sparse vector.

4.3.2   Alternative moment tensors

Let $X \in \mathcal{X}$, where $\mathcal{X} = \mathcal{X}_1 \times \cdots \mathcal{X}_m$ and $\mathcal{X}_i = \{0, \ldots, r_i\}$. By the moment aliasing principle of Proposition 3.12, the probability distribution p of X can be uniquely represented by the moments $M = [\mu_{\mathcal{A}(x)}]_{x \in \mathcal{X}}$ or central moments $M' = [\mu'_{\mathcal{A}(x)}]_{x \in \mathcal{X}}$. The problem is that in general the representation in terms of moments is more complicated than in the binary case. In particular, the conditional expectation does not have a linear form anymore (c.f. (3.10)). In this section we present a simple alternative. We show how appropriately organized marginal distributions can replace moments. This alternative representation resembles moments, it is simple to compute using linear transformations, and it specializes to moments in the binary case. Moreover, it enables us to work efficiently with conditional expectations.

Suppose temporarily that $\mathcal{X} = \mathcal{X}_1 \times \mathcal{X}_2$, where $\mathcal{X}_i = \{0, 1, \ldots, r_i\}$. Denote $p_{i+} = p_1(i)$ and $p_{+j} = p_2(j)$. By adding all rows of p to its first row and all its columns to the first column we obtain a new matrix

$$\boldsymbol{\mu}_{12} \quad = \quad \begin{bmatrix} 1 & p_{+1} & \cdots & p_{+r_2} \\ \hline p_{1+} & p_{11} & \cdots & p_{1r_2} \\ \vdots & & \cdots & \vdots \\ p_{r_1+} & p_{r_1 1} & \cdots & p_{r_1 r_2} \end{bmatrix}. \qquad (4.29)$$

We denote the top left corner of this matrix by $\mu_\emptyset = 1 \in \mathbb{R}$, the bottom left block by $\mu_1 \in \mathbb{R}^{r_1}$, the top right block by $\mu_2 \in \mathbb{R}^{r_2}$, and the remaining block by $\mu_{12} \in \mathbb{R}^{r_1 \times r_2}$.

To extend this construction in a notationally compact way to the general case, we can represent the row and column operations with elementary matrices that we will call $M(r)$ for $r \in \mathbb{N}$, defined as follows

$$M(r) = \left[\begin{array}{c|c} 1 & \mathbf{1}^T \\ \hline \mathbf{0} & I_r \end{array}\right], \quad \text{so} \quad M(r)^{-1} = \left[\begin{array}{c|c} 1 & -\mathbf{1}^T \\ \hline \mathbf{0} & I_r \end{array}\right]. \tag{4.30}$$

Here $\mathbf{0}, \mathbf{1} \in \mathbb{R}^r$ denotes the vector of zeros and ones, respectively, and I_r is the $r \times r$ identity matrix. The matrix $\dot{\mu}_{12}$ in (4.29) is equal to $(M(r_1), M(r_2)) \cdot p$ (c.f. Definition 2.51). For the case of general m where p is a higher-order tensor, for each multiset $A \subseteq [m]$ we let $M(A) := \otimes_{i \in A} M(r_i)$, and define

$$\dot{\mu}_A \quad := \quad M(A) \cdot p_A. \tag{4.31}$$

Note that $p_A = M(A)^{-1} \cdot \dot{\mu}_A$, where $M(A)^{-1} = \otimes_{i \in A} M(r_i)^{-1}$. Moreover, μ_A is defined as the block of $\dot{\mu}_A$ corresponding to entries with no 0 in their index.

Example 4.42. Suppose $m = 2$, $\mathcal{X}_1 = \{0, 1, 2\}$, and $\mathcal{X}_2 = \{0, 1\}$. Writing p_{i+} for $p_1(i)$, p_{+1} for $p_2(1)$, and p_{ij} for $p(i, j)$ we obtain

$$\dot{\mu}_{12} = \left[\begin{array}{c|c} 1 & p_{+1} \\ \hline p_{1+} & p_{11} \\ p_{2+} & p_{21} \end{array}\right] = \left[\begin{array}{c|c} 1 & \mu_2 \\ \hline \mu_1 & \mu_{12} \end{array}\right], \quad \dot{\mu}_1 = \left[\begin{array}{c} 1 \\ \hline p_{1+} \\ p_{2+} \end{array}\right] = \left[\begin{array}{c} 1 \\ \hline \mu_1 \end{array}\right], \quad \text{and}$$

$$\dot{\mu}_{11} = \left[\begin{array}{c|cc} 1 & p_{1+} & p_{2+} \\ \hline p_{1+} & p_{1+} & 0 \\ p_{2+} & 0 & p_{2+} \end{array}\right] = \left[\begin{array}{c|c} 1 & \mu_1 \\ \hline \mu_1 & \mu_{11} \end{array}\right], \quad \text{so}$$

$$\mu_{12} = \left[\begin{array}{c} p_{11} \\ p_{21} \end{array}\right], \quad \mu_1 = \left[\begin{array}{c} p_{1+} \\ p_{2+} \end{array}\right] \quad \text{and} \quad \mu_{11} = \left[\begin{array}{cc} p_{1+} & 0 \\ 0 & p_{2+} \end{array}\right].$$

Note that $X_1 \perp\!\!\!\perp X_2$ if and only if $\mu_{12} = \mu_1 \otimes \mu_2$. As in Example 4.39 we can write the transformation from p to $\dot{\mu}_{12}$ using the vec-mat notation. We obtain

$$\text{vec}(\dot{\mu}_{12}) = \text{mat}(M(2) \otimes M(1)) \cdot \text{vec}(p).$$

By explicitly computing the Kronecker product $\text{mat}(M(2) \otimes M(1))$, this can be written as

$$\left[\begin{array}{c} 1 \\ p_2(1) \\ p_1(1) \\ p(1,1) \\ p_1(2) \\ p(2,1) \end{array}\right] = \left[\begin{array}{cccccc} 1 & 1 & 1 & 1 & 1 & 1 \\ 0 & 1 & 0 & 1 & 0 & 1 \\ 0 & 0 & 1 & 1 & 0 & 0 \\ 0 & 0 & 0 & 1 & 0 & 0 \\ 0 & 0 & 0 & 0 & 1 & 1 \\ 0 & 0 & 0 & 0 & 0 & 1 \end{array}\right] \cdot \left[\begin{array}{c} p(0,0) \\ p(0,1) \\ p(1,0) \\ p(1,1) \\ p(2,0) \\ p(2,1) \end{array}\right].$$

Tensors $\dot{\boldsymbol{\mu}}_A$ for $A \subseteq [m]$ generalize regular moments in two ways. First, if $\mathcal{X} = \{0,1\}^m$, then $\boldsymbol{\mu}_A$ is a number that coincides with $\mu_A = \mathbb{E}[\prod_{i \in A} X_i]$ for every $A \subseteq [m]$. Moreover, for any discrete data, $\boldsymbol{\mu}_A$ is indeed a moment of a multivariate random variable.

Definition 4.43. Let $\mathbf{X}_i$ be a random variable with values in $\{0, e_1, \ldots, e_{r_i}\} \subset \mathbb{R}^{r_i}$, where e_i are unit vectors in $\mathbb{R}^{r_i}$. We fix the joint distribution of all $\mathbf{X}_i$ to be equal to the joint distribution of X, that is, the event $\{X_1 = x_1, \ldots, X_m = x_m\}$ corresponds to $\{\mathbf{X}_1 = \mathbf{x}_1, \ldots, \mathbf{X}_m = \mathbf{x}_m\}$, where $\mathbf{x}_i = \mathbf{0}$ if $x_i = 0$ and $\mathbf{x}_i = e_{x_i}$ otherwise. Define $\dot{\mathbf{X}}_i = (1, \mathbf{X}_i)$, so that $\dot{\mathbf{X}}_i \in \{e_0, e_0 + e_1, \ldots, e_0 + e_{r_i}\} \subseteq \mathbb{R}^{\mathcal{X}_i}$ and for every $A \subseteq [m]$ let

$$\mathbf{X}_A := \bigotimes_{i \in A} \mathbf{X}_i, \qquad \dot{\mathbf{X}}_A := \bigotimes_{i \in A} \dot{\mathbf{X}}_i.$$

The following result has a straightforward proof.

Proposition 4.44. *The tensor moment $\dot{\boldsymbol{\mu}}_A$ is equal to $\mathbb{E}[\dot{\mathbf{X}}_A]$. Moreover, $\boldsymbol{\mu}_A = \mathbb{E}[\mathbf{X}_A]$.*

Marginal independence implies factorization of joint moments.

Proposition 4.45. *Let $B_1 | \cdots | B_r$ be a partition of A. Then*

$$B_1 \perp\!\!\!\perp \cdots \perp\!\!\!\perp B_r \quad \text{if and only if} \quad \dot{\boldsymbol{\mu}}_A = \dot{\boldsymbol{\mu}}_{B_1} \otimes \cdots \otimes \dot{\boldsymbol{\mu}}_{B_r}.$$

This is equivalent to

$$\boldsymbol{\mu}_{A \cap C} = \boldsymbol{\mu}_{B_1 \cap C} \otimes \cdots \otimes \boldsymbol{\mu}_{B_r \cap C} \qquad \text{for all } C \subseteq A, |C| \geq 2.$$

Proof. By definition in (4.26), $B_1 \perp\!\!\!\perp \cdots \perp\!\!\!\perp B_r$ is equivalent to $p_A = p_{B_1} \otimes \cdots \otimes p_{B_r}$. Applying $M(A)$ on both sides yields $\dot{\boldsymbol{\mu}}_A = \dot{\boldsymbol{\mu}}_{B_1} \otimes \cdots \otimes \dot{\boldsymbol{\mu}}_{B_r}$. □

Next, we show how to generalize *central moments* in this setting. For every $i \in [m]$ let

$$M'(i) := \left[\begin{array}{c|c} 1 & \mathbf{0}^T \\ \hline -\boldsymbol{\mu}_i & I_{r_i} \end{array} \right] \cdot M(r_i), \qquad \text{so} \qquad (M'(i))^{-1} = M(r_i)^{-1} \cdot \left[\begin{array}{c|c} 1 & \mathbf{0}^T \\ \hline \boldsymbol{\mu}_i & I_{r_i} \end{array} \right]$$

and let $M'(A) := \bigotimes_{i \in A} M'(i)$. Now define

$$\dot{\boldsymbol{\mu}}'_A := M'(A) \cdot p_A. \tag{4.32}$$

We define $\boldsymbol{\mu}'_A$ as the block of $\dot{\boldsymbol{\mu}}'_A$ entries with no 0 in their index. To get a sense of how these quantities behave, observe that $\boldsymbol{\mu}'_i = 0$ for $i = 1, \ldots, m$. For any $i, j \in [m]$, we have

$$\dot{\boldsymbol{\mu}}_{ij} = \left[\begin{array}{c|c} 1 & \boldsymbol{\mu}_j \\ \hline \boldsymbol{\mu}_i & \boldsymbol{\mu}_{ij} \end{array} \right] \qquad \text{and} \qquad \dot{\boldsymbol{\mu}}'_{ij} = \left[\begin{array}{c|c} 1 & 0 \\ \hline 0 & \boldsymbol{\mu}_{ij} - \boldsymbol{\mu}_i \otimes \boldsymbol{\mu}_j \end{array} \right]$$

and hence $\boldsymbol{\mu}'_{ij} = \boldsymbol{\mu}_{ij} - \boldsymbol{\mu}_i \otimes \boldsymbol{\mu}_j$, and in particular,

$$\text{rank}(\dot{\boldsymbol{\mu}}_{ij}) = \text{rank}(\dot{\boldsymbol{\mu}}'_{ij}) = 1 + \text{rank}(\boldsymbol{\mu}'_{ij}).$$

This is convenient, because $\text{rank}(\dot{\boldsymbol{\mu}}_{ij}) = \text{rank}(p_{ij}) \leq 1$ if and only if $i \perp\!\!\!\perp j$. We record this as a proposition:

Proposition 4.46. *For any two $i, j \in [m]$, we have $i \perp\!\!\!\perp j \iff \boldsymbol{\mu}'_{ij} = \mathbf{0}$.*

4.3.3   Kronecker products and binary moments

In this section, $X \in \{0,1\}^m$ is a binary distribution with probability tensor p. By (4.31) the moment tensor $\dot{\mu}_{1\cdots m}$ is given by

$$\dot{\mu}_{1\cdots m} \;\;=\;\; A^{\otimes m} \cdot p,$$

where

$$A = \begin{bmatrix} 1 & 1 \\ 0 & 1 \end{bmatrix}.$$

By (4.32) the tensor of central moments is obtained using a multilinear transformation

$$\dot{\mu}'_{1\cdots m} \;\;=\;\; (B_1 \otimes \cdots \otimes B_m) \cdot p,$$

where

$$B_i = \begin{bmatrix} 1 & 1 \\ -\mu_i & 1 \end{bmatrix} \begin{bmatrix} 1 & 1 \\ 0 & 1 \end{bmatrix} = \begin{bmatrix} 1 & 1 \\ -p_i(1) & p_i(0) \end{bmatrix}.$$

These formulas give very efficient ways of computing moments for binary variables. In practice, we compute the Kronecker products $\boldsymbol{A} = \mathrm{mat}(A^{\otimes m})$ and $\boldsymbol{B} = \mathrm{mat}(B_1 \otimes \cdots \otimes B_m)$ and use the identities (see also [Teugels, 1990, Theorem 1])

$$\begin{aligned}
\mathrm{vec}(\dot{\mu}_{1\cdots m}) &= \boldsymbol{A}\,\mathrm{vec}(p), \\
\mathrm{vec}(\dot{\mu}'_{1\cdots m}) &= \boldsymbol{B}\,\mathrm{vec}(p).
\end{aligned} \tag{4.33}$$

The inverse expressions are easily obtained because $\boldsymbol{A}^{-1} = \mathrm{mat}((A^{-1})^{\otimes m})$ and $\boldsymbol{B}^{-1} = \mathrm{mat}(B_1^{-1} \otimes \cdots \otimes B_m^{-1})$, where

$$A^{-1} = \begin{bmatrix} 1 & -1 \\ 0 & 1 \end{bmatrix}, \qquad B_i^{-1} = \begin{bmatrix} p_i(0) & -1 \\ p_i(1) & 1 \end{bmatrix}.$$

It is also not hard to give an explicit form of $\boldsymbol{A}$ and $\boldsymbol{B}$. We provide such a formula for $\boldsymbol{A}$.

Lemma 4.47. *We have* $\boldsymbol{A}(\boldsymbol{x}, \boldsymbol{y}) = 1$ *if* $\boldsymbol{x} \leq \boldsymbol{y}$ *and* $\boldsymbol{A}(\boldsymbol{x}, \boldsymbol{y}) = 0$ *otherwise. By (4.1),* $\boldsymbol{A}$ *represents the zeta function on the Boolean lattice. By Möbius inversion,*

$$\boldsymbol{A}^{-1}(\boldsymbol{x}, \boldsymbol{y}) \;\;=\;\; \begin{cases} (-1)^{|\boldsymbol{y}-\boldsymbol{x}|} & \textit{if } \boldsymbol{x} \leq \boldsymbol{y}, \\ 0 & \textit{otherwise.} \end{cases}$$

4.4   Submodularity and log-supermodularity

4.4.1   Basic definitions

We say that a lattice L is *distributive* if for all $x, y, z \in L$,

$$x \wedge (y \vee z) = (x \wedge y) \vee (x \wedge z).$$

An important example is the Boolean lattice. In this section we fix a finite distributive lattice and consider functions $f : L \to \mathbb{R}$. We say that such a function is *submodular* if

$$f(x) + f(y) \;\geq\; f(x \vee y) + f(x \wedge y) \qquad \text{for all } x, y \in L.$$

Two alternative definitions are popular when L is the Boolean lattice of subsets of $[m]$.

Proposition 4.48. *A function* $f : 2^{[m]} \to \mathbb{R}$ *is submodular if and only if one of the following holds:*

1. *for all* $A \subseteq B$, $i \in [m] \setminus B$

$$f(A \cup \{i\}) - f(A) \geq f(B \cup \{i\}) - f(B)$$

2. *for all* $A \subseteq [m]$ *and* $i, j \in [m] \setminus A$

$$f(A \cup \{i\}) + f(A \cup \{j\}) \geq f(A \cup \{i, j\}) - f(A).$$

In this book we are going to use a notion closely related to submodularity.

Definition 4.49. A nonnegative set function $f : L \to [0, \infty)$ is *log-supermodular* if

$$f(x)f(y) \leq f(x \vee y)f(x \wedge y) \qquad \text{for all } x, y \in L.$$

Remark 4.50. A strictly positive function f is log-supermodular if and only if $-\log(f)$ is submodular.

Example 4.51. Let $L = \{0, 1\}^m$, where $\boldsymbol{x} \vee \boldsymbol{y}$ and $\boldsymbol{x} \wedge \boldsymbol{y}$ are given by componentwise maximum and minimum. The log-supermodular functions $p : \{0, 1\}^m \to [0, \infty)$ that satisfy the normalizing equation $\sum_{\boldsymbol{x} \in \{0,1\}^m} p(\boldsymbol{x}) = 1$ are said to satisfy the MTP$_2$ (multivariate total positivity of order 2) constraints. If $m = 2$, then there is only one constraint

$$p(0, 1)p(1, 0) \;\leq\; p(1, 1)p(0, 0). \tag{4.34}$$

If $m = 3$, then there are 9 constraints

$$p(0,0,1)p(1,1,0) \leq p(0,0,0)p(1,1,1) \quad p(0,1,0)p(1,0,1) \leq p(0,0,0)p(1,1,1)$$
$$p(1,0,0)p(0,1,1) \leq p(0,0,0)p(1,1,1) \quad p(0,1,1)p(1,0,1) \leq p(0,0,1)p(1,1,1)$$
$$p(0,1,1)p(1,1,0) \leq p(0,1,0)p(1,1,1) \quad p(1,0,1)p(1,1,0) \leq p(1,0,0)p(1,1,1)$$
$$p(0,0,1)p(0,1,0) \leq p(0,0,0)p(0,1,1) \quad p(0,0,1)p(1,0,0) \leq p(0,0,0)p(1,0,1)$$
$$p(0,1,0)p(1,0,0) \leq p(0,0,0)p(1,1,0).$$

Note that the inequality in (4.34) is equivalent to the covariance being nonnegative. There is a remarkable connection between the MTP$_2$ property and positive dependence, which we are going to present in the next section.

4.4.2   The Ahlswede–Daykin inequality

For any collections $\mathcal{C}, \mathcal{C}'$ of elements of L define

$$
\begin{aligned}
\mathcal{C} \vee \mathcal{C}' &= \{x \vee y : x \in \mathcal{C}, y \in \mathcal{C}'\} \\
\mathcal{C} \wedge \mathcal{C}' &= \{x \wedge y : x \in \mathcal{C}, y \in \mathcal{C}'\}.
\end{aligned}
$$

Theorem 4.52 (Four Function Theorem, Ahlswede and Daykin [1978]). *If $\alpha, \beta, \gamma, \delta$ are non-negative functions defined on a finite distributive lattice L satisfying*

$$
\alpha(x)\beta(y) \quad \leq \quad \gamma(x \vee y)\delta(x \wedge y) \qquad \text{for all } x, y \in L,
$$

and if $\mathcal{C}, \mathcal{C}'$ are any two collections of elements of L, then

$$
\alpha(\mathcal{C})\beta(\mathcal{C}') \quad \leq \quad \gamma(\mathcal{C} \vee \mathcal{C}')\delta(\mathcal{C} \wedge \mathcal{C}'),
$$

where we use short-hand notation $f(\mathcal{C}) = \sum_{x \in \mathcal{C}} f(x)$.

Theorem 4.52 has a couple of useful corollaries. For example, take $\alpha(x) = p(x)f(x)$, $\beta(x) = p(x)g(x)$, $\gamma(x) = p(x)$, and $\delta(x) = p(x)f(x)g(x)$, where f, g are decreasing functions ($f(x) \geq f(y)$ if $x \leq y$ in L) and where p is a log-supermodular function on L. If $\alpha(x)\beta(y) \leq \gamma(x \vee y)\delta(x \wedge y)$, then equivalently

$$
p(x)p(y)f(x)g(y) \quad \leq \quad p(x \vee y)p(x \wedge y)f(x \wedge y)g(x \wedge y).
$$

If p is a probability distribution, that is, $\sum_{x \in L} p(x) = 1$ holds, then for any collection $\mathcal{C}$

$$
\alpha(\mathcal{C}) = \sum_{x \in \mathcal{C}} p(x)f(x),
$$

$$
\beta(\mathcal{C}) = \sum_{x \in \mathcal{C}} p(x)g(x),
$$

$$
\gamma(\mathcal{C}) = \sum_{x \in \mathcal{C}} p(x) = p(\mathcal{C}),
$$

$$
\delta(\mathcal{C}) = \sum_{x \in \mathcal{C}} p(x)f(x)g(x).
$$

Note that $\frac{1}{p(\mathcal{C})} \sum_{x \in \mathcal{C}} p(x) = 1$ and hence

$$
p(x; \mathcal{C}) = \frac{1}{p(\mathcal{C})} p(x) \qquad \text{for } x \in \mathcal{C}
$$

is a probability distribution. Denote

$$
\mathbb{E}_{\mathcal{C}}[f] := \sum_{x \in \mathcal{C}} p(x; \mathcal{C})f(x), \qquad \operatorname{cov}_{\mathcal{C}}(f, g) := \mathbb{E}_{\mathcal{C}}[fg] - \mathbb{E}_{\mathcal{C}}[f]\mathbb{E}_{\mathcal{C}}[g].
$$

We obtain the following corollary.

Proposition 4.53 (The FKG inequality). *If p is a log-supermodular (MTP$_2$) probability distribution on L and if f and g are both decreasing (or both increasing) functions on L, then for any collection C of elements of L*

$$\operatorname{cov}_C(f, g) \geq 0.$$

Example 4.54. Consider again Example 4.51. Every monomial $\boldsymbol{x}^a$, where $a \in \mathbb{N}^m$, defines an increasing function on $\{0,1\}^m$. Let $m = 3$ and take C be all points in $\{0,1\}^3$ with the last coordinate 0. Then taking $f(\boldsymbol{x}) = x_1$, $g(\boldsymbol{x}) = x_2$ we obtain

$$\mathbb{E}_C[f] = \frac{1}{p(C)}(p(1,0,0) + p(1,1,0))$$

$$\mathbb{E}_C[g] = \frac{1}{p(C)}(p(0,1,0) + p(1,1,0))$$

$$\mathbb{E}_C[fg] = \frac{1}{p(C)}p(1,1,0)$$

and therefore for every probability distribution over $\{0,1\}^3$ that satisfies the MTP$_2$ constraints,

$$p(C)p(1,1,0) - (p(1,0,0) + p(1,1,0))(p(0,1,0) + p(1,1,0)) \geq 0.$$

The above quantity is just the conditional covariance of random variables X_1 and X_2 given $X_3 = 0$.

We use our standard notation $\mathcal{X} = \mathcal{X}_1 \times \cdots \times \mathcal{X}_m$, where $\mathcal{X}_i = \{0, \ldots, r_i\}$. A tensor $p = [p(\boldsymbol{x})] \in \mathbb{R}^{\mathcal{X}}$ is log-supermodular if

$$p(\boldsymbol{x}) \cdot p(\boldsymbol{y}) \quad \leq \quad p(\boldsymbol{x} \vee \boldsymbol{y}) \cdot p(\boldsymbol{x} \wedge \boldsymbol{y}) \quad \text{for any two } \boldsymbol{x}, \boldsymbol{y} \in \mathcal{X}, \qquad (4.35)$$

where $\vee, \wedge$ are the coordinate maximum and minimum, respectively. If L is a set is equal to $\mathcal{X}$, we prefer to use the following more direct result, which follows from Theorem 4.52 by taking all four functions to be equal.

Proposition 4.55. *Fix $m \geq 2$ and a set $\mathcal{X}$ as above and let p be a tensor in $\mathbb{R}^{\mathcal{X}}$. If p is log-supermodular, then for any two collections C, C', we have*

$$p(C) \cdot p(C') \leq p(C \vee C') \cdot p(C \wedge C').$$

In particular, the computation in Example 4.54 can also be easily derived from Proposition 4.55 by taking $\wedge, \vee$ to be the coordinatewise minimum and maximum, and by taking $C = \{(1,0,0), (1,1,0)\}$, $C' = \{(0,1,0), (1,1,0)\}$.

Theorem 4.56. *Let $\mathcal{X} = \prod_{i=1}^{m} \mathcal{X}_i$ and $\mathcal{X}' = \prod_{i=1}^{m-1} \mathcal{X}_i$. If $p \in \mathbb{R}^{\mathcal{X}}$ is log-supermodular, then the marginalization p' of p defined by*

$$p' = (\boldsymbol{I}, \ldots, \boldsymbol{I}, \boldsymbol{1}) \cdot p \in \mathbb{R}^{\mathcal{X}'}$$

is also log-supermodular. In particular, if p is a probability distribution of $\boldsymbol{X} = (X_1, \ldots, X_m)$, then every marginal distribution p_A is also log-supermodular.

Proof. Let $x, y \in \mathcal{X}'$ and $\mathcal{C}_x$ and $\mathcal{C}_y$ two collections of elements of $\mathcal{X}$ such that they agree with x and y over the first $m-1$ coordinates. By Proposition 4.55, $p(\mathcal{C}_x)p(\mathcal{C}_y) \leq p(\mathcal{C}_x \vee \mathcal{C}_y)p(\mathcal{C}_x \wedge \mathcal{C}_y)$. The result follows because $p(\mathcal{C}_x) = p'(x)$, $p(\mathcal{C}_y) = p'(y)$, $p(\mathcal{C}_x \vee \mathcal{C}_y) = p'(x \vee y)$, and $p(\mathcal{C}_x \wedge \mathcal{C}_y) = p'(x \wedge y)$. $\quad\square$

4.5   Bibliographical notes

Section 4.1 is mostly based on Chapter 3 in Stanley [2002]. Other good references are Davey and Priestley [2002] and Aigner [1997]. A special case of Lemma 4.19 was first formulated by Weisner [1935] and then the result was generalized by Rota et al. [1973]. Our proof strategy follows the proof of a similar result in [Stanley, 2002, Section 3.9]. The basic examples of partition lattices are motivated by the free probability theory in Lehner [2002], Speicher [1997]. Non-crossing partitions are analyzed in Kreweras [1972], Krawczyk and Speicher [2000], Speicher [1994, 1997]. The interval partitions are discussed for example in Speicher and Woroudi [1997]. Cumulants were first defined and studied by Danish scientist T.N. Thiele [1899] (cited after Hald [2000]) who called them semi-invariants. They were first called cumulants in Fisher and Wishart [1932]. In the definition of cumulants we underline the links with combinatorics (see Rota and Shen [2000], Speed [1983]) and computational algebraic geometry (see Pistone and Wynn [1999], Pistone and Wynn [2006]). The general references are Barndorff–Nielsen and Cox [1989], McCullagh [1987]. Binary cumulants were defined and used in algebraic geometry in Sturmfels and Zwiernik [2012] and Pistone and Wynn [2006]. The generalization to L-cumulants has been developed in Zwiernik [2012]. The application of these ideas in algebraic geometry is given in Ciliberto et al. [2014], Manivel and Michałek [2014], Michałek et al. [2015]. Our exposition on Ahlswede–Daykin inequality comes mostly from [Anderson, 1987, Section 6.2]. For the original result, see Ahlswede and Daykin [1978]. For the FKG inequality, see Fortuin et al. [1971]. For more on submodularity, see, for example, Bach [2011], Fujishige [2005].

Part II

Latent tree graphical models

Chapter 5

Phylogenetic trees and their models

[]
In this chapter, we introduce the main object of study of this book, which are trees and their statistical models. In the beginning, the main motivation will be to provide first examples of how various geometric spaces are associated to trees. We discuss the space of tree metrics, the space of phylogenetic oranges, and the latent tree model. Some necessary combinatorics, including tree splits and the Tuffley poset, provides us a language to describe these spaces and their special points. What we will try to indicate in this chapter and prove in the following ones is that these various tree spaces have many features in common and they should be studied together.

We start with some standard definitions.

5.1 Trees

5.1.1 Phylogenetic trees and semi-labeled trees

In Section 3.4.1 we introduced some basic graph-theoretic concepts. In this section we extend this material specializing to trees. A *tree* $T = (V, E)$ is a connected undirected graph with no cycles. In particular, for any two $u, v \in V$ there is a unique path between them, which we denote by $\overline{uv}$. A vertex of T that has only one neighbor is called a *leaf*. A vertex of T that is not a leaf is called an *inner vertex*. An edge of T is *inner* if both of its ends are inner vertices, otherwise it is called *terminal*. A connected subgraph of T is called a *subtree* of T.

A *rooted tree* T^r is a directed graph whose underlying undirected graph is a tree that has one distinguished vertex r, called the *root*, and all the edges are directed away from r. For every vertex v of a rooted tree T^r such that $v \in V \setminus r$, the set pa(v) of parents of v is a singleton. As an example consider the quartet tree in Figure 5.1 with one of its rooted versions, where the root is given by an inner vertex.

Figure 5.1: *A quartet tree and a rooted quartet tree.*

We often constrain ourselves to *binary trees*.

Definition 5.1. A tree T is *binary* if every inner vertex has exactly three neighbors. A rooted tree T^r is binary if every inner vertex has two children.

For example, the undirected quartet tree is a binary tree but its rooted version on the right in Figure 5.1 is not because the root has three children. In our analysis, leaves of a tree are always labeled and represented by solid vertices. More generally, we consider so-called semi-labeled trees.

Definition 5.2. A *semi-labeled tree* on $[m]$ is an ordered pair $\mathcal{T} = (T; \phi)$, where T is a tree with vertex set V and $\phi : [m] \to V$ is a map such that, for each $v \in V$ that has at most two neighbors, $v \in \phi([m])$. We say that $\mathcal{T} = (T; \phi)$ is a *phylogenetic tree* if ϕ is a bijection from $[m]$ to the set of leaves of T. In a similar way, we define a *rooted semi-labeled tree* and a *rooted phylogenetic tree*.

If $v \in \phi([m])$, then we say that v is *labeled* and we depict it by a solid vertex. A vertex that is not labeled is called *unlabeled*. Note that multiple labels at a vertex are allowed. The map ϕ, called the labeling map, is always implicit and hence typically it is omitted in our notation. The tree T of $\mathcal{T}$ is called the underlying tree of $\mathcal{T}$. The set of vertices of the underlying tree T is denoted by $V(\mathcal{T})$ and its set of edges by $E(\mathcal{T})$. An example of a semi-labeled tree is given in Figure 5.2.

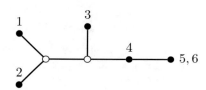

Figure 5.2: *A semi-labeled tree with the labeling set* $\{1, 2, 3, 4, 5, 6\}$.

Although we mainly consider geometric spaces associated to phylogenetic trees, the main reason to introduce semi-labeled trees is that they parameterize some special subspaces. More generally we need to consider *semi-labeled forests* that are disjoint unions of semi-labeled trees.

Definition 5.3 (Semi-labeled forest). Given a partition $\pi = B_1 | \cdots | B_r$ of $[m]$, consider a collection of semi-labeled trees $\mathcal{T}_{B_i}$ with labeling sets B_i. The union of these trees is called a *semi-labeled forest*

$$\mathcal{F} \;=\; \{\mathcal{T}_B : B \in \pi\}.$$

The partition π is called the partition of $\mathcal{F}$. Often we write $\mathcal{F}$ as a pair (F, ϕ), where F is the underlying forest and ϕ the labeling map.

As an example, consider the graph in Figure 5.3, which is a semi-labeled forest with partition $123|4|5|6$.

As we show below, the connection between set partitions and trees is

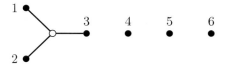

Figure 5.3: *A semi-labeled forest with the partition* $123|4|5|6$.

much deeper. We now define two basic operations on semi-labeled forests called *edge deletion* and *edge contraction*. This enables us later to introduce a suitable partial ordering on the set of semi-labeled forests.

Definition 5.4 (Edge deletion). Let $\mathcal{F} = (F; \phi)$ be a semi-labeled forest and let e be an edge of F. Then by $\mathcal{F} \setminus e$, denote the semi-labeled forest obtained from $\mathcal{F}$ by

1. removing from F the edge e,

2. *suppressing* all the resulting unlabeled vertices with only two neighbors.

Here in step 2, by suppressing we mean the following: if step 1 introduces a subgraph $u - v - w$ such that v is unlabeled and has no other neighbors apart from u and w, we remove v together with both its edges from F and add edge $u - w$.

It is easy to see that $\mathcal{F} \setminus e$ is indeed a semi-labeled forest. More generally, for a subset E' of edges of $\mathcal{F}$, by $\mathcal{F} \setminus E'$ denote the semi-labeled forest obtained from $\mathcal{F}$ by deleting edges of E' and then

(i) recursively deleting all degree-one unlabeled vertices

(ii) suppressing degree-two unlabeled vertices.

This will be illustrated by the following example.

Example 5.5. Consider the semi-labeled tree $\mathcal{T}$ in Figure 5.4. Let e_4, e_5, e_6 be edges incident with the labeled vertices $4, 5, 6$. Then the semi-labeled forest $\mathcal{T} \setminus \{e_4, e_5, e_6\}$ is given in Figure 5.3. It is obtained by first deleting the given edges. The resulting forest on the left in Figure 5.5 is then corrected by recursively deleting unlabeled degree-one vertices as in step (i). We then apply step (ii) to the resulting forest, which is given on the right in Figure 5.5.

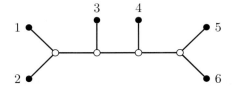

Figure 5.4: *A binary tree with six leaves.*

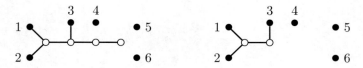

Figure 5.5: *An illustration for Example 5.5.*

Example 5.6. Consider the quartet tree in Figure 5.1. After removing the edges incident with the leaves 1 and 4 we obtain the following semi-labeled forest with no unlabeled vertices

$$
\begin{array}{cccc}
1 & 2 & 3 & 4 \\
\bullet & \bullet\!\!-\!\!\bullet & & \bullet
\end{array}
$$

Definition 5.7 (Edge contraction). Let $\mathcal{F} = (F; \phi)$ be a semi-labeled forest and let e be an edge of F. The semi-labeled forest obtained from $\mathcal{F}$ by identifying the ends of e and then deleting e, denoted $\mathcal{F}/e$, is said to be obtained from $\mathcal{F}$ by *contracting* e. The semi-labeled forest obtained from $\mathcal{F}$ by contracting each of the edges of E' is denoted by $\mathcal{F}/E'$.

Example 5.8. Consider again the tree $\mathcal{T}$ in Figure 5.4. Let e_4, e_5, e_6 be like in Example 5.5. Then the semi-labeled forest $\mathcal{T}/\{e_4, e_5, e_6\}$ is given in Figure 5.2.

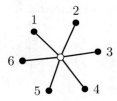

Figure 5.6: *A star tree with $m = 6$ leaves.*

Remark 5.9. More generally, any semi-labeled tree can be obtained in this way from a binary phylogenetic tree. For example, the *star tree* in Figure 5.6 is obtained from the binary tree in Figure 5.4 by contracting all inner edges.

If either $\mathcal{F}' = \mathcal{F} \setminus E'$ or $\mathcal{F}' = \mathcal{F}/E'$ for some set of edges E', then we say that $\mathcal{F}'$ is a *subforest* of $\mathcal{F}$. We discuss this concept and its link to set partitions in the next section.

5.1.2 Splits and $\mathcal{T}$-partitions

A *split* of $[m]$ is a partition of $[m]$ into two non-empty sets. Let now $\mathcal{T} = (T; \phi)$ be a semi-labeled tree with the labeling set $[m]$. A *tree split* (or a split *induced by* $\mathcal{T}$) is a split of the labeling set $[m]$ obtained by removing an edge of the underlying tree T and considering two connected components of $\mathcal{T} \setminus e$. The set of all tree splits induced by $\mathcal{T}$ is denoted by $\mathcal{S}(\mathcal{T})$. The labeling set $[m]$ is implicit in this notation.

Example 5.10. The splits of the quartet tree in Figure 5.1 are 1|234, 2|134, 3|124, 4|123, and 12|34.

Each split in $\mathcal{S}(\mathcal{T})$ is a partition of $[m]$ and thus an element of the lattice $\mathbf{\Pi}_m$ of all set partitions introduced in Section 4.1.3. Recall from Section 4.1.3 that if $\pi = B_1|\cdots|B_r$ and $\nu = B_1'|\cdots|B_s'$ are two partitions in $\mathbf{\Pi}_m$, then $\pi \wedge \nu$ is the common refinement of π and ν, that is, the partition with blocks given by all the non-empty sets of the form $B_i \cap B_j'$ for $i = 1, \ldots, r$, $j = 1, \ldots, s$.

Definition 5.11. Let $\mathcal{T}$ be a semi-labeled tree with the labeling set $[m]$ and let $\mathcal{S}(\mathcal{T})$ be the corresponding set of tree splits of $[m]$. We say that $\pi \in \mathbf{\Pi}_m$ is a $\mathcal{T}$-*partition* if $\pi = \pi_1 \wedge \cdots \wedge \pi_k$ for some splits $\pi_1, \ldots, \pi_k \in \mathcal{S}(\mathcal{T})$. The one-block partition $\hat{1} = [m]$ by convention is a $\mathcal{T}$-partition. The set of all $\mathcal{T}$-partitions is denoted by $\mathbf{\Pi}(\mathcal{T})$.

The set of all $\mathcal{T}$-partitions forms a meet-semilattice and hence $\mathbf{\Pi}(\mathcal{T})$ is a lattice by Lemma 4.12. For a simple example, consider the phylogenetic tree in Figure 5.1. The corresponding poset of $\mathcal{T}$-partitions is given in Figure 5.7. Note that $\mathbf{\Pi}(\mathcal{T})$ shares the meet operation with $\mathbf{\Pi}_m$ but the join operation is not the same. For example, from the Hasse diagram in Figure 5.7 we see that the join of 1|4|23 and 14|2|3 in $\mathbf{\Pi}(\mathcal{T})$ is 1234, whereas in $\mathbf{\Pi}_m$ it is 14|23; see Figure 4.1. In other words, $\mathbf{\Pi}(\mathcal{T})$ is a lattice and it is a subposet of $\mathbf{\Pi}_m$ but it is not a sublattice of $\mathbf{\Pi}_m$.

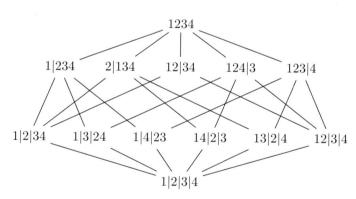

Figure 5.7 *The Hasse diagram of the poset of $\mathcal{T}$-partitions for the quartet tree in Figure 5.1.*

Remark 5.12. Both the lattice of interval partitions and the lattice of one-cluster partitions in Definition 4.15 are $\mathcal{T}$-partitions. The lattice of interval partitions is obtained as a $\mathcal{T}$-partition lattice, when $\mathcal{T}$ is a chain graph like in Figure 5.8. The lattice of one-cluster partitions is a $\mathcal{T}$-partition lattice, when $\mathcal{T}$ is a phylogenetic star tree model like in Figure 5.6.

Given a subset of splits $\mathcal{S} \subset \mathbf{\Pi}_m$ we may wonder if there exists a semi-

Figure 5.8: *A chain graph.*

labeled tree $\mathcal{T}$ such that $\mathcal{S} = \mathcal{S}(\mathcal{T})$. The answer to this question is classically known thanks to Buneman [1971].

Definition 5.13. A pair of different splits $\pi_1 = A_1|B_1$ and $\pi_2 = A_2|B_2$ of $[m]$ are *compatible* if exactly one of the sets $A_1 \cap A_2$, $A_1 \cap B_2$, $A_2 \cap B_1$, and $B_1 \cap B_2$ is the empty set. Equivalently, the partition $\pi_1 \wedge \pi_2$ has three blocks.

For example, 12|34 and 13|24 are not compatible and 1|234 and 12|34 are.

Theorem 5.14. *Let $\mathcal{S}$ be a collection of splits of $[m]$. Then, there is a semi-labeled tree $\mathcal{T}$ such that $\mathcal{S} = \mathcal{S}(\mathcal{T})$ if and only if the splits in $\mathcal{S}$ are pairwise compatible. Moreover, if such a semi-labeled tree exists, then it is unique.*

If $\mathcal{T}$ is a *phylogenetic* tree, then $\mathcal{S}(\mathcal{T})$ contains the set $\mathcal{S}_0$ of trivial splits (see Definition 4.15) that correspond to removing terminal edges of the underlying tree T. Moreover, to every semi-labeled tree $\mathcal{T}'$ with splits $\mathcal{S}(\mathcal{T}')$ we can associate a phylogenetic tree with splits $\mathcal{S}(\mathcal{T}') \cup \mathcal{S}_0$. This construction is correct, which follows from Theorem 5.14 and the fact that trivial splits are compatible with all other splits of $[m]$ and hence the splits of $\mathcal{S}(\mathcal{T}') \cup \mathcal{S}_0$ are pairwise compatible.

The fact that $\mathcal{S}_0 \subseteq \mathcal{S}(\mathcal{T})$ for every phylogenetic tree $\mathcal{T}$ also implies that the whole topological information that distinguishes $\mathbf{\Pi}(\mathcal{T})$ and $\mathbf{\Pi}(\mathcal{T}')$ for two phylogenetic trees $\mathcal{T}, \mathcal{T}'$ is contained in the set of non-trivial splits. The corresponding posets are denoted by $\mathbf{\Pi}'(\mathcal{T})$ and $\mathbf{\Pi}'(\mathcal{T}')$. Consider, for example, two different binary trees $\mathcal{T}$ and $\mathcal{T}'$, both with six leaves, given in Figure 5.9. Their associated posets $\mathbf{\Pi}'(\mathcal{T})$ and $\mathbf{\Pi}'(\mathcal{T}')$ are given in Figure 5.10.

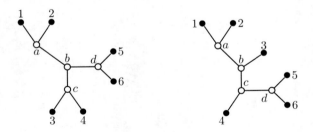

Figure 5.9: *Two non-isomorphic binary phylogenetic trees with six leaves.*

Definition 5.15. Let $\mathcal{T} = (T; \phi)$ be a semi-labeled tree with a labeling set $[m]$. For any $B \subseteq [m]$, by $\mathcal{T}(B)$ denote the semi-labeled tree whose labeling map is the restriction of ϕ to B. The underlying tree of $\mathcal{T}(B)$ is the smallest subtree of T containing $\phi(B)$ (spanned over B) with all unlabeled vertices that have two neighbors suppressed.

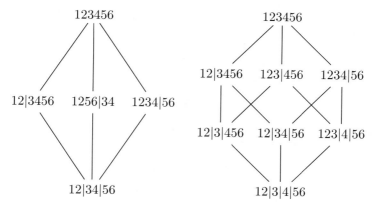

Figure 5.10 *The Hasse diagrams of* $\mathbf{\Pi}'(\mathcal{T})$ *and* $\mathbf{\Pi}'(\mathcal{T}')$, *where* $\mathcal{T}$, $\mathcal{T}'$ *are trees in Figure 5.9.*

Figure 5.11: *The tripod tree.*

Example 5.16. Suppose that $\mathcal{T}$ is the quartet tree in Figure 5.1. Then $\mathcal{T}(\{1,2\}) = \overset{1}{\bullet} - \overset{2}{\bullet}$, $\mathcal{T}(\{1,3\}) = \overset{1}{\bullet} - \overset{3}{\bullet}$, and $\mathcal{T}(\{1,2,3\}$ is the tripod tree in Figure 5.11.

Lemma 5.17. *Define* $\mathbf{\Pi}(\mathcal{T}, B)$ *as the induced subposet of* $\mathbf{\Pi}(\mathcal{T})$ *obtained by restricting each partition* $\pi \in \mathbf{\Pi}(\mathcal{T})$ *to* B. *Then*

$$\mathbf{\Pi}(\mathcal{T}, B) \quad = \quad \mathbf{\Pi}(\mathcal{T}(B)).$$

Proof. If $\pi_1, \ldots, \pi_k$ are splits generating $\mathbf{\Pi}(\mathcal{T})$, then the poset $\mathbf{\Pi}(\mathcal{T}, B)$ is a semi-lattice generated by $\pi_1(B), \ldots, \pi_k(B)$, where we disregard π_i such that $\pi_i(B) = B$. This is because the meet operation $\wedge$ and restriction to B commute. Also, the splits in $\mathcal{S}(\mathcal{T}(B))$ are precisely the non-trivial elements of $\pi_1(B), \ldots, \pi_k(B)$. $\qquad\qquad\square$

By $\mathfrak{m}_B$ and $\mathfrak{m}$ we denote the Möbius functions on $\mathbf{\Pi}(\mathcal{T}, B)$ and $\mathbf{\Pi}(\mathcal{T})$. Their minimal and maximal elements are denoted by $\hat{0}_B, \hat{0}$ and $\hat{1}_B \equiv B$, $\hat{1} \equiv [m]$, respectively. Recall that, by construction, the maximal element always corresponds to the one-block partition B or $[m]$.

For any partition $\pi \in \mathbf{\Pi}(\mathcal{T})$, the interval $[\hat{0}, \pi]$ has a natural structure of a product of posets $\prod_{B \in \pi} \mathbf{\Pi}(\mathcal{T}, B)$ and so $[\nu, \pi] \simeq \prod_{B \in \pi}[\nu_B, \hat{1}_B]$, where

$\nu_B \in \mathbf{\Pi}(\mathcal{T}, B)$ is the restriction of $\nu \in \mathbf{\Pi}(\mathcal{T})$ to the block containing only elements from $B \subset [m]$. By Proposition 4.7, the Möbius function on the product of posets $\prod_{B \in \pi} \mathbf{\Pi}(\mathcal{T}, B)$ can be written as the product of Möbius functions for each of the posets $\mathbf{\Pi}(\mathcal{T}, B)$. Thus for $\nu \leq \pi$ in $\mathbf{\Pi}(\mathcal{T})$

$$\mathfrak{m}(\nu, \pi) = \prod_{B \in \pi} \mathfrak{m}_B(\nu_B, B). \tag{5.1}$$

Example 5.18. Consider the semi-labeled forest in Figure 5.13 and take $\pi = 12|34|56$, $\nu = 12|3|4|56$. We have

$$\mathfrak{m}(\nu, \pi) = \mathfrak{m}_{12}(12, 12)\mathfrak{m}_{34}(3|4, 34)\mathfrak{m}_{56}(56, 56).$$

Now directly from definition in (4.2) we have $\mathfrak{m}_{12}(12, 12) = \mathfrak{m}_{56}(56, 56) = 1$ and $\mathfrak{m}_{34}(3|4, 34) = -1$ and therefore $\mathfrak{m}(\nu, \pi) = -1$.

5.1.3 Tree-based metrics

In this section we follow [Semple and Steel, 2003, Chapter 7] introducing tree-based metrics, which is the first geometric object that we relate to semi-labeled trees. Understanding this space is very important because many of its features are recurrent also in the context of other spaces we are going to discuss later.

Definition 5.19. A *metric* on a set S is a function $d : S \times S \to \mathbb{R}_\geq$ such that, for all $x, y, z \in S$, the following hold

(i) $d(x, y) = 0$ if and only if $x = y$,

(ii) $d(x, y) = d(y, x)$,

(iii) $d(x, z) \leq d(x, y) + d(y, z)$ (triangle inequality).

The pair (S, d) is called a *metric space*.

In our case, S is always a finite set.

We consider metrics on $[m]$ induced by semi-labeled trees in the following sense. Let $T = (V, E)$ be a tree and suppose that $w : E \to \mathbb{R}_+$ is a map that assigns lengths to the edges of T. Recall that for any pair $u, v \in V$ by $\overline{uv}$ we denote the path in T joining u and v. We now define the map $d_{T,w} : V \times V \to \mathbb{R}$ by setting, for all $u, v \in V$,

$$d_{T,w}(u, v) = \begin{cases} \sum_{e \in \overline{uv}} w(e), & \text{if } u \neq v, \\ 0, & \text{otherwise.} \end{cases}$$

Suppose now we are interested only in the distances between labeled vertices.

Definition 5.20. An arbitrary function $d : [m] \times [m] \to \mathbb{R}$ is called a *tree metric* if there exists a semi-labeled tree $\mathcal{T} = (T; \phi)$ $(T = (V, E))$ with the labeling set $[m]$ and a positive length assignment $w : E \to \mathbb{R}_+$ such that for all $i, j \in [m]$

$$d(i, j) = d_{T,w}(\phi(i), \phi(j)).$$

We will call d a $\mathcal{T}$-metric if we want to make $\mathcal{T}$ explicit.

Example 5.21. Consider a quartet tree with edge lengths as indicated in Figure 5.12. The distance between vertices 1 and 3 is $d(1,3) = 9.5$ and the whole distance matrix is

$$\begin{bmatrix} 0 & 5.5 & 9.5 & 8 \\ \cdot & 0 & 11 & 9.5 \\ \cdot & \cdot & 0 & 3.5 \\ \cdot & \cdot & \cdot & 0 \end{bmatrix},$$

where the dots indicate that this matrix is symmetric.

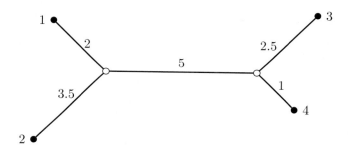

Figure 5.12: *A metric on a quartet tree.*

It is easy to describe the set of all possible tree metrics.

Definition 5.22. We say that a map $d : [m] \times [m] \to \mathbb{R}$ satisfies the *four-point condition* if for every four (not necessarily distinct) elements $i, j, k, l \in [m]$,

$$d(i,j) + d(k,l) \leq \max \begin{cases} d(i,k) + d(j,l) \\ d(i,l) + d(j,k). \end{cases}$$

Since the elements $i, j, k, l \in [m]$ in Definition 5.22 need not be distinct, every such map is a metric on $[m]$ given that $d(i,i) = 0$ and $d(i,j) = d(j,i)$ for all $i, j \in [m]$. The following fundamental theorem links tree metrics with the four-point condition.

Theorem 5.23 (Tree-Metric Theorem, Buneman [1974]). *Suppose that $d : [m] \times [m] \to \mathbb{R}$ is such that $d(i,i) = 0$ and $d(i,j) = d(j,i)$ for all $i, j \in [m]$. Then, d is a tree metric on $[m]$ if and only if it satisfies the four-point condition. Moreover, a tree metric defines the defining semi-labeled tree uniquely.*

Some intuition on how the proof of this result should work comes from the following example.

Example 5.24. Let $m = 4$ and let

$$\Delta = d(1,3) + d(2,4) - d(1,2) - d(3,4),$$

$$\Delta' = d(1,4) + d(2,3) - d(1,2) - d(3,4).$$

There are three possible phylogenetic trees on four leaves. On one of them $\Delta > 0, \Delta' > 0$, on another $\Delta < 0, \Delta' = 0$, and on the third $\Delta = 0, \Delta' < 0$:

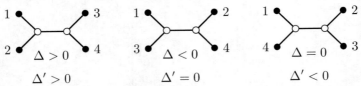

In particular, for each tree either $\Delta \geq 0$ or $\Delta' \geq 0$ and the values of Δ and Δ' enable us to distinguish between all possible trees with four leaves.

We can extract from Theorem 5.23 the constraints on the space of $\mathcal{T}$-metrics for a fixed phylogenetic tree $\mathcal{T}$.

Proposition 5.25. *Let $\mathcal{T}$ be a fixed phylogenetic tree. The set of all $\mathcal{T}$-metrics is the set of all metrics d on $[m]$ satisfying*

$$d(i,j) + d(k,l) \quad \leq \quad d(i,k) + d(j,l) \quad = \quad d(i,l) + d(j,k)$$

for all disjoint $i, j, k, l \in [m]$ such that the split $\{i,j\}|\{k,l\}$ lies in $\Pi(\mathcal{T}, \{i,j,k,l\})$.

The proof can be completed by using an idea like in Example 5.24.

Whenever we write inequalities like those above, we want to think about constrained geometric spaces. There are many ways to embed a metric space in an affine space. For statistical reasons we implicitly always embed metric spaces in the space of $|S| \times |S|$ symmetric matrices with zeros on the diagonal (c.f. Example 5.21). If d is a metric, then $\rho := e^{-d}$ corresponds to a matrix whose (x,y)-element is $\rho(x,y) = e^{-d(x,y)}$. Note that ρ has ones on the diagonal and all its elements lie in $[0,1]$. The triangle inequality implies sign constraints on certain minors

$$\rho(x,z)\rho(y,y) \quad \geq \quad \rho(x,y)\rho(y,z).$$

If $\mathcal{T}$ is fixed, then the space of all $\mathcal{T}$-metrics forms a polyhedral cone described in Proposition 5.25. By Theorem 5.23, the space of all tree metrics on $[m]$ can be identified by a subset obtained by gluing together all these various cones; see Billera et al. [2001]. For a fixed $\mathcal{T}$, the analysis of $\rho = e^{-d}$ leads to simple rank constraints on the space of $\mathcal{T}$-metrics. Directly from Proposition 5.25 it follows that if $A|B$ is a tree split, then the submatrix of ρ with rows in A and columns in B has rank one.

In the next section we take a closer look at the space of all $\rho = e^{-d}$, where d is a tree metric. More precisely, we are going to study its compactification, called the *space of phylogenetic oranges*.

5.1.4 Phylogenetic oranges and the Tuffley poset

An object very closely related to the space of all tree metrics, described in Theorem 5.23, is the space of phylogenetic oranges. It was first defined by

Kim [2000] and then further studied by Moulton and Steel [2004], Gill et al. [2008]. Let $\mathcal{T}$ be a semi-labeled tree with the labeling set $[m]$ and edge set $E(\mathcal{T})$. Given a map $\lambda : E(\mathcal{T}) \to [0, 1]$ define a map $p_{(\mathcal{T},\lambda)} : [0, 1]^{E(\mathcal{T})} \to \mathbb{R}^{\binom{m}{2}}$ defined for any two labeled vertices i, j by

$$p_{(\mathcal{T},\lambda)}(i,j) \quad = \quad \prod_{e \in \overline{ij}} \lambda(e), \qquad (5.2)$$

where $\overline{ij}$ denotes the set of edges on the path between i and j. The image of this map is denoted by $\mathcal{E}(\mathcal{T})$. The union of all $\mathcal{E}(\mathcal{T})$ for all binary phylogenetic trees $\mathcal{T}$ on $[m]$ is called the *edge product space* or the *space of phylogenetic oranges* and is denoted by $\mathcal{E}(m)$.

The connection between the space of tree metrics and the space of phylogenetic oranges is immediate. Taking $-\log(\cdot)$ on the both sides of (5.2) gives the tree metric because

$$-\log p_{(\mathcal{T},\lambda)}(i,j) \quad = \quad \sum_{e \in \overline{ij}} -\log \lambda(e),$$

and $-\log \lambda(e) \in (0, +\infty)$. A subtlety arises when some of $\lambda(e)$ are zero. Here we set $-\log 0 = \infty$, which also shows that $\mathcal{E}(\mathcal{T})$ is a compactification of the space of all $\mathcal{T}$-metrics.

In our applications it is important to understand the geometry of $\mathcal{E}(m)$. Directly by Definition 2.43, for any phylogenetic tree $\mathcal{T}$, the space $\mathcal{E}(\mathcal{T})$ is a toric cube. In particular, by Theorem 2.45, $\mathcal{E}(\mathcal{T})$ is a basic semialgebraic set given by binomial inequalities and equal to the closure of its interior. The space of strictly positive points of $\mathcal{E}(m)$ and $\mathcal{E}(\mathcal{T})$ for every $\mathcal{T}$ is in one-to-one correspondence with the space of tree metrics. Therefore, the inequalities describing $\mathcal{E}(m)$ and $\mathcal{E}(\mathcal{T})$ can be derived from the four-point condition in Definition 5.22 and from Proposition 5.25.

Proposition 5.26. *A point $x = [x_{ij}]$ with nonnegative coordinates lies in $\mathcal{E}(m) \subset \mathbb{R}^{\binom{m}{2}}$ if and only if for every four (not necessarily distinct) leaves i, j, k, l in V,*

$$x_{ij}x_{kl} \quad \geq \quad \min \begin{cases} x_{ik}x_{jl} \\ x_{il}x_{jk}. \end{cases}$$

Moreover, if this holds, then $(\mathcal{T}, \lambda)$ such that $x_{ij} = p_{(\mathcal{T},\lambda)}(i,j)$ is defined uniquely.

Proof. Consider the set of strictly positive points in $\mathcal{E}(m)$. The space is isomorphic with the space of tree metrics via $\rho \mapsto -\log(\rho)$. Thus, constraints on this open space can be derived directly from the four-point condition and are exactly the constraints above. Now note that these constraints describe a larger closed set, the smallest closed set containing this open space. By Theorem 2.45 this closed set is precisely $\mathcal{E}(m)$.   □

A proof of the next result is essentially the same.

Proposition 5.27. *If $\mathcal{T}$ is given, then the space $\mathcal{E}(\mathcal{T})$ has dimension $|E(\mathcal{T})|$ and is described by all points with nonnegative coordinates satisfying the following set of constraints. For any four (not necessarily distinct) leaves i, j, k, l such that the split $\{i, j\}|\{k, l\}$ is compatible with $\mathcal{T}$ we have*

$$x_{ik}x_{jl} \;=\; x_{il}x_{jk} \;\leq\; x_{ij}x_{kl}. \qquad (5.3)$$

Note that (5.3) implies that for any three distinct leaves i, j, k

$$x_{ik}x_{jk} \;\leq\; x_{ij}.$$

By Theorem 2.47, $\mathcal{E}(\mathcal{T})$ forms a CW-complex, with the simplest instance given by Example 2.44. This CW-complex structure was analyzed in detail by Moulton and Steel [2004] and it is closely linked to the so-called Tuffley poset denoted by $\mathrm{Tuff}(m)$. The poset structure of $\mathrm{Tuff}(m)$ is defined as follows. For two forests $\mathcal{F}, \mathcal{F}'$ we have $\mathcal{F}' \preceq \mathcal{F}$ in $\mathrm{Tuff}(m)$ if:

(O1) $\pi' \leq \pi$ in $\mathbf{\Pi}_m$.

(O2) Given $A \in \pi$ and $\nu = \pi'(A) = B_1|\cdots|B_r$ (restriction of π' to A):
  (i) for every $B \in \nu$, $\mathbf{\Pi}(\mathcal{T}'_B) \subseteq \mathbf{\Pi}(\mathcal{T}_A, B)$,
  (ii) for any two distinct $B_i, B_j \in \nu$ there exists a split $F_i|F_j \in \mathbf{\Pi}(\mathcal{T}_A)$ with $B_i \subseteq F_i$ and $B_j \subseteq F_j$.

For a semi-labeled tree $\mathcal{T}$ by $\mathrm{Tuff}(\mathcal{T})$ we denote the poset of all $\mathcal{F} \preceq \mathcal{T}$.

We now describe how the Tuffley poset is related to the CW-complex structure of $\mathcal{E}(\mathcal{T})$ (c.f. Definition 2.46). To a semi-labeled tree $\mathcal{T}$ we associate a closed ball $\boldsymbol{B}(\mathcal{T}) = [0, 1]^{E(\mathcal{T})}$. Further, for a semi-labeled forest $\mathcal{F} = \{\mathcal{T}_A : A \in \pi\}$ we define

$$\boldsymbol{B}(\mathcal{F}) = \prod_{A \in \pi} \boldsymbol{B}(\mathcal{T}_A).$$

We can extend the map in (5.2) to forests as follows. If $\mathcal{F} = \{\mathcal{T}_A : A \in \pi\}$ is a semi-labeled forest, then let $\psi_{\mathcal{F}} : \boldsymbol{B}(\mathcal{F}) \to [0, 1]^{\binom{m}{2}}$ be defined by setting, for $\lambda = (\lambda_A : A \in \pi)$,

$$\psi_{\mathcal{F}}(\lambda)(i, j) \;=\; \begin{cases} p_{(\mathcal{T}_A, \lambda_A)}(i, j) & \text{if } i, j \in A \text{ for } A \in \pi, \\ 0 & \text{otherwise.} \end{cases}$$

The following example should make clear that the image of $\psi_{\mathcal{F}}(\lambda)$ can be realized as a subset of $\mathcal{E}(\mathcal{T})$ for any phylogenetic tree $\mathcal{T}$ such that $\mathcal{F}$ is obtained from $\mathcal{T}$ by removing and contracting its edges.

Example 5.28. Let $\mathcal{F}$ be a semi-labeled forest $\overset{1}{\bullet} - \overset{2}{\bullet}\quad\overset{3}{\bullet} - \overset{4}{\bullet}$. Then

$$\psi_{\mathcal{F}}(\lambda)(1, 2) = \lambda_{12}, \qquad \psi_{\mathcal{F}}(\lambda)(3, 4) = \lambda_{34}, \qquad \psi_{\mathcal{F}}(\lambda)(i, j) = 0 \text{ otherwise,}$$

where $\lambda_{12}, \lambda_{34} \in [0, 1]$ are some parameters of the edges of $\mathcal{F}$. Let $\mathcal{T}$ be the quartet tree in Figure 5.1. Consider a point in $\mathcal{E}(\mathcal{T})$ parameterized by

$$(\lambda'_0, \lambda'_1, \lambda'_2, \lambda'_3, \lambda'_4) = (0, 1, \lambda_{12}, 1, \lambda_{34}),$$

where λ'_0 is the parameter of the inner edge, and λ'_i for $i = 1, 2, 3, 4$ are the parameters of the terminal edges of the corresponding leaves. With this choice of parameters, we have $p_{(\mathcal{T}, \lambda')}(i, j) = \psi_{\mathcal{F}}(\lambda)(i, j)$ for all i, j. Note that setting $\lambda'_e = 0$ or $\lambda'_e = 1$ corresponds to edge deletion or edge contraction, respectively.

We formulate the following theorem without proof; see [Moulton and Steel, 2004, Theorem 3.3].

Theorem 5.29. $\mathcal{E}(m)$ *is a CW-complex with cell decomposition*

$$\{(\boldsymbol{B}(\mathcal{F}), \psi_{\mathcal{F}}) : \mathcal{F} \in \mathrm{Tuff}(m)\}.$$

Furthermore, $\mathrm{Tuff}(m)$ *is isomorphic to the face poset of* $\mathcal{E}(m)$ *under the map that sends* $\mathcal{F}$ *to* $\psi_{\mathcal{F}}(\boldsymbol{B}(\mathcal{F}))$. *Similarly, for every phylogenetic tree* $\mathcal{T}$, $\mathrm{Tuff}(\mathcal{T})$ *is isomorphic to the face poset of* $\mathcal{E}(\mathcal{T})$.

To better understand the Tuffley poset it is convenient to reformulate the partial ordering given by (O1) and (O2). Let $\mathcal{F}$ be a semi-labeled forest and π the corresponding partition of $[m]$, $\mathcal{F} = \{\mathcal{T}_B : B \in \pi\}$ (c.f. Definition 5.3). Define $\boldsymbol{\Pi}(\mathcal{F})$ to be the set of induced partitions defined as a product of $\boldsymbol{\Pi}(\mathcal{T}_{B_i})$ for all components $\mathcal{T}_{B_i}$ of $\mathcal{F}$:

$$\boldsymbol{\Pi}(\mathcal{F}) \quad = \quad \prod_{B \in \pi} \boldsymbol{\Pi}(\mathcal{T}_B). \tag{5.4}$$

In particular, π is the maximal element of $\boldsymbol{\Pi}(\mathcal{F})$ and we consider $\boldsymbol{\Pi}(\mathcal{F})$ as a subset of $\boldsymbol{\Pi}(\mathcal{T})$ in a natural way. It is clear that $\boldsymbol{\Pi}(\mathcal{F} \setminus e) \subset \boldsymbol{\Pi}(\mathcal{F})$ and $\boldsymbol{\Pi}(\mathcal{F}/e) \subset \boldsymbol{\Pi}(\mathcal{F})$ for all $e \in E(\mathcal{F})$.

Now we are ready to define a poset structure on the set of all semi-labeled forests with labeling set $[m]$.

Definition 5.30. Given two semi-labeled forests $\mathcal{F}, \mathcal{F}'$, both with the labeling set $[m]$, we say that $\mathcal{F}'$ is a *semi-labeled subforest* of $\mathcal{F}$ if $\boldsymbol{\Pi}(\mathcal{F}') \subseteq \boldsymbol{\Pi}(\mathcal{F})$.

Example 5.31. Consider again a semi-labeled tree $\mathcal{T}$ in Figure 5.2. Let $\mathcal{F}$ be the semi-labeled forest in Figure 5.13. Every element in $\boldsymbol{\Pi}(\mathcal{F})$ lies in $\boldsymbol{\Pi}(\mathcal{T})$ and hence $\mathcal{F}$ is a semi-labeled subforest of $\mathcal{T}$.

Figure 5.13: *A semi-labeled forest.*

The product structure of $\boldsymbol{\Pi}(\mathcal{F})$ given by (5.4) implies the following lemma.

Lemma 5.32. *Let* $\mathcal{F}, \mathcal{F}'$ *be two semi-labeled forests with the same labeling set* $[m]$ *and* π, π' *the corresponding partitions of* $[m]$. *Then* $\mathcal{F}'$ *is a semi-labeled subforest of* $\mathcal{F}$, *or equivalently* $\boldsymbol{\Pi}(\mathcal{F}') \subseteq \boldsymbol{\Pi}(\mathcal{F})$, *if and only if:*

(O1) $\pi' \leq \pi$ *in* $\boldsymbol{\Pi}_m$,

(O2') for every block $A \in \pi$

$$\prod_{B \in \pi'(A)} \mathbf{\Pi}(\mathcal{T}_B') \quad \subseteq \quad \mathbf{\Pi}(\mathcal{T}_A),$$

where $\pi'(A)$ denotes restriction of π' to elements of A.

In other words, $\mathcal{F}'$ was obtained from $\mathcal{F}$ by some edge deletions and contractions.

Proof. We prove the "if" direction by contradiction. If $\pi' \not\leq \pi$, then $\mathbf{\Pi}(\mathcal{F}')$ cannot be contained in $\mathbf{\Pi}(\mathcal{F})$ because π' is the maximal element of $\mathbf{\Pi}(\mathcal{F}')$ and every element of $\mathbf{\Pi}(\mathcal{F})$ is less than or equal to π. So suppose that (O1) holds. Now if (O2') fails to hold, then there must be a block $A \in \pi$ and an element $\nu \in \mathbf{\Pi}(\mathcal{F}')$ such that $\nu(A) \notin \mathbf{\Pi}(\mathcal{T}_A)$, which is a contradiction with the fact that $\mathbf{\Pi}(\mathcal{F}') \subseteq \mathbf{\Pi}(\mathcal{F})$. The "only if" part of the proof follows from (5.4). $\square$

The following result shows that $\mathrm{Tuff}(\mathcal{T})$ is equal to the set of all semi-labeled forests with the partial order given by Definition 5.30. This provides a clearer description of the ordering in terms of the corresponding partition lattices.

Theorem 5.33. *The Tuffley poset $\mathrm{Tuff}(m)$ is the set of all semi-labeled forests with $[m]$ as the labeling set. We have $\mathcal{F}' \preceq \mathcal{F}$ if and only if $\mathbf{\Pi}(\mathcal{F}') \subseteq \mathbf{\Pi}(\mathcal{F})$.*

Proof. It suffices to show that (O1) and (O2') are equivalent to conditions (O1) and (O2) in Lemma 5.32. Suppose first that (O1) and (O2) hold. Condition (O2')(i) follows directly. If $B_i, B_j \in \nu$, then by (O2)

$$\mathbf{\Pi}(\mathcal{T}_{B_i}') \times \mathbf{\Pi}(\mathcal{T}_{B_j}') \subseteq \mathbf{\Pi}(\mathcal{T}_A, B_i \cup B_j),$$

which implies that in $\mathbf{\Pi}(\mathcal{T}_A)$ there must exist an element $F_i|F_j$ that constrained to $B_i \cup B_j$ becomes $B_i|B_j$ and hence (O2')(ii) also holds. Now suppose (O1) and (O2') hold. Because $\pi' \leq \pi$, to show (O2), it suffices to show that for every $A \in \pi$, $\prod_{B \in \pi'(A)} \mathbf{\Pi}(\mathcal{T}_B') \subseteq \mathbf{\Pi}(\mathcal{T}_A)$. Let $\hat{0}$ be the minimal element of $\mathbf{\Pi}(\mathcal{T}_A)$. Suppose that $\nu \in \mathbf{\Pi}(\mathcal{T}_A)$, then by (O2')(i) the interval $[\hat{0}, \nu]$ is precisely $\prod_{B \in \pi'(A)} \mathbf{\Pi}(\mathcal{T}_B')$ and hence (O2) holds. Thus, it is enough to show that $\nu \in \mathbf{\Pi}(\mathcal{T}_A)$. By (O2') for every two $B_i, B_j \in \nu$ there exists an edge e_{ij} in $\mathcal{T}_A$ separating B_i from B_j. We can require that e_{ij} does not split any other $B_k \in \nu$. Indeed, suppose that for every $F_i|F_j$ like in (O2')(ii) there exists $k \neq i, j$ such that $B_k \cap F_i \neq \emptyset$ and $B_k \cap F_j \neq \emptyset$. In addition to a fixed $F_i|F_j$, consider any split $F_i'|F_k'$ in $\mathbf{\Pi}(\mathcal{T}_A)$ such that $B_i \subseteq F_i'$ and $B_k \subseteq F_k'$, which exists by (O2')(ii). These two splits are necessarily compatible and hence

$$\emptyset \in \{F_i \cap F_i', F_i \cap F_k', F_j \cap F_i', F_j \cap F_k'\}.$$

However, $B_i \subseteq F_i \cap F_i'$ and hence $F_i \cap F_i'$ cannot be empty. By assumption,

$B_k \cap F_i \neq \emptyset$ and $B_k \cap F_j \neq \emptyset$ and hence also $F_i \cap F_k' \neq \emptyset$ and $F_j \cap F_k' \neq \emptyset$. This implies that necessarily $F_j \cap F_i' = \emptyset$ and hence $F_i = \boldsymbol{f}_i'$, $F_j = \boldsymbol{f}_k'$ and finally $B_k \subseteq F_j$, which leads to contradiction. It follows that there exists a set of edges separating all B_i from each other. Removing all these edges induces the partition ν. $\qquad\square$

For every $\pi \in \mathbf{\Pi}(\mathcal{T})$ we define two semi-labeled forests:

$$\mathcal{F}^\pi \quad := \quad \{\mathcal{T}(B) : B \in \pi\}, \tag{5.5}$$

where $\mathcal{T}(B)$ is the subtree of $\mathcal{T}$ spanned over B (see Definition 5.15) and

$$\mathcal{F}_\pi \quad := \quad \{\mathcal{T}_B : B \in \pi\}, \tag{5.6}$$

such that each $\mathcal{T}_B$ is a single vertex labeled with B. Note that $\mathcal{F}^\pi$ depends both on $\mathcal{T}$ and π, whereas $\mathcal{F}_\pi$ depends only on π. For example, if $\mathcal{T}$ is the phylogenetic tree in Figure 5.4 and $\pi = 123|4|5|6$, then $\mathcal{F}^\pi$ is the semi-labeled forest in Figure 5.3 and $\mathcal{F}_\pi$ is given in Figure 5.14.

$$\underset{\bullet}{1,2,3} \qquad \underset{\bullet}{4} \qquad \underset{\bullet}{5} \qquad \underset{\bullet}{6}$$

Figure 5.14: *A semi-labeled forest $\mathcal{F}_\pi$ induced by the partition $\pi = 123|4|5|6$.*

Proposition 5.34. *The maximal elements of* $\mathrm{Tuff}(m)$ *correspond to binary phylogenetic trees. The minimal elements are forests $\mathcal{F}_\pi$ for $\pi \in \bigcup_\mathcal{T} \mathbf{\Pi}(\mathcal{T})$, where the union is taken over all binary phylogenetic trees.*

The Tuffley poset is very big even for very small m. The simplest case, when $m = 3$, is depicted in Figure 5.15.

5.2 Markov process on a tree

5.2.1 Graphical model formulation

Let $T^r = (V, E)$ be a rooted tree and $\boldsymbol{Y} = (Y_v)_{v \in V}$ be a vector of random variables such that every Y_v takes $k_v + 1$ values in $\mathcal{Y}_v := \{0, \ldots, k_v\}$. Denote the state space of $\boldsymbol{Y}$ by $\mathcal{Y} = \prod_{v \in V} \mathcal{Y}_v$.

Definition 5.35. The *Markov process* on a rooted tree T^r is the space of all distributions for $\boldsymbol{Y} = (Y_v)_{v \in V}$ of the form

$$p(\boldsymbol{y}; \theta, T) \quad = \quad \theta_r(y_r) \prod_{u \to v} \theta_{v|u}(y_v|y_u) \qquad \text{for } \boldsymbol{y} = (y_v)_{v \in V} \in \mathcal{Y}, \tag{5.7}$$

for some nonnegative parameters $\theta_r \in \mathbb{R}^{k_r+1}$ and $\theta_{v|u} \in \mathbb{R}^{(k_u+1) \times (k_v+1)}$ for every $u \to v$ in T^r. We assume that $\sum_{y=0}^{k_r} \theta_r(y) = 1$ and $\sum_{x=0}^{k_v} \theta_{v|u}(x|y) = 1$ for every y.

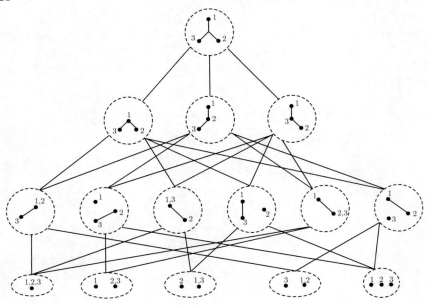

Figure 5.15: *The Tuffley poset for the tripod tree.*

The marginal distribution p_r induced from (5.7) is equal to θ_r and so θ_r is the *root distribution*. For each $u \to v \in E$, the matrix $\theta_{v|u}$ is called a *transition matrix* — it is a stochastic matrix, in which each *row* sums to 1. If the marginal distribution p_u induced from (5.7) contains no zeros, then $p_{v|u} = \theta_{v|u}^T$, in which case we interpret $\theta_{v|u}$ as the conditional distribution of Y_v given Y_u. This interpretation may break down if some of the parameters in (5.7) are zero.

It is convenient to think about a Markov process on a tree as a generalization of a (non-homogeneous) Markov chain, which is a Markov process on a tree $1 \to 2 \to \cdots \to m$ such that $k_1 = \ldots = k_m = k$ for some k.

Remark 5.36. Comparing (5.7) to (3.26) shows that the Markov process on T^r is a DAG model on T^r. Since T^r has no immoralities, the essential graph of T^r (see Section 3.4.3) is equal to the undirected version T of T^r. By Corollary 3.58 the Markov process on T^r is equal to the undirected graphical model on T. In particular, the model does not depend on the rooting and we denote it by $\mathbf{N}(T)$ (or $\mathbf{N}(T, \mathbf{k} + \mathbf{1})$ if $\mathbf{k} = (k_i)$ needs to be made explicit).

The model $\mathbf{N}(T)$ is completely specified by setting all transition matrices $\theta_{v|u}$ for $u \to v \in E$ together with the root distribution θ_r. Since $\sum_{y=0}^{k_r} \theta_r(y) = 1$ and $\sum_{x=0}^{k_v} \theta_{v|\operatorname{pa}(v)}(x|y) = 1$ for all $v \in V \setminus \{r\}$ and $y \in \{0, \ldots, k_{\operatorname{pa}(v)}\}$, then the set of parameters consists of exactly $d = (m+1)m|E| + m$ free parameters. We denote the parameter space by $\Theta_T = [0, 1]^d$. For instance in the binary case, $k_v = 1$ for all $v \in V$, we have two *free* parameters: $\theta_{v|u}(1|0)$, $\theta_{v|u}(1|1)$ for each edge $u \rightarrow v \in E$ and one parameter $\theta_r(1)$ for the root.

By Remark 5.36 we can freely switch between the parametric and the implicit representation of $\mathbf{N}(T)$, where the latter is given by Markov properties.

Proposition 5.37. *Since T is decomposable, by Proposition 3.49, all different Markov properties on T are equivalent. By the global Markov properties, $A \perp\!\!\!\perp B | C$ as long as C separates A and B in T. In particular, taking $A = \{u\}$, $B = \{v\}$ and $C = V \setminus \{u, v\}$, Lemma 3.10 gives the defining equations of $\mathbf{N}(T)$.*

The undirected factorization in (3.21) shows that $\mathbf{N}(T)$, when restricted to positive distributions, forms a discrete exponential family. In the two-state case this exponential family formulation links $\mathbf{N}(T, (2, \ldots, 2))$ to the Ising model on T widely studied in statistical physics. Then the joint distribution in $\mathbf{N}(T)$ has the following exponential form (c.f. (3.12))

$$p(\boldsymbol{x}; \omega, T) = \exp \left\{ \sum_{v \in V(T)} \omega_v x_v + \sum_{(u,v) \in E(T)} \omega_{uv} x_u x_v - \log Z(\omega) \right\},$$

where $Z(\omega)$ is the normalizing constant and ω_v, ω_{uv} are real parameters. Thus there are $|V| + |E|$ functions forming the sufficient statistics: x_v for $v \in V(T)$ and $x_u x_v$ for $u - v$ in $E(T)$. The matrix A representing the sufficient statistics has $2^{|V|}$ columns and $|V| + |E|$ rows. If $\hat{p}$ are sample proportions, then $A\hat{p}$ computes the sample moments of order ≤ 2 given by the marginal sample probabilities $\hat{p}_v(1)$ for $v \in V(T)$ and $\hat{p}_{uv}(1, 1)$ for $(u, v) \in E(T)$.

In biology there is a special interest in models where $k_v = k$ for all v and some $k \in \mathbb{N}$. This model is denoted by $\mathbf{N}(T, k)$. In the phylogenetic context we usually make the following simplifying assumptions for $\mathbf{N}(T, k)$:

(M1) $\theta_r(x) > 0$, for all $x \in \{0, \ldots, k\}$, and

(M2) $\det(\theta_{v|u}) \neq 0$, for each $u \rightarrow v \in E$.

Using the fact that $p_r = \theta_r$ and $p_v = p_{v|u} p_u$ for every $u \rightarrow v$, conditions (M1) and (M2) imply the following weaker condition

$$p_v(x) > 0, \quad \text{for all } v \in V \text{ and all } x \in \{0, \ldots, k\}. \tag{5.8}$$

Indeed, let $u \rightarrow v$, then by a recursive argument we can assume $p_u > 0$. By (M2) each row of $p_{v|u}$ is nonzero and hence $p_v = p_{v|u} p_u > 0$.

5.2.2 Edge contraction and deletion

We generalize the Markov process on a tree to a *Markov process on a forest* by defining it as a Cartesian product of its tree components. Formally, if F is a forest and $T_1, \ldots, T_s$ are its tree components with vertices $V_1, \ldots, V_s$, then

$$p(\boldsymbol{y}; \theta, F) \quad = \quad p(\boldsymbol{y}_1; \theta, T_1) \cdots p(\boldsymbol{y}_s; \theta, T_s), \tag{5.9}$$

where $\boldsymbol{y}$ is a vector indexed by V_i, $\boldsymbol{y} = (\boldsymbol{y}_1, \ldots, \boldsymbol{y}_s)$ and each $p(\boldsymbol{y}_i; \theta, T_i)$ is of the form (5.7). The Markov process on a forest F is denoted by $\boldsymbol{N}(F)$.

By a similar argument as in the case of the phylogenetic oranges (c.f. Example 5.28) model $\boldsymbol{N}(F)$ can be realized as a submodel of $\boldsymbol{N}(T)$ for any tree T such that F can be obtained from T by edge contraction and deletion.

Lemma 5.38. *Let T be a tree and $\boldsymbol{N}(T)$ the corresponding Markov process on T. If $e = u \to v$ is an edge of T, then the model of the forest $T \setminus e$ is obtained as a submodel of $\boldsymbol{N}(T)$, where the parameters are constrained to satisfy* rank$(\theta_{v|u}) = 1$.

Proof. A matrix has rank one if and only if each row is a multiple of the first row. Since $\theta_{v|u}$ is a stochastic matrix of rank one, each row has to be equal. However, directly from (5.7), if $\theta_{v|u}(y_v|y_u)$ does not depend on y_u, y_v, then $p(\mathbf{y}; \theta, T)$ factorizes into $p(\boldsymbol{y}_1; \theta, T_1)p(\boldsymbol{y}_2; \theta, T_2)$, where T_1, T_2 are two components of $T \setminus e$. The fact that it factorizes is immediate. The fact that the components are themselves probabilities follows from the fact that marginalizing over any of $\boldsymbol{y}_i$ yields a probability distribution. Since $p(\mathbf{y}; \theta, T) = p(\boldsymbol{y}_1; \theta, T_1)p(\boldsymbol{y}_2; \theta, T_2)$, it factorizes $\boldsymbol{N}(T \setminus e)$. $\square$

Lemma 5.39. *Let T be a tree and $\boldsymbol{N}(T)$ the corresponding Markov process on T. If $e = u \to v$ is an edge of T, then the model of the forest T/e is obtained as a submodel of $\boldsymbol{N}(T)$, where the parameters are constrained to satisfy $\theta_{v|u}(x|y) = 1$ if $x = y$ and $\theta_{v|u}(x|y) = 0$ otherwise, that is, $\theta_{v|u}$ is the identity matrix.*

Proof. This again follows directly from (5.7). $\square$

5.2.3   The Chow–Liu algorithm

Let T be a tree and $\boldsymbol{N}(T)$ the corresponding tree Markov process on T. Let

$$\boldsymbol{Y} = (Y_v) \in \mathcal{Y} = \prod_{v \in V(T)} \mathcal{Y}_v$$

be a vector of discrete random variables represented by vertices of T. Suppose that a random sample of length n was observed. We report it in the form of an observed counts table u or the sample proportions $\hat{p} = u/n$, which we assume have only strictly positive entries. Since $\boldsymbol{N}(T)$ is a graphical model of a decomposable graph, the maximum likelihood estimate is easily obtained for any fixed T (see Proposition 3.52)

$$\hat{\theta}_r(y_r) = \hat{p}_r(y_r).$$

Similarly, for any edge $u \to v$ the maximum likelihood estimate of $\theta_{v|u}$ is given by

$$\hat{\theta}_{v|u}(y_v|y_u) = \frac{\hat{p}_{uv}(y_u, y_v)}{\hat{p}_u(y_u)}.$$

Suppose now that the tree T is also considered to be a (discrete) parameter of our model. In this case, at least in principle, we should search over the space of all possible trees with a given number of vertices and find the one that gives the highest value of the likelihood function. The Chow–Liu algorithm, proposed by Chow and Liu [1968], is a remarkably simple algorithm, which gives an efficient way to find the best tree approximation that maximizes the likelihood.

For two random variables Y_i, Y_j with joint distribution p_{ij} we define their *mutual information* $I(Y_i, Y_j)$ as the Kullback–Leibler divergence between p_{ij} and the product of marginals $p_i p_j$

$$\boldsymbol{I}_p(Y_i, Y_j) \;\; = \;\; \sum_{y_i, y_j} p_{ij}(y_i, y_j) \log \frac{p_{ij}(y_i, y_j)}{p_i(y_i) p_j(y_j)}.$$

The mutual information satisfies $\boldsymbol{I}_p(Y_i, Y_j) \geq 0$ and is zero precisely when Y_i and Y_j are independent.

Fix a tree T. The model $\boldsymbol{N}(T)$ is a closure of an exponential family. By Theorem 3.33 and Remark 3.34, maximizing the likelihood function of the model $\boldsymbol{N}(T)$ under data $\hat{p}$ is equivalent to finding the unique distribution q such that

(i) $q \in \boldsymbol{N}(T)$,

(ii) $q_{ij}(y_i, y_j) = \hat{p}_{ij}(y_i, y_j)$ for all $i, j \in V$ and $y_i \in \mathcal{Y}_i, y_j \in \mathcal{Y}_j$.

If $q \in \boldsymbol{N}(T)$, then

$$q(\boldsymbol{y}) \;\; = \;\; q_r(y_r) \prod_{u \to v \in T} q_{v|u}(y_v|y_u) = \prod_v q_v(y_v) \prod_{u \to v \in E(T)} \frac{q_{uv}(y_u, y_v)}{q_u(y_u) q_v(y_v)}.$$
$$(5.10)$$

Further, using (ii) we obtain

$$q(\boldsymbol{y}) \;\; = \;\; \prod_v \hat{p}_v(y_v) \prod_{u \to v \in E(T)} \frac{\hat{p}_{uv}(y_u, y_v)}{\hat{p}_u(y_u) \hat{p}_v(y_v)}.$$

The log-likelihood function $n \sum_{\boldsymbol{y} \in \mathcal{Y}} \log(q(\boldsymbol{y})) \hat{p}(\boldsymbol{y})$ becomes

$$n \sum_{\boldsymbol{y}} \sum_v \hat{p}(\boldsymbol{y}) \log \hat{p}_v(y_v) + n \sum_{\boldsymbol{y}} \sum_{u \to v \in E(T)} \hat{p}(\boldsymbol{y}) \log \frac{\hat{p}_{uv}(y_u, y_v)}{\hat{p}_u(y_u) \hat{p}_v(y_v)}.$$

Note that for a fixed v the sum $\sum_{\boldsymbol{y}} \hat{p}_v(\boldsymbol{y}) \log \hat{p}_v(y_v)$ is just $\sum_{y_v} \hat{p}(y_v) \log \hat{p}_v(y_v)$. A similar simplification holds for the second term. Hence, by changing the order of summation in both terms above, we can rewrite the log-likelihood function as

$$n \sum_v \sum_{y_v} \hat{p}(y_v) \log \hat{p}_v(y_v) + n \sum_{u \to v \in E(T)} \boldsymbol{I}_{\hat{p}}(Y_u, Y_v). \qquad (5.11)$$

The value of that expression yields the maximum value of the log-likelihood function over the model $\boldsymbol{N}(T)$ for a fixed tree T. Now to find the highest value over all possible trees we note that the first term of the above sum does not depend on the tree topology. It follows that the maximum log-likelihood over all possible models $\boldsymbol{N}(T)$ will be attained for a tree T such that the second summand in (5.11) is maximized.

Since all $\boldsymbol{I}_{\hat{p}}(Y_u, Y_v)$ are nonnegative, we interpret them as edge weights. With this interpretation, the link to minimum-cost spanning trees is immediate. More precisely, let K_V be the complete graph on the set V. For each edge $u - v$ in K_V, let $\boldsymbol{I}_{\hat{p}}(Y_u, Y_v)$ be its weight. The Chow–Liu algorithm is essentially the same as the minimum-cost spanning tree algorithm on K_V proposed by Kruskal [1956] and Prim [1957]:

- Compute all mutual informations of $\hat{p}$ and order them from the largest to the smallest.

- Move along this ordered sequence adding subsequently the corresponding edges unless adding an edge introduces a cycle.

If all weights are different, then there is a unique best solution. If some of the weights are equal, then multiple solutions are possible, but as we showed above they will all give the same value of the likelihood function.

Example 5.40. Suppose that four binary random variables are given with the joint distribution that comes from the tripod tree model in Figure 5.16. Fixing parameter values, $q_4(1) = 0.6$ and

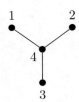

Figure 5.16: *The tripod tree with all vertices observed.*

$$q_{1|4} = \begin{bmatrix} 0.7 & 0.3 \\ 0.4 & 0.6 \end{bmatrix}, \quad q_{2|4} = \begin{bmatrix} 0.8 & 0.2 \\ 0.5 & 0.5 \end{bmatrix}, \quad q_{3|4} = \begin{bmatrix} 0.6 & 0.4 \\ 0.4 & 0.6 \end{bmatrix}$$

we can quickly obtain the true distribution q, for example, using the following R code

```
q <- array(0,c(2,2,2,2))
p4 <- c(0.4,0.6)
p14 <- matrix(c(0.7,0.4,0.3,0.6),2,2)
p24 <- matrix(c(0.8,0.5,0.2,0.5),2,2)
p34 <- matrix(c(0.6,0.4,0.4,0.6),2,2)
for (i in 1:2){
  for (j in 1:2){
```

```
for (k in 1:2){
  for (l in 1:2){
    q[i,j,k,l] <- p4[l]*p14[l,i]*p24[l,j]*p34[l,k]}}}}
```

Now generate a random sample of size n from q. Note that our computations will make sense even if $\hat{p}$ has zeros as long as two-way margins are strictly positive. We use the following R code

```
n <- 200
dat <- sample(1:16,n,replace=TRUE,c(q))
phat <- tabulate(dat)/n
dim(phat) <- c(2,2,2,2)
```

Given sample proportions $\hat{p}$ we compute mutual informations

```
MI <- matrix(0,4,4)
for (i in 1:3){
  for (j in (i+1):4){
    pi <- apply(phat,c(i),sum)
    pj <- apply(phat,c(j),sum)
    pij <- apply(phat,c(i,j),sum)
    MI[i,j]=sum(log(pij/outer(pi,pj))*pij)}}
```

The six possible mutual informations truncated to three decimal points are given in the matrix below

$$
\begin{bmatrix}
\cdot & 0.000 & 0.003 & 0.043 \\
\cdot & \cdot & 0.004 & 0.027 \\
\cdot & \cdot & \cdot & 0.045 \\
\cdot & \cdot & \cdot & \cdot
\end{bmatrix}
$$

We start building the maximum likelihood tree by first adding the edge $3-4$ corresponding to the mutual information value 0.045. Then, the next two biggest values of the mutual information correspond to edges $1-4$ and $2-4$. No cycles are formed by adding these edges and so we can include them. Since no more edges can be added, we stop recovering the original tree.

5.2.4 Markov tree models in biology

The models used in phylogenetics are usually defined via continuous time Markov processes evolving on a tree. In this formulation, to each edge e we associate a *rate matrix* $Q_e = [q(x,y)]$. We assume that the state space of each random variable is the same and equal to $k \geq 2$. Thus Q is a $k \times k$ matrix whose (x,y)-th entry is the instantaneous rate of substitution from state x to state y. By conservation, each row of Q_e sums to 0. The product $q(x,y)\Delta t$ gives the probability that x changes to y in an infinitely small time interval Δt. On each edge $e = u \to v$ the rate matrix Q_e operates with intensity λ_e for time t_e. The triple (Q_e, λ_e, t_e) corresponds to the transition matrix $\theta_{v|u}$ via

$$\theta_{v|u} = \exp(Q_e \alpha_e), \tag{5.12}$$

where $\alpha_e = \lambda_e t_e$ and $\exp(A) := \sum_{k \geq 0} \frac{1}{k!} A^k$ is the matrix exponential.

In this section, we want to show how the continuous time Markov process on a tree relates to the graphical model on T. By Jacobi's identity, for every square matrix A,

$$\det(\exp(A)) \;\; = \;\; e^{\mathrm{tr}(A)}.$$

Note that because

$$\det(\theta_{v|u}) \;\; = \;\; \det(\exp(Q_e \alpha_e)) \;\; = \;\; \exp(\mathrm{Tr}(Q_e)\alpha_e),$$

in particular $\det(\theta_{v|u}) > 0$. It turns out that for *binary* data this is the only constraint that distinguishes the two models.

Proposition 5.41. *A continuous time Markov process on T with two states corresponds to the submodel of the Markov process $\boldsymbol{N}(T,2)$ such that $\det(\theta_{v|u}) > 0$ (or equivalently $\mathrm{Tr}(\theta_{u|v}) > 1$) for all edges $u \to v$ of T^r.*

Proof. It is a classical result (see [Kingman, 1962, Proposition 2]) that $\det(\theta_{u|v}) > 0$ if and only if there exists a 2×2 matrix Q such that $\theta_{u|v} = \exp(Q)$, or equivalently if $\theta_{u|v}$ has representation (5.12). $\qquad\square$

For general state spaces this result does not hold. By [Kingman, 1962, Proposition 3] if $k \geq 3$, then the space of transition matrices that have an exponential representation as in (5.12) is relatively closed (in the classical topology) in the space of all transition matrices with positive determinant. It is an interesting open question to list all constraints that assure a continuous time representation.

5.3   The general Markov model

The model we want to study in this book is given as a collection of marginal distributions of the Markov process $\boldsymbol{N}(T)$ on T. Of course such a marginal model is necessarily more complicated. In this section we introduce this model and show some of its basic properties. This helps to set up the scene for the coming chapters.

5.3.1   Definition and examples

Let $\mathcal{T} = (T; \phi)$ be a semi-labeled tree with the labeling set $[m]$. In this section we consider the marginal model of $\boldsymbol{N}(T)$ over the random variables represented by the labeled vertices of $\mathcal{T}$. The parameterization of this new model is easily induced from the parameterization of $\boldsymbol{N}(T)$ by summing over all possible values of unlabeled vertices. More precisely, the vector $\boldsymbol{Y} \in \mathcal{Y} = \prod_{v \in V} \{0, \ldots, k_v\}$ has as its components all variables Y_v for $v \in V(T)$, both those that are observed and those that are hidden. The observed vector $\boldsymbol{X} = (X_1, \ldots, X_n)$ is identified with a subvector of $\boldsymbol{Y}$ by $X_i = Y_{\phi(i)}$. The subvector of the remaining (hidden) variables in $\boldsymbol{Y}$ is denoted by $\boldsymbol{H}$. The corresponding state spaces are denoted by $\mathcal{X}$ and $\mathcal{H}$.

Definition 5.42. The *latent tree model* on $\mathcal{T}$, denoted by $M(\mathcal{T})$, is the marginal model for X induced from probability distributions over $Y = (X, H)$ in $N(T)$ by

$$p(x; \theta, \mathcal{T}) = \sum_{v \notin \phi([m])} \sum_{h_v \in \mathcal{Y}_v} p((x, h); \theta, T), \qquad (5.13)$$

where $p(y; \theta, \mathcal{T})$ for $y = (x, h)$ is given by (5.7). In other words $M(\mathcal{T}) = f_{\mathcal{T}}(\Theta_T)$, where $f_{\mathcal{T}} : \Theta_T \to \Delta_{\mathcal{X}}$ is the parameterization given by (5.13).

When we want $k = (k_v)_{v \in V}$ explicit we write $M(\mathcal{T}, k+1)$, and when all k_v are equal, we write $M(\mathcal{T}, k+1)$. In the case where all k_i are equal, the corresponding latent tree model is called a *general Markov model*. We pay special attention to the two-state general Markov model $M(\mathcal{T}, 2)$.

Remark 5.43. Parameterization (5.13) also makes sense if $\mathcal{T}$ has vertices with multiple labels. However, if $\phi(i) = \phi(j)$ for some i, j, then the support of distributions in the resulting model is concentrated over points x such that $x_i = x_j$, which corresponds to setting $X_i = X_j$.

Example 5.44. Let T^r be a rooted quartet tree in Figure 5.1, where we denote the root by r and the other inner vertex by a. Assume that all the variables in this system have two states, that is, $k_v = 1$ for all v. Then the general Markov model $M(\mathcal{T}, 2)$ is given as the image of $f_{\mathcal{T}} : \Theta_T \to \Delta_{\mathcal{X}}$, where $\mathcal{X} = \{0, 1\}^4$ and where the parameter space Θ_T has exactly 11 free parameters: $\theta_r(1)$ and $\theta_{v|\text{pa}(v)}(1|0)$, $\theta_{v|\text{pa}(v)}(1|1)$ for each $v \in V \setminus r$. For every $x \in \mathcal{X}$

$$p(x; \theta, T) = \sum_{h_r, h_a = 0}^{1} \theta_r(h_r)\theta_{a|r}(h_a|h_r)\theta_{1|r}(x_1|h_r)\theta_{2|r}(x_2|h_r)\theta_{3|a}(x_3|h_a)\theta_{4|a}(x_4|h_a).$$

5.3.2 Reduction to non-degenerate vertices

In our general setting we always assume (5.8). In the two-state case this means that all variables in the system are non-degenerate, that is, that $\text{Var}(Y_v) > 0$ for all v. It is important to note that this assumption does not change the model space as long as the observed vector X is non-degenerate, which we can always assume with no loss of generality because a binary random variable is degenerate if and only if it is constant. The following proposition makes this statement more precise.

Proposition 5.45. *Suppose that Θ_T^* is the set of all parameters in Θ_T such that $\text{var}(Y_v) = p_v(0)p_v(1) > 0$ for all inner vertices v. Then $f_{\mathcal{T}}(\Theta_T) = f_{\mathcal{T}}(\Theta_T^*)$.*

Proof. It is clear that $f_{\mathcal{T}}(\Theta_T^*) \subseteq f_{\mathcal{T}}(\Theta_T)$. For the opposite inclusion it is enough to show that if $\theta \in \Theta_T$ is such that the induced variance of Y_v is zero for some inner vertex v, then there exists $\theta' \in \Theta_T^*$ such that $f_{\mathcal{T}}(\theta) = f_{\mathcal{T}}(\theta')$. Suppose that θ is such that the variance of Y_r is zero, where r is the root, which we can assume to be an inner vertex. This means that $\theta_r(0)\theta_r(1) = 0$

and thus θ_r is a unit vector $e_i \in \mathbb{R}^2$, where i is either 1 or 2. Let $\theta'_r = (1/2, 1/2)$ be the uniform distribution and for every u such that $r \to u$ in T define $\theta'_{v|r}$ as the matrix in which every row is equal to the i-th row of $\theta_{v|r}$. Suppose that all other coordinates of the parameter vector θ' coincide with the corresponding coordinates of θ. It is easy to check that $\boldsymbol{f}_{\mathcal{T}}(\theta) = \boldsymbol{f}_{\mathcal{T}}(\theta')$. If v is any inner vertex, we proceed in exactly the same way — if $u \to v$ in T, then $\theta_{v|u}$ is zero apart from the i-th column in which all entries are 1. We can replace it with a new matrix $\theta'_{v|u}$ in which all entries are equal. Now we need only adjust $\theta_{w|v}$ in exactly the same way as above. $\square$

5.3.3   Reduction to binary phylogenetic trees

Binary phylogenetic trees form an important subclass of semi-labeled trees in the study of latent tree models. We show that assuming that each inner vertex has exactly three neighbors and only the leaves are labeled, we still keep the full structure of the problem. More precisely, any model for an arbitrary semi-labeled tree is a submodel of a tree model of a binary phylogenetic tree. To show this we formulate the following definition.

Definition 5.46. Let $\mathcal{T}$ be a semi-labeled tree. A *binary expansion* of $\mathcal{T}$, denoted by $\mathcal{T}^*$, is *any* phylogenetic tree $\mathcal{T}^*$ whose underlying tree is binary and there exists $E' \subseteq E(\mathcal{T}^*)$ such that $\mathcal{T} = \mathcal{T}^*/E'$, that is, $\mathcal{T}$ is obtained from $\mathcal{T}^*$ by contracting edges in E'.

To see that such $\mathcal{T}^*$ always exists, consider the following procedure on $\mathcal{T} = (T, \phi)$, which gives a (non-unique) construction of $\mathcal{T}^*$. For simplicity, assume that $\mathcal{T}$ has no multiple labels but a general procedure is very similar.

Definition 5.47 (A construction of $\mathcal{T}^*$). Let $\mathcal{T} = (T, \phi)$ be a semi-labeled tree with the labeling set $[m]$. The proposed recursive procedure gives a binary phylogenetic tree with the same labeling set such that $\mathcal{T} = \mathcal{T}^*/E'$ for some $E' \subseteq E(\mathcal{T}^*)$. Set $\mathcal{T}_0 := \mathcal{T}$. The following two basic operations will be used. Each takes as the input a semi-labeled tree $\mathcal{T}_j = (T_j, \phi_j)$ and outputs a semi-labeled tree $\mathcal{T}_{j+1} = (T_{j+1}, \phi_{j+1})$:

1. If v is a labeled inner vertex with label i, that is, $v = \phi_j(i)$ for some $i \in [m]$, then $\mathcal{T}_{j+1}$ is obtained as follows: Extend $V(\mathcal{T}_j)$ and $E(\mathcal{T}_j)$ by adding another vertex v' and an edge $v - v'$. Now define ϕ_{j+1} so that $\phi_j(i') = \phi_{j+1}(i')$ for all $i' \neq i$, $\phi_{j+1}(i) = v'$.

2. Suppose v is an unlabeled inner vertex of $\mathcal{T}_j$ with $d > 3$ neighbors $u_1, \ldots, u_d$. Then define $V(\mathcal{T}_{j+1})$ extending $V(\mathcal{T}_j)$ by adding vertices $v_1, \ldots, v_{d-3}$ and $E(\mathcal{T}_{j+1})$ by modifying $E(\mathcal{T}_j)$ in such a way that the induced subgraph of $\mathcal{T}_{j+1}$ over $u_1, \ldots, u_d, v, v_1, \ldots, v_{d-3}$ is a binary tree with inner vertices given by $v, v_1, \ldots, v_{d-3}$; see Figure 5.17.

First, use Step 1 recursively until the output is a semi-labeled tree $\mathcal{T}'$ such that only its leaves are labeled. Now use Step 2 for each inner vertex of $\mathcal{T}'$ with more than three neighbors. The procedure will stop when all inner vertices have exactly three neighbors. In this way we obtain a binary phylogenetic tree that is a binary expansion of $\mathcal{T}$.

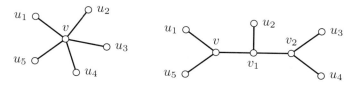

Figure 5.17 *An illustration of the second step of the procedure in Definition 5.47 of constructing a binary expansion.*

Lemma 5.48. *Let $\mathcal{T}$ be a semi-labeled tree and $\mathcal{T}^*$ its binary expansion with $E' \subseteq E(\mathcal{T}^*)$ such that $\mathcal{T} = \mathcal{T}^*/E'$. Then $\boldsymbol{M}(\mathcal{T}) \subseteq \boldsymbol{M}(\mathcal{T}^*)$.*

Proof. This follows from Theorem 5.49 below. □

We can extend this analysis to a semi-labeled forest $\mathcal{F}$ defining $\boldsymbol{M}(\mathcal{F})$ as the Cartesian product of $\boldsymbol{M}(\mathcal{T}_A)$, where $\mathcal{T}_A$ are the tree components of $\mathcal{F}$. More precisely, like in (5.13) we parameterize $\boldsymbol{M}(\mathcal{F})$ by

$$p(x; \theta, \mathcal{F}) = \sum_{v \notin \phi([m])} \sum_{h_v \in \mathcal{Y}_v} p((x, h); \theta, F), \tag{5.14}$$

where F is the underlying forest of $\mathcal{F}$ and $p((x, h); \theta, F)$ is a Markov process on a forest given in (5.9). We have the following result.

Theorem 5.49. *Let $\mathcal{F}, \mathcal{F}'$ be two semi-labeled forests with common labeling set $[m]$. If $\mathcal{F} \preceq \mathcal{F}'$, then $\boldsymbol{M}(\mathcal{F}) \subseteq \boldsymbol{M}(\mathcal{F}')$.*

Proof. Since $\boldsymbol{M}(\mathcal{F})$ is a product of $\boldsymbol{M}(\mathcal{T}_B)$ for all its tree components $\mathcal{T}_B$, without loss of generality we can assume that $\mathcal{F}'$ is a semi-labeled tree, that is, $\mathcal{F}'$ is connected. By Lemma 5.32, $\mathcal{F} \preceq \mathcal{F}'$ if and only if $\mathcal{F}$ is obtained from $\mathcal{F}'$ by some edge deletions and contractions. Thus, without loss of generality we can assume that there exists $e = u \to v \in E(\mathcal{F}')$ such that either $\mathcal{F} = \mathcal{F}' \setminus e$ or $\mathcal{F} = \mathcal{F}'/e$. If $\mathcal{F} = \mathcal{F}'/e$, then $\boldsymbol{M}(\mathcal{F})$ is a submodel of $\boldsymbol{M}(\mathcal{F}')$, where $\theta_{u|v}$ is set to be the identity matrix, which follows from Lemma 5.39. Similarly, by Lemma 5.38, $\mathcal{F} = \mathcal{F}' \setminus e$, then $\boldsymbol{M}(\mathcal{F})$ is a submodel of $\boldsymbol{M}(\mathcal{F}')$, where $\theta_{u|v}$ is set to be the matrix with all rows equal. □

5.3.4 Degree-two vertices

We say that a vertex v in a graph has *degree* d if it has exactly d neighbors. In the next chapter it will be important for us to consider latent tree models $\boldsymbol{M}(\mathcal{T})$ in the case when $\mathcal{T}$ is allowed to have unlabeled vertices of degree two. In that case, the parameterization in (5.13) still makes sense. The following result shows that from a model selection point of view allowing degree-two unlabeled vertices does not change anything.

Lemma 5.50. *Suppose that $\mathcal{T}$ is allowed to have degree-two unlabeled vertices. Let v be such a vertex and let $\mathcal{T}'$ be the tree obtained from $\mathcal{T}$ by suppressing v. Then $\boldsymbol{M}(\mathcal{T})$ and $\boldsymbol{M}(\mathcal{T}')$ are equal.*

Proof. Let u and w be neighbors of v. By Remark 5.36, the rooting of $\mathcal{T}$ does not matter and we can assume that $u \to v$, $v \to w$ in $E(\mathcal{T})$. The product $\theta_{w|v}\theta_{v|u}$ defines a transition matrix which we denote by $\theta_{w|u}$. From this it is clear that $\boldsymbol{M}(\mathcal{T}) \subseteq \boldsymbol{M}(\mathcal{T}')$. The opposite inclusion is also immediate. Let p be a point in $\boldsymbol{M}(\mathcal{T}')$ for a given parameter vector in $\Theta_{T'}$. Now in Θ_T we can set $\theta_{v|u} = \theta_{w|u}$ and $\theta_{w|v}$ equal to the identity matrix, which also implies that $p \in \boldsymbol{M}(\mathcal{T})$. $\qquad\square$

Although degree-two unlabeled vertices can be ignored in the sense of Lemma 5.50, including this case in our analysis is still useful in some situations. Let $\mathcal{T}$ be a semi-labeled tree and $B \subseteq [m]$. Recall Definition 5.15 of the induced semi-labeled tree $\mathcal{T}(B)$ and consider the following related object.

Definition 5.51. Let $\mathcal{T} = (T; \phi)$ be a semi-labeled tree with a labeling set $[m]$. For any $B \subseteq [m]$, by $\mathcal{T}^0(B)$ denote the pair $(T(B), \phi|_B)$, where $T(B)$ is the smallest subtree of T containing $\phi(B)$ and $\phi|_B$ is the restriction of ϕ to B.

Note that $\mathcal{T}^0(B)$ in general is not a semi-labeled tree because it is allowed to have degree-two unlabeled vertices. For example, consider the tree on the left of Figure 5.18. If i, j, k are three leaves, then $\mathcal{T}^0(\{i, j, k\})$ is the boldfaced subtree with four degree-two vertices and one degree-three vertex. The corresponding phylogenetic tree $\mathcal{T}(\{i, j, k\})$ is obtained from $\mathcal{T}^0(\{i, j, k\})$ by suppressing all degree-two vertices.

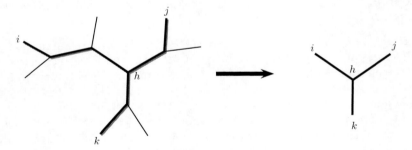

Figure 5.18 *The boldfaced tree on the left is* $\mathcal{T}^0(\{i, j, k\})$, *which defines the same general Markov model as the tree* $\mathcal{T}(\{i, j, k\})$ *on the right.*

Proposition 5.52. *Suppose that* $\boldsymbol{X}$ *has distribution in* $\boldsymbol{M}(\mathcal{T})$ *for some phylogenetic tree* $\mathcal{T}$. *If* $B \subseteq [m]$, *then the space of all possible marginal distributions of the subvector* $\boldsymbol{X}_B$ *is* $\boldsymbol{M}(\mathcal{T}^0(B))$, *which by Lemma 5.50 is equal to* $\boldsymbol{M}(\mathcal{T}(B))$.

5.3.5   Continuous time formulation

In this section we consider again the Markov process on a tree induced by a continuous time Markov process as discussed in Section 5.2.4. Proposition 5.41 motivates the following definition.

Definition 5.53. Let $\mathcal{T}$ be a semi-labeled tree. By $M_{++}(\mathcal{T}, 2)$, denote the *positive part of* $M(\mathcal{T}, 2)$, that is the model parameterized like $M(\mathcal{T}, 2)$ but with assumption that $\det(\theta_{u|v}) > 0$ for every $u \to v$. By Proposition 5.41 the positive part $M_{++}(\mathcal{T}, 2)$ represents exactly the distributions in $M(\mathcal{T}, 2)$ that come from a continuous Markov process on $\mathcal{T}$.

By $M_+(\mathcal{T}, 2)$ we denote the *nonnegative part of* $M(\mathcal{T}, 2)$, that is the model parameterized like $M(\mathcal{T}, 2)$ but with assumption that $\det(\theta_{u|v}) \geq 0$ for every $u \to v$. The set $M_+(\mathcal{T}, 2)$ is the (Euclidean) closure of $M_{++}(\mathcal{T}, 2)$. We extend this definition on semi-labeled forests. If $\mathcal{F} = \{\mathcal{T}_B : B \in \pi\}$, then $M_{++}(\mathcal{F}, 2)$ is the Cartesian product of $M_{++}(\mathcal{T}_B, 2)$.

The models used in phylogenetic analysis typically have some additional constraints. The most general of those constrained models is the *General Time Reversible (GTR) model*; see Tavaré [1986].

Definition 5.54 (GTR model). Let T^r be a rooted tree. The General Time-Reversible (GTR) model on T^r is a Markov process on T^r satisfying:

   (i) there exists Q such that for every $e = u \to v$ in T^r, $\theta_{v|u} = \exp(Q\alpha_e)$,

   (ii) there exists π such that $\pi Q = 0$ and $\theta_r = \pi$,

   (iii) $\mathrm{diag}(\pi)Q$ is symmetric (time reversibility).

Some of the reasons why the GTR model became popular in phylogenetic analysis is its computational tractability, identifiability, and interpretability. However, recently some bad features of this model class have been pointed out by Sumner et al. [2012a,b], who showed that the model is not multiplicatively closed. That is, the result of multiplying of two GTR matrices is not necessarily GTR, which has important consequences for estimation. Also it is important to note that time reversibility has no biological reason and is just a mathematical convenience; see [Yang, 2006, Section 1.5.2].

Other models popular in biology are submodels of the GTR model with some additional constraints on the matrix Q. From a mathematical point of view, these models have been studied in algebraic statistics as general *group-based models*; see Eriksson et al. [2005], Sturmfels and Sullivant [2005] and references therein. It is important to realize the modeling assumptions underlying all these models. The Markov process on a tree, being a graphical model, is itself a fairly constrained model, which assumes a number of conditional independence statements. The GTR model assumes further constraints, which, as we pointed out above, are problematic. All these constraints seem very restrictive given that we expect our model only to conveniently represent evolution along a tree. Constraining this model further produces low dimensional models with good computational tractability but possibly with some robustness issues. The geometric viewpoint of this book enables us to obtain a unique perspective on the above problem as we show in the next sections.

5.4   Phylogenetic invariants

Given a phylogenetic tree $\mathcal{T}$ and the general Markov model $M(\mathcal{T})$ on $\mathcal{T}$, we associate to it an ideal $\mathcal{I}_{\mathcal{T}}$ of all polynomials vanishing on $M(\mathcal{T})$ and we call it the *phylogenetic ideal*. Every polynomial $f \in \mathcal{I}_{\mathcal{T}}$ is called a *phylogenetic invariant*. Some authors call *phylogenetic invariants* only the polynomials that lie in $\mathcal{I}_{\mathcal{T}}$ but do not lie in $\mathcal{I}_{\mathcal{T}'}$ for any other tree $\mathcal{T}'$. In this short section we present briefly some aspects of the theory of phylogenetic invariants giving references to the existing literature.

5.4.1   The basic idea

One motivation for investigating phylogenetic invariants is that they provide techniques for developing new methods for tree reconstruction. The basic idea behind application of phylogenetic invariants is as follows. Let $M \subseteq \Delta_{m-1}$ be an algebraic parametric model for a discrete random variable X with m possible values and let $\mathcal{I}_M$ be the corresponding ideal. Given n independent observations of X we compute the sample proportions $\hat{p}$. If the data were generated from $p^* = p(\theta^*) \in M$ for some $\theta^* \in \Theta$, then $\hat{p}$ converges almost surely to p^* by the law of large numbers. It follows that for large sample sizes $\hat{p}$ should be close to a point in M; see Figure 5.19. Because $f(p^*) = 0$ for every $f \in \mathcal{I}_M$, for large n also $f(\hat{p}) \approx 0$. Note that this procedure will never require parameter estimation.

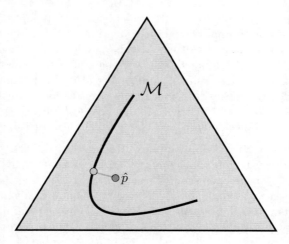

Figure 5.19: *The basic idea behind the application of phylogenetic invariants.*

To give an idea of how we could make inference based on phylogenetic invariants we provide the following example. Let X, Y be two binary variables. The independence model is given by a simple equation

$$p_{00}p_{11} - p_{01}p_{10} \quad = \quad 0. \tag{5.15}$$

Given sample proportions $\hat{p}$ for sample size n, to test if (5.15) holds when evaluated at the unknown true data generating distribution, we can equivalently test whether the *log-odds ratio* satisfies

$$\mathrm{LOR}(p) := \log \frac{p_{00}p_{11}}{p_{01}p_{10}} \;\; = \;\; 0.$$

The reason why the log-odds ratio may be preferred here is that its sample distribution is easier to understand. More specifically, if a random sample of size n from the true distribution p^* is observed, then $\mathrm{LOR}(\hat{p})$ has approximately normal distribution with mean $\mathrm{LOR}(p^*)$ and variance, which is approximately

$$\frac{1}{n}\left(\frac{1}{\hat{p}_{00}} + \frac{1}{\hat{p}_{01}} + \frac{1}{\hat{p}_{10}} + \frac{1}{\hat{p}_{11}}\right).$$

For a more detailed analysis, see [Agresti, 2002, Section 3.1.1].

5.4.2   The practice of phylogenetic invariants

The methods proposed in the phylogenetic literature are mainly simple diagnostic tests. A nice recent overview is given by Allman and Rhodes [2007]. The first papers developing the theory of phylogenetic invariants focused on linear invariants; see Cavender [1989], Lake [1987], and Navidi et al. [1993]. These were used to distinguish between different tree topologies under various simplified phylogenetic tree models. This method was showed by Huelsenbeck [1995] to produce rather bad results. However, using recent developments in the field, this simplified method was extended leading to powerful and robust algorithms, as shown for example by Fernández–Sánchez and Casanellas [2014].

The basic setting is as follows. Given a collection of phylogenetic trees $\mathcal{T}, \ldots, \mathcal{T}'$ and their corresponding models, we provide the set of defining phylogenetic ideals $\mathcal{I}_{\mathcal{T}}, \ldots, \mathcal{I}_{\mathcal{T}'}$. Until very recently, the complete ideals were not available so the main idea was to list as many phylogenetic invariants as possible. Considering simple quadratic invariants allows us to generalize directly asymptotic chi-square tests for independence in a contingency table; see Sankoff [1990]. General quadratic invariants are considered in Drolet and Sankoff [1990].

One of the problems with this method is that the number of phylogenetic invariants grows exponentially with the tree size. Moreover, for phylogeny reconstruction we need to check invariants for each possible tree. A basic idea of Eriksson [2007] is to use the phylogenetic invariants that are related with rank constraints on various flattenings of the observed frequencies tensor; see Section 7.1.4 for details. These so-called *edge invariants* can be shown to be the only invariants relevant from the phylogeny reconstruction point of view; see Casanellas and Fernández–Sánchez [2011] for details. We state it more precisely below.

Theorem 5.55. *Let $\mathcal{C} = \{\mathcal{T}, \ldots, \mathcal{T}'\}$ be a collection of phylogenetic trees on* *m leaves with corresponding general Markov models $\boldsymbol{M}(\mathcal{T}),\ldots,\boldsymbol{M}(\mathcal{T}')$. For each tree $\mathcal{T}$ in this collection there exists an open set $U_{\mathcal{T}}$ such that if p lies in $\bigcup_{\mathcal{T}\in\mathcal{C}} U_{\mathcal{T}}$, then p belongs to the Zariski closure of $\boldsymbol{M}(\mathcal{T})$ for some $\mathcal{T}$ if and only if p belongs to the zero set of the edge invariants of $\mathcal{T}$.*

To test the edge invariants Eriksson [2007] uses the singular value decomposition and the Frobenius norm to compute the distance of a matrix to the set of matrices of certain rank. Recently this method has been further improved by Fernández–Sánchez and Casanellas [2014]. In its current form the method is robust and simulations show that it outperforms most of the commonly used methods.

A significant deficiency of the method of phylogenetic invariants is that they all ignore additional constraints on models that can be used to improve estimation. This issue will be addressed later in the book. The basic idea can be, however, easily described by a picture. Let $\mathbf{M}$ be an algebraic parametric model. By construction it is a semialgebraic set. The phylogenetic invariants of the model define the Zariski closure $\overline{\mathbf{M}}$ of $\mathbf{M}$. Informally speaking, all the diagnostic tests based on phylogenetic invariants analyze how far a given data point is from $\overline{\mathbf{M}}$. Given two different data sets imagine that, with a fixed metric, $\hat{p}_2$ is "closer" than $\hat{p}_1$ to $\overline{\mathbf{M}}$. We have to be very careful to draw conclusions based solely on this fact. In particular, as it is shown in Figure 5.20, it is theoretically possible for $\hat{p}_1$ to lie actually closer to model $\mathbf{M}$. Analogously, we can consider two different tree topologies with two different

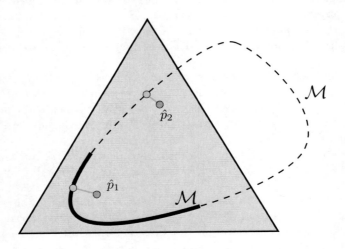

Figure 5.20: *Inequality constrains matter.*

phylogenetic ideals. Again, the data may support one of the ideals better but still come from another topology. The method of phylogenetic invariants may perform poorly in picking the correct tree if a model is misspecified, that is, if the true distribution does not lie in any of $\boldsymbol{M}(\mathcal{T})$.

5.4.3   Toward basic statistical analysis

The method of phylogenetic invariants gives a way to select the best tree under a given criterion. It does not, however, give any way of quantifying how well the chosen tree fits the data. There is currently no literature on testing the semialgebraic structures in the phylogenetic context. In simplistic situations, tests can be developed using a more general approach for constrained multinomial models; see for example Davis–Stober [2009], Shapiro [1985, 1988].

In view of Theorem 5.55 we can constrain our analysis only to edge invariants. These are given as minors of flattenings of the tensor of sample proportions. Thus, every such minor is a polynomial in sample proportions, and thus as a random variable, it has its mean and variance. This information can then be used to construct simple diagnostic statistics. An alternative route is to understand the distribution of singular values of every flattening. We expect this line of research to become increasingly active.

5.5   Bibliographical notes

A large part of this chapter is based on Semple and Steel [2003] and we also use their combinatorial approach as the starting point for the geometric analysis. A good introduction to trees and tree models can be also found in Felsenstein [2004], Yang [2006]. Another book on combinatorics of tree models is Dress et al. [2012]. Some of the corresponding links to geometry are exploited in Billera et al. [2001, 2002] and Zwiernik and Smith [2012].

In the biology literature, the general Markov model is also known as Barry Hartigan's model; see Barry and Hartigan [1987]. Recently this model is becoming increasingly more popular in biological analysis; see for example Allman and Rhodes [2008], Jayaswal et al. [2005, 2007].

Chapter 6

The local geometry

[]

Our strategy is to define a new parameterization for the tree models defined in Section 5.3. The new coordinate system is based on moments rather than conditional probabilities. This helps us to exploit various invariance properties of tree models, which in turn enables us to express the dependence structure implied by the tree more elegantly. In most of this chapter we focus on the two-state case $M(\mathcal{T}, 2)$. We show that using special L-cumulants gives a great understanding of the model $M(\mathcal{T}, 2)$. Throughout this chapter we assume that (5.8) holds, that is, the marginal distribution over each variable in the system has only positive entries. By Proposition 5.45 this assumption does not change the model.

6.1 Tree cumulant parameterization

In this section we present tree cumulants. The motivation is given by a simple example with three leaves, which we treat in detail below.

6.1.1 The tripod tree

Let $\mathcal{T}$ be the tripod tree in Figure 6.1 and let $M(\mathcal{T}, 2)$ be the corresponding general Markov model. Here the black vertices represent three observed variables X_1, X_2, X_3, and the white vertex indicates a variable H that remains *hidden*. Since all variables in the system are binary, the parameterization is

$$p(x) = \sum_{y=0}^{1} \theta_h(y)\theta_{1|h}(x_1|y)\theta_{2|h}(x_2|y)\theta_{3|h}(x_3|y) \qquad \text{for all } x \in \{0, 1\}^3. \quad (6.1)$$

There are seven free parameters needed to specify $p(x)$: $\theta_h(1)$ together with $\theta_{1|h}(1|y)$, $\theta_{2|h}(1|y)$ and $\theta_{3|h}(1|y)$ for $y = 0, 1$. The Zariski closure of this model corresponds exactly to the first Secant $\mathrm{Sec}(1, 1, 1)$ as defined in Section 2.4; see also Example 2.60.

Following Section 3.1.3, define

$$\bar{X}_i := \frac{X_i - \mathbb{E}X_i}{\mathrm{var}(X_i)}, \qquad \bar{H} := \frac{H - \mathbb{E}H}{\mathrm{var}(H)}$$

and $\rho_{ij} = \mathbb{E}(\bar{X}_i \bar{X}_j)$, $\rho_{ih} = \mathbb{E}(\bar{X}_i \bar{H})$, $\rho_{123} = \mathbb{E}(\bar{X}_1 \bar{X}_2 \bar{X}_3)$, $\rho_{hhh} = \mathbb{E}(\bar{H}^3)$. We

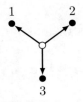

Figure 6.1: *The tripod tree.*

always assume that (5.8) holds and therefore all the above quantities are well defined. By Proposition 3.12 there is a one-to-one correspondence between the probabilities $p(x)$ for $x \in \{0, 1\}^3$ and the four central moments μ'_{ij} for $i, j = 1, 2, 3$ and μ'_{123} supplemented by the three means μ_i for $i = 1, 2, 3$. Thus, probabilities are in one-to-one correspondence with μ_1, μ_2, μ_3, ρ_{12}, ρ_{13}, ρ_{23} and ρ_{123}. What we will show later explicitly, is that the original parameters $\theta_h(1)$ and $\theta_{i|h}(1|0), \theta_{i|h}(1|1)$ are in one-to-one correspondence with seven new parameters ρ_{111}, ρ_{222}, ρ_{333}, ρ_{hhh} and ρ_{1h}, ρ_{2h}, ρ_{3h}.

Because $X_1 \perp\!\!\!\perp X_2 \perp\!\!\!\perp X_3 | H$, by Lemma 3.19

$$\mu_i \quad \in \quad [0, 1],$$

$$\rho_{ij} \quad = \quad \rho_{ih}\rho_{jh} \qquad \text{for all } i < j \text{ and} \qquad (6.2)$$

$$\rho_{123} \quad = \quad \rho_{hhh}\rho_{1h}\rho_{2h}\rho_{3h}$$

and in particular there are no non-trivial constraints on the means μ_i or the corresponding skewnesses ρ_{iii}. Moreover, there are no equality constraints in this model.

The product form of this parameterization enables us to derive constraints on the observed vertices. For example, (6.2) implies that

$$\rho_{12}\rho_{13}\rho_{23} \quad = \quad (\rho_{1h}\rho_{2h})(\rho_{1h}\rho_{3h})(\rho_{2h}\rho_{3h}) \quad = \quad \rho_{1h}^2\rho_{2h}^2\rho_{3h}^2$$

and hence $\rho_{12}\rho_{13}\rho_{23} \geq 0$ must hold for every distribution in $M(\mathcal{T}, 2)$. It also allows us to find explicit formulas for the parameters of the model for every distribution in $M(\mathcal{T}, 2)$ such that $\rho_{12}\rho_{13}\rho_{23} > 0$. Indeed, from (6.2) it follows that

$$\rho_{hhh}^2 \quad = \quad \frac{\rho_{123}^2}{\rho_{12}\rho_{13}\rho_{23}}, \qquad \rho_{hi}^2 \quad = \quad \frac{\rho_{ij}\rho_{ik}}{\rho_{jk}}, \qquad (6.3)$$

which, up to a sign, identifies parameters in terms of the distribution of (X_1, X_2, X_3).

Remark 6.1. Given formulas (6.3), the corresponding formulas for the original parameters $\theta_h(1)$ and $\theta_{i|h}(1|j)$ can be given; see Theorem 7.6.

6.1.2   *The general tree case*

Let $\mathcal{T}$ be a semi-labeled tree. Consider the general Markov model $M(\mathcal{T}, 2)$. The lattice of $\mathcal{T}$-partitions $\Pi(\mathcal{T})$ defined in Section 5.1.2 leads to the definition of *tree cumulants* (or $\mathcal{T}$-cumulants if we want to make $\mathcal{T}$ explicit). We follow the general construction of L-cumulants given in Definition 4.23, where L now stands for $\mathbf{\Pi}(\mathcal{T})$. We denote tree cumulants by t and their standardized version by $\bar{\mathsf{t}}$. Thus, for any subset of leaves A we have

$$\mathsf{t}_A = \sum_{\pi \in \mathbf{\Pi}(\mathcal{T}, A)} \mathfrak{m}_A(\pi) \prod_{B \in \pi} \mu_B, \qquad (6.4)$$

where $\mathfrak{m}_A(\pi) := \mathfrak{m}_A(\pi, \hat{1})$ is the Möbius function on $\mathbf{\Pi}(\mathcal{T}, A)$. The expression for $\bar{\mathsf{t}}_A$ is the same with μ_B replaced by $\rho_B = \mathbb{E}[\prod_{i \in B} \bar{X}_i]$. In the formula for $\bar{\mathsf{t}}$, by Proposition 4.38, the poset $\mathbf{\Pi}(\mathcal{T}, A)$ can be replaced with a much smaller poset $\mathbf{\Pi}'(\mathcal{T}, A)$ generated by non-trivial splits of $\mathcal{T}$.

The Möbius inversion formula (c.f. Proposition 4.4) gives the expression for moments in terms of cumulants

$$\mu_A = \sum_{\pi \in \mathbf{\Pi}(\mathcal{T}, A)} \prod_{B \in \pi} \mathsf{t}_B. \qquad (6.5)$$

Remark 6.2. The map between the probabilities $p(x)$ for $x \in \{0, 1\}^m$ and moments μ_B for $B \subseteq [m]$ is a simple linear map in (4.23). It follows that the change from the raw probabilities to tree cumulants is a polynomial map. Hence, the map to standardized tree cumulants is an algebraic map.

Remark 6.3. For any phylogenetic tree $\mathcal{T}$ with at most three leaves, the corresponding $\mathbf{\Pi}(\mathcal{T})$ coincides with the lattice of all partitions. It follows that t_A for $|A| \leq 3$ coincides with the corresponding cumulant (or central moment) $k_A = \mu'_A$. For the standardized tree cumulants we have $\bar{\mathsf{t}}_A = \bar{k}_A = \rho_A$.

Example 6.4. Let $\mathcal{T}$ be a quartet phylogenetic tree from Example 5.44. The $\mathcal{T}$-cumulants are given by 15 coordinates t_A for all non-empty $A \subseteq \{1, 2, 3, 4\}$. By Remark 6.3 we have $\mathsf{t}_i = \mu_i$, $\mathsf{t}_{ij} = \mu'_{ij} = k_{ij}$ for $1 \leq i < j \leq 4$ and $\mathsf{t}_{ijk} = \mu'_{ijk} = k_{ijk}$ for all $1 \leq i < j < k \leq 4$. Finally, we have three ways of computing t_{1234}. One way is to use Proposition 4.25, which gives

$$\mathsf{t}_{1234} = k_{1234} + k_{13}k_{24} + k_{14}k_{23},$$

where the formula for cumulants k_A in terms of moments is given by (4.16). Another way is to compute directly the Möbius function and use (6.4). There is also a third simpler way, which generalizes to bigger caterpillar trees; see Section 6.3.3. Let $L = \mathbf{\Pi}(\mathcal{T})$ with the Hasse diagram given in Figure 5.7. By K denote the partition lattice defined by interval splits: $1|234$, $12|34$, $123|4$, which is also a tree partition for the semi-labeled tree $\overset{1}{\bullet}-\overset{2}{\bullet}-\overset{3}{\bullet}-\overset{4}{\bullet}$. Both K and L have only one non-trivial split $12|34$ and we easily check that assumptions of Proposition 4.38 are satisfied, which gives

$$\mathsf{t}_{1234} = \sum_{\pi \in K} \mathfrak{m}_K(\pi) \prod_{B \in \pi} \mu'_B = \mu'_{1234} - \mu'_{12}\mu'_{34},$$

where the second equation follows from the fact that K is isomorphic to the Boolean lattice on $\{1, 2, 3\}$ and hence $\mathfrak{m}_K(\pi) = (-1)^{|\pi|-1}$. The formula for μ'_{1234}, μ'_{12} and μ'_{34} can be derived from Proposition 4.37.

Definition 6.5 (Moment parameters). Consider the change of parameters from $\theta_r(1)$, $\theta_{v|u}(1|y)$ for $y = 0, 1$ to a new set of parameters given by *skewness* $\rho_{vvv} = \mathbb{E}\bar{Y}_v^3$ for all $v \in V(\mathcal{T})$ and *edge correlation* $\rho_{uv} = \mathbb{E}(\bar{Y}_u\bar{Y}_v)$ for all $u - v$ in $\mathcal{T}$. This new parameter space is denoted by $\Omega_{\mathcal{T}}$.

To write the map to moment parameters explicitly, first note that $\mu_v = \mathbb{E}Y_v = \mathbb{P}(Y_v = 1)$. We first compute $\mu_r = \theta_r(1)$ and the remaining means are computed recursively for each v assuming that the formula for $\mu_{\mathrm{pa}(v)}$ is given. The recursive formula is

$$\mu_v \quad = \quad (1 - \mu_{\mathrm{pa}(v)})\theta_{v|\mathrm{pa}(v)}(1|0) + \mu_{\mathrm{pa}(v)}\theta_{v|\mathrm{pa}(v)}(1|1). \qquad (6.6)$$

Recall that the variance μ'_{vv} of Y_v is given by $\sigma_v = \mu_v(1 - \mu_v)$ and the formula for the skewness of Y_v by

$$\rho_{vvv} \quad = \quad \frac{1 - 2\mu_v}{\sqrt{\sigma_v}} \qquad \text{for all } v \in V(\mathcal{T}), \qquad (6.7)$$

which can be obtained by rewriting $(Y_v - \mu_v)^3$ as $Y_v(1 - 3\mu_v + 2\mu_v^2) - \mu_v^3$ using the fact that $Y_v^k = Y_v$ for all $k \geq 1$. The formula for an edge correlation ρ_{uv} is

$$\rho_{uv} \quad = \quad (\theta_{v|u}(1|1) - \theta_{v|u}(1|0))\sqrt{\frac{\sigma_u}{\sigma_v}}, \qquad (6.8)$$

which for simplicity we write partly in terms of the means μ_u, μ_v, whose explicit expressions in terms of the θ parameters is given in (6.6).

Simple linear constraints defining Θ_T become slightly more complicated when expressed in the new parameters $\omega \in \Omega_{\mathcal{T}}$.

Lemma 6.6 (Constraints on $\Omega_{\mathcal{T}}$). *Suppose that the values of means* $\mu_v \in [0, 1]$ *are fixed for every vertex* v *(equivalently, by (6.7),* $\rho_{vvv} \in \mathbb{R}$ *are fixed). Then the correlations are constrained to satisfy*

$$-\min\{t_ut_v, \frac{1}{t_ut_v}\} \quad \leq \quad \rho_{uv} \quad \leq \quad \min\{\frac{t_u}{t_v}, \frac{t_v}{t_u}\},$$

where $t_u = \sqrt{\frac{1 - \mu_u}{\mu_u}}$.

Proof. We first fix means μ_v in the interval $(0, 1)$. Given this choice, edge correlations ρ_e cannot now take any value in $[-1, 1]$. To obtain the induced constraints on ρ_{uv} induced by this choice note that

$$\max\{0, p_u(1) + p_v(1) - 1\} \quad \leq \quad p_{uv}(1, 1) \quad \leq \quad \min\{p_u(1), p_v(1)\},$$

where $p_u(1) = \mu_u$ and $p_{uv}(1, 1) = \mu_{uv}$. Since $\mu'_{uv} = \mu_{uv} - \mu_u\mu_v$, we further obtain that

$$\max\{-\mu_u\mu_v, -(1 - \mu_u)(1 - \mu_v)\} \leq \mu'_{uv} \leq \min\{\mu_u(1 - \mu_v), (1 - \mu_u)\mu_v\},$$

which after normalizing by $\sqrt{\sigma_u\sigma_v}$ gives the final constraint. $\qquad \square$

The following result provides a partial understanding of why tree cumulants are well suited for general Markov models.

Proposition 6.7. *Let $\mathcal{T}$ be a phylogenetic tree with the labeling set $[m]$ and let $\boldsymbol{X}$ be a binary vector with distribution in $\boldsymbol{M}(\mathcal{T}, 2)$. There exists a tree split $B_1|B_2 \in \boldsymbol{\Pi}(\mathcal{T})$ such that $X_{B_1} \perp\!\!\!\perp X_{B_2}$ if and only if $\mathsf{t}_A = 0$ for all $A \subseteq [m]$ unless either $A \subseteq B_1$ or $A \subseteq B_2$. Moreover, this in general is not true if $B_1|B_2 \notin \boldsymbol{\Pi}(\mathcal{T})$.*

Proof. Because for every phylogenetic tree $\mathcal{T}$ the $\mathcal{T}$-partition lattice is saturated (see Definition 4.29), Theorem 4.31 and Remark 4.32 apply as long as the vector $\boldsymbol{X}$ is non-degenerate. $\qquad\square$

A generic distribution in $\boldsymbol{M}(\mathcal{T}, 2)$ does not satisfy any marginal independencies. Suppose that $q \in \boldsymbol{M}(\mathcal{T}, 2)$ satisfies $\boldsymbol{X}_{B_1} \perp\!\!\!\perp \boldsymbol{X}_{B_2}$ but no other marginal independences. This can happen if and only if $B_1|B_2 \in \boldsymbol{\Pi}(\mathcal{T})$ and $Y_u \perp\!\!\!\perp Y_v$, where $u - v$ is the edge inducing the tree split $B_1|B_2$. This in turn is equivalent to $\rho_{uv} = 0$ and by Proposition 6.7 to $\mathsf{t}_A = 0$ for all A such that $u - v \in E(\mathcal{T}^0(A))$ (c.f. Definition 5.51). This informally shows that the parameter ρ_e can be a factor in all t_A such that $u - v \in E(\mathcal{T}^0(A))$. The following result, which also generalizes (6.2), shows that this is true.

Theorem 6.8. *Let $\mathcal{T} = (T, \phi)$ be a binary phylogenetic tree with the labeling set $[m]$. Then $\boldsymbol{M}(\mathcal{T}, 2)$ in the space of standardized tree cumulants is given by*

$$\bar{\mathsf{t}}_A = \prod_{v \in V(\mathcal{T}^0(A))} \rho_{vvv}^{\alpha_v} \prod_{e \in E(\mathcal{T}^0(A))} \rho_e \quad \text{for } |A| \geq 2, \tag{6.9}$$

where $\mathcal{T}^0(A)$ is the tree in Definition 5.51, $\alpha_v := \max\{0, \deg(v) - 2\}$ and the degree of v is taken in $\mathcal{T}^0(A)$. Moreover, there are no constraints on the means $\mu_i \in (0, 1)$ for $i = 1, \ldots, m$.

Proof. We prove a more general formulation of this theorem when $\mathcal{T}$ is allowed to have unlabeled degree-two vertices. In this case all the inner vertices of $\mathcal{T}$ have degree 2 or 3 and the same holds true for $\mathcal{T}^0(B)$ for any $B \subseteq [m]$. It is then sufficient to show that the theorem is true for $A = [m]$. The proof proceeds by induction with respect to m. Let $m = 2$. Since by definition $\bar{\mathsf{t}}_{12} = \rho_{12}$, (6.9) becomes

$$\rho_{12} = \prod_{e \in E(\mathcal{T})} \rho_e. \tag{6.10}$$

By Proposition 5.37, X_1 and X_2 are independent given any vertex, say v, separating them. By Lemma 3.19 $\rho_{12} = \rho_{1v}\rho_{2v}$. To prove (6.10) we imply use this result recursively.

Now assume that $m \geq 3$ and the theorem is true for all $k \leq m - 1$. We can always find two leaves separated from all the other leaves by an inner vertex. Denote the leaves by $1, 2$ and the separating inner vertex by a. The global Markov properties in (3.20) give that for each $C \subseteq A := \{3, \dots, n\}$ and any distribution in $\boldsymbol{M}(\mathcal{T}, 2)$ we have

$$\rho_{12C} \;=\; \rho_{12}\rho_C + \rho_{12}\rho_{aaa}\rho_{aC}. \tag{6.11}$$

To prove (6.11) use Proposition 3.18 and the fact that $X_1 \perp\!\!\!\perp X_2 \perp\!\!\!\perp \boldsymbol{X}_C | H_a$ to write

$$\rho_{12C} \;=\; \mathbb{E}(\bar{X}_1 \bar{X}_2 \prod_{i \in C} \bar{X}_i) \;=\; \mathbb{E}\big(\mathbb{E}(\bar{X}_1 | \bar{H}_a)\mathbb{E}(\bar{X}_2 | \bar{H}_a)\mathbb{E}(\prod_{i \in C} \bar{X}_i | \bar{H}_a)\big).$$

Because H_a is binary, we can use (3.10) to write

$$\mathbb{E}(\bar{X}_1 | \bar{H}_a) = \rho_{1a}\bar{H}_a, \quad \mathbb{E}(\bar{X}_2 | \bar{H}_a) = \rho_{2a}\bar{H}_a, \quad \mathbb{E}(\prod_{i \in C} \bar{X}_i | \bar{H}_a) = \rho_C + \rho_{C\cup a}\bar{H}_a.$$

Since $\rho_{12} = \rho_{1a}\rho_{2a}$, (6.11) follows.

The inner vertex a separates three sets $\{1\}$, $\{2\}$, A and it has degree ≤ 3. Therefore its degree must be exactly 3. Let $E' = \overline{12} \subset E(\mathcal{T})$. The tree $\mathcal{T}/E'$ is obtained from $\mathcal{T}$ by contracting all edges between 1 and 2, and hence it contains a multiple labeled vertex a with label $\{1, 2\}$. Define $\mathcal{T}_a$ as the tree with labeling set $\{a\} \cup A$ obtained from $\mathcal{T}/E'$ by replacing the multiple label $\{1, 2\}$ with $\{a\}$. Let $\pi_0 = 12 | \hat{0}_A \in \boldsymbol{\Pi}(\mathcal{T})$. The *trimming map* with respect to $\{1, 2\}$ is the map

$$[\pi_0, \hat{1}] = \boldsymbol{\Pi}(\mathcal{T}/E') \;\to\; \boldsymbol{\Pi}(\mathcal{T}_a), \qquad \pi \mapsto \widetilde{\pi}$$

that changes the block $12C$ in $\pi \in [\pi_0, \hat{1}]$ to aC. Note that the trimming map constitutes an isomorphism of posets between $[\pi_0, \hat{1}]$ and $\boldsymbol{\Pi}(\mathcal{T}_a)$.

The definition of $\mathcal{T}$-cumulants implies that

$$\bar{\mathfrak{t}}_{[m]} \;=\; \sum_{\pi \in [\pi_0, \hat{1}]} \mathfrak{m}(\pi) \prod_{B \in \pi} \rho_B + \sum_{\pi \notin [\pi_0, \hat{1}]} \mathfrak{m}(\pi) \prod_{B \in \pi} \rho_B. \tag{6.12}$$

The second summand in (6.12) is zero since every $\pi \in \boldsymbol{\Pi}(\mathcal{T})$ such that $\pi \notin [\pi_0, \hat{1}]$ necessarily contains either 1 or 2 as one of the blocks and $\rho_1 = \rho_2 = 0$. Let $\pi_1 = 12 | A$. Applying (6.11) to each ρ_{12C} for each $\pi \in [\pi_0, \hat{1}]$ we obtain

$$\prod_{B \in \pi} \rho_B \;=\; \prod_{B \in \pi \wedge \pi_1} \rho_B + \rho_{12}\rho_{aaa} \prod_{B \in \widetilde{\pi}} \rho_B,$$

where $\widetilde{\pi}$ denotes the map of the partition π under the trimming map described above. We can rewrite

$$\bar{\mathfrak{t}}_{[m]} = \sum_{\pi \in [\pi_0, \hat{1}]} \mathfrak{m}(\pi) \prod_{B \in \pi \wedge \pi_1} \rho_B + \rho_{12}\rho_{aaa} \sum_{\pi \in [\pi_0, \hat{1}]} \mathfrak{m}(\pi) \prod_{B \in \widetilde{\pi}} \rho_B. \tag{6.13}$$

The first summand in (6.13) can be rewritten as

$$\sum_{\delta \in [\pi_0, \pi_1]} \left[\left(\sum_{\pi \wedge \pi_1 = \delta} \mathfrak{m}(\pi) \right) \prod_{B \in \delta} \rho_B \right]. \tag{6.14}$$

However, from Lemma 4.19, since $\pi_1 \neq \hat{1}$, for each δ the sum $\sum_{\pi \wedge \pi_1 = \delta} \mathfrak{m}(\pi)$ in (6.14) is zero. It follows that

$$\bar{t}_{[m]} \quad = \quad \rho_{12}\rho_{aaa} \sum_{\pi \in [\pi_0, \hat{1}]} \mathfrak{m}(\pi) \prod_{B \in \bar{\pi}} \rho_B.$$

By Proposition 4.11 the Möbius function of $[\pi_0, \hat{1}]$ is equal to the restriction of the Möbius function on $\mathbf{\Pi}(\mathcal{T})$ to the interval $[\pi_0, \hat{1}]$. The trimming map constitutes an isomorphism between $[\pi_0, \hat{1}]$ and $\mathbf{\Pi}(\mathcal{T}_a)$. Consequently, the Möbius function on $[\pi_0, \hat{1}]$ is equal to the Möbius function on $\mathbf{\Pi}(\mathcal{T}_a)$. It follows that

$$\bar{t}_{[m]} \quad = \quad \rho_{12}\rho_{aaa}\bar{t}_{aA}$$

and by induction we can write $\bar{t}_{aA}$ in the form (6.9), which already implies the final result because $\rho_{12} = \prod_{e \in \overline{12}} \rho_e$ by case $m = 2$.   □

Recall from Definition 5.46 a binary expansion $\mathcal{T}^*$ of a semi-labeled tree $\mathcal{T}$. By Lemma 5.48 $\mathbf{M}(\mathcal{T}, 2) \subseteq \mathbf{M}(\mathcal{T}^*, 2)$ and we can use this fact to obtain the parameterization of $\mathbf{M}(\mathcal{T}, 2)$ for any semi-labeled tree $\mathcal{T}$ using the parameterization for its binary expansion $\mathcal{T}^*$. Let $E' \subset E(\mathcal{T}^*)$ such that $\mathcal{T}^*/E' = \mathcal{T}$. For a vertex $w \in V(\mathcal{T})$, let $\lfloor w \rfloor$ denote the subset of vertices in $V(\mathcal{T}^*)$ that get identified with w in the process of going from $\mathcal{T}^*$ to $\mathcal{T}$.

The following theorem generalizes Theorem 6.8. Here we first conveniently extend the definition of the degree.

Definition 6.9. Let $\mathcal{T} = (T, \phi)$ be a semi-labeled tree and let $v \in V(\mathcal{T})$ of degree $\deg(v)$. The $\mathcal{T}$-degree of v is defined as

$$\deg(v; \mathcal{T}) \quad := \quad \deg(v) + |\phi^{-1}(v)|.$$

In particular, the degree of v is equal to its $\mathcal{T}$-degree if v is unlabeled but is strictly lower otherwise.

Theorem 6.10. *Let $\mathcal{T}$ be a semi-labeled tree and let $\mathcal{T}^*$ be its binary expansion. If $\bar{t}_A^*$ for $A \subseteq [m]$ are $\mathcal{T}^*$-cumulants, then for all distributions in $\mathbf{M}(\mathcal{T}, 2)$, $\mu_i \in (0, 1)$ for $i \in [m]$ and*

$$\bar{t}_A^* = \prod_{v \in V(\mathcal{T}^0(A))} \rho_{vvv}^{\beta_v} \prod_{e \in E(\mathcal{T}^0(A))} \rho_e \quad \text{for } |A| \geq 2, \qquad (6.15)$$

where $\beta_v := \max\{0, \deg(v; \mathcal{T}^0(A)) - 2\}$ and the degree is taken in $\mathcal{T}^0(A)$.

Remark 6.11. Theorem 6.8 is a special case of this result. Indeed, if $\mathcal{T}$ is a binary phylogenetic tree, which is allowed to have degree-two unlabeled vertices, then

$$\max\{0, \deg(v) - 2\} = \max\{0, \deg(v; \mathcal{T}) - 2\}.$$

This is elementary for unlabeled vertices and for labeled vertices this follows from the fact that only the leaves are labeled.

Proof of Theorem 6.10. Let $\mathcal{T}^*$ be a binary expansion of $\mathcal{T}$ obtained from $\mathcal{T}$ by the procedure in Definition 5.47, and let $E' \subseteq E(\mathcal{T}^*)$ be such that $\mathcal{T} = \mathcal{T}^*/E'$. By the proof of Theorem 5.49, $\boldsymbol{M}(\mathcal{T}, 2)$ is given as a submodel of $\boldsymbol{M}(\mathcal{T}^*, 2)$ by setting $\theta^*_{v|u}$ to be the identity matrix for every edge $(u, v) \in E'$. In the new parameters, $\Omega_{\mathcal{T}}$ is isomorphic to the subset of $\Omega_{\mathcal{T}^*}$ given by

$$
\begin{aligned}
\rho^*_e &= \rho_e \quad \text{for all } e \notin E', \\
\rho^*_e &= 1 \quad \text{for all } e \in E', \text{ and} \\
\rho^*_{vvv} &= \rho_{www} \quad \text{for all } w \in V(\mathcal{T}) \text{ and } v \in \lfloor w \rfloor.
\end{aligned} \tag{6.16}
$$

Indeed, if $\theta^*_{v|u}$ is the identity matrix, then by (6.6) $\mu^*_u = \mu^*_v$, that is $\rho^*_{uuu} = \rho^*_{vvv}$. By (6.8) also $\rho^*_{uv} = 1$.

We first show (6.15) for $A = [m]$. By Theorem 6.8 the model $\boldsymbol{M}(\mathcal{T}^*)$ is parameterized by (c.f. Remark 6.11)

$$
\bar{\mathfrak{t}}^*_{[m]} = \prod_{v \in V(\mathcal{T}^*)} \rho^{* \, \max\{0, \deg(v; \mathcal{T}^*) - 2\}}_{vvv} \prod_{e \in E(\mathcal{T}^*)} \rho^*_e. \tag{6.17}
$$

Since $E(\mathcal{T}^*) = E(\mathcal{T}) \cup E'$, applying (6.16) to $\prod_{e \in E(\mathcal{T}^*)} \rho^*_e$ gives:

$$
\prod_{e \in E(\mathcal{T}^*)} \rho^*_e \quad \overset{(6.16)}{\longrightarrow} \quad \prod_{e \in E(\mathcal{T})} \rho_e.
$$

If $w \in V(\mathcal{T})$ is an inner vertex, then we consider three cases. Case 1: If w is labeled and $\mathcal{T}$-degree is $\deg(w; \mathcal{T}) = 3$, then w has degree 2 and $|\phi^{-1}(w)| = 1$. In that case, by construction in Definition 5.47, $\lfloor w \rfloor$ consists of one leaf ($\mathcal{T}^*$-degree 2) and one unlabeled inner vertex of $\mathcal{T}^*$-degree 3. Therefore,

$$
\prod_{v \in \lfloor w \rfloor} \rho^{* \, \max\{0, \deg(v; \mathcal{T}^*) - 2\}}_{vvv} \quad \overset{(6.16)}{\longrightarrow} \quad \rho^{\deg(w; \mathcal{T}) - 2}_{www}. \tag{6.18}
$$

Case 2: If w has $\mathcal{T}$-degree ≥ 4, the set $\lfloor w \rfloor$ has $|\phi^{-1}(w)| + \deg(w; \mathcal{T}) - 2$ elements out of which $|\phi^{-1}(w)|$ are leaves of $\mathcal{T}^*$ and the rest are inner vertices of $\mathcal{T}^*$-degree 3. Indeed, going from $\mathcal{T}$ to $\mathcal{T}^*$ in Definition 5.47, in Step 1 we introduce $|\phi^{-1}(w)|$ new leaves that are all in $\lfloor w \rfloor$, and in Step 2 we introduce further $\deg(w; \mathcal{T}) - 2$ inner vertices. Therefore, (6.18) still applies if $\deg(w; \mathcal{T}) \geq 4$, which implies that after applying (6.16) to (6.17) we obtain exactly (6.15) with $A = [m]$. The proof for a general $A \subset [m]$ is similar but greater care must be paid to degree-two unlabeled vertices. □

Remark 6.12. Note that the theorem holds also in the case when $\mathcal{T}$ has vertices with multiple labels.

Example 6.13. Consider a simple model given by a semi-labeled tree $\mathcal{T}$ of the form $\overset{1}{\bullet} - \overset{2}{\bullet} - \overset{3}{\bullet}$. Since $\deg(2; \mathcal{T}) = 3$, we have

$$
\bar{\mathfrak{t}}^*_{123} = \rho_{222} \rho_{12} \rho_{23}.
$$

However, the $\deg(2; \mathcal{T}^0(\{1,3\})) = 2$ because in this tree the vertex 2 is unlabeled. This gives

$$\bar{t}^*_{13} = \rho_{12}\rho_{23}.$$

Here we use $\mathcal{T}^*$-cumulants, where $\mathcal{T}^*$ is the tripod tree model. By Remark 6.3, $\mathcal{T}^*$-cumulants are equal to cumulants and so in particular $\bar{t}^*_{123} = \rho_{123}$ and $\bar{t}^*_{13} = \rho_{13}$.

Example 6.14. Suppose now that $\mathcal{T}$ is a phylogenetic tree with one inner vertex and m labeled leaves like in Figure 5.6. We denote the inner vertex with letter h. In this case any phylogenetic binary tree with the same number of leaves is a binary expansion of $\mathcal{T}$. Computations are particularly simple if $\mathcal{T}^*$ is a caterpillar tree; see Section 6.3.3. In tree cumulants, the model $M(\mathcal{T}, 2)$ is parameterized by $\mu_i \in (0, 1)$ for $i \in [m]$ and

$$\bar{t}^*_A \;=\; \rho_{hhh}^{|A|-2} \prod_{i\in A} \rho_{ih} \qquad \text{for } |A| \geq 2.$$

It may be tempting to think that it is enough to consider $\mathcal{T}$-cumulants instead of $\mathcal{T}^*$-cumulants. However, for example, if $\mathcal{T}$ is a star tree, then $\mathbf{\Pi}(\mathcal{T})$ is the lattice of one-cluster partitions (c.f. Remark 5.12) and hence by Proposition 4.36 the $\mathcal{T}$-cumulants are equal to central moments. This set of coordinates gives some insight into the geometry of $M(\mathcal{T}, 2)$ but it is not sufficient for our purposes; see the appendix in Geiger et al. [2001].

Recall the model $M_+(\mathcal{T}, 2)$ from Definition 5.53.

Lemma 6.15. *The constraint on the parameter space Θ_T to satisfy $\det(\theta_{u|v}) > 0$ for every $u \to v$ translates to $\rho_e > 0$ for all $e \in \mathcal{T}$ in $\Omega_{\mathcal{T}}$.*

Proof. We have $\det \theta_{v|u} = \theta_{v|u}(1|1) - \theta_{v|u}(1|0)$ and hence by (6.8) $\det(\theta_{u|v}) > 0$ if and only if $\rho_{uv} > 0$. □

6.2   Geometry of unidentified subspaces

In this section we analyze the geometry of q-fibers as defined in Definition 3.23 for $q \in M(\mathcal{T}, 2)$. Here the q-fibers are taken assuming that $\mathbf{Y} = (Y_v)$ is a non-degenerate vector, which excludes some points of the original parameter space; see Proposition 5.45. Although, in some applications (see Rusakov and Geiger [2005], Zwiernik [2011]) it is important to understand fibers of the parameterization of $M(\mathcal{T}, 2)$ without assuming non-degeneracy (so that the parameter space is compact); in this book we restrict to the non-degenerate case.

For a distribution $q \in M(\mathcal{T}, 2)$ let μ^q_B, ρ^q_B, $\bar{t}^q_B$ for $B \subseteq [m]$ denote the corresponding moments, standardized moments, and standardized $\mathcal{T}^*$-cumulants, where $\mathcal{T}^*$ is any binary expansion of $\mathcal{T}$. The q-fiber is the set of all points of $\Omega_{\mathcal{T}}$ such that $\rho_{iii} = \rho^q_{iii}$ for all $i \in [m]$ and for every $A \subseteq [m]$ such that $|A| \geq 2$

$$\prod_{v\in V(\mathcal{T}^0(A))} \rho_{vvv}^{\beta_v} \prod_{e\in E(\mathcal{T}^0(A))} \rho_e \;=\; \bar{t}^q_A, \qquad (6.19)$$

where $\beta_v = \max\{0, \deg(v; \mathcal{T}^0(A)) - 2\}$. Since the second-order standard-ized tree cumulants are always equal to corresponding correlations, by Theorem 6.10 for any $\omega = ((\rho_{vvv}), (\rho_e))$ in the q-fiber we have that

$$\prod_{e \in \overline{ij}} \rho_e = \rho_{ij}^q. \tag{6.20}$$

Let $\Sigma^q = [\rho_{ij}^q]$ be the corresponding *correlation matrix*, which is a symmetric positive semi-definite $m \times m$ matrix. We show that the geometry of the q-fiber, denoted by $\Omega_q \subseteq \Omega_{\mathcal{T}}$, is determined by zeros in Σ^q.

Definition 6.16. An edge $e \in E(\mathcal{T})$ is *isolated relative to* q if $\rho_{ij}^q = 0$ for all $i, j \in [m]$ such that $e \in \overline{ij}$. We denote the set of all edges of $\mathcal{T}$ which are isolated relative to q by $E^q \subseteq E(\mathcal{T})$. We define the *$q$-forest $\mathcal{F}^q$* as the forest obtained from $\mathcal{T}$ by removing edges in E^q but keeping all the vertices. Often, $\mathcal{F}^q$ will not be a semi-labeled forest because we allow it to have unlabeled vertices of degree less than three.

We illustrate this construction in the example below.

Example 6.17. Let $\mathcal{T}$ be the phylogenetic tree given in Figure 6.2 (ignore at first that some edges are dashed) and assume that the correlation matrix contains zeros given in the provided 7×7-matrix where the asterisks mean any non-zero values such that the matrix is positive semi-definite. We have

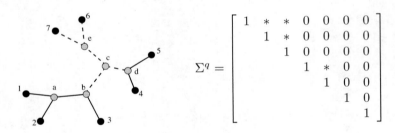

Figure 6.2: *An example of a phylogenetic tree and a correlation matrix Σ^q.*

$E^q = \{b - c, c - d, c - e, e - 6, e - 7\}$ and these edges are depicted in Figure 6.2 as dashed lines. The q-forest $\mathcal{F}^q$ is obtained by removing the edges in E^q. It is not a semi-labeled forest because it contains degree zero and two unlabeled vertices.

Lemma 6.18. *All the unlabeled vertices of $\mathcal{T}$ have either degree zero in $\mathcal{F}^q$ or the degree is strictly greater than one.*

Proof. Suppose that v is an unlabeled vertex of $\mathcal{T}$ whose degree in $\mathcal{F}^q$ is one. Denote by e the edge of v that is not isolated relative to q. Let i, j be *any* two labeled vertices of $\mathcal{T}$ such that $e \in \overline{ij}$. Since all other edges of v are isolated relative to q, then $\overline{ij}$ necessarily contains such an edge and therefore $\rho_{ij}^q = 0$. This gives a contradiction. $\square$

Lemma 6.19. *Let u, v be two distinct vertices of $\mathcal{T}$ such that each: is labeled or unlabeled with degree ≥ 3 in $\mathcal{F}^q$. Then the monomial $\prod_{e \in \overline{uv}} \rho_e^2$ is a constant function on the q-fiber Ω_q. Moreover, if there are no degree-zero vertices on the path between u and v, then the value of this monomial is nonzero.*

Proof. It is enough to prove this result only in the case when all vertices on the path between u and v are unlabeled of degree ≤ 2 in $\mathcal{F}^q$. If both u and v are labeled, then there is nothing to show. Thus consider first the case when both u and v are unlabeled of degree ≥ 3 in $\mathcal{F}^q$. We can find four labeled vertices i, j, k, l such that u separates i from j in $\mathcal{F}^q$, v separates k and l, and $\{u, v\}$ separates $\{i, j\}$ from $\{k, l\}$. In Figure 6.3 we depict this situation in the case when there are no degree-zero vertices on the path between u and v. By construction we can require ρ_{ij}^q, ρ_{kl}^q to be nonzero. Also ρ_{ik}^q, ρ_{jl}^q can be

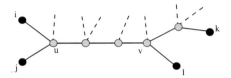

Figure 6.3: *An illustration for the proof of Lemma 6.19.*

required to be nonzero if all the vertices on the path between u and v have degree exactly two in $\mathcal{F}^q$. Consider the quantity $\frac{\rho_{ik}^q \rho_{jl}^q}{\rho_{ij}^q \rho_{kl}^q}$, which has a fixed value in the q-fiber, and substitute (6.20) for each of the terms. A simple rearrangement now gives that

$$\prod_{e \in \overline{uv}} \rho_e^2 \;=\; \frac{\rho_{ik}^q \rho_{jl}^q}{\rho_{ij}^q \rho_{kl}^q}.$$

If u is labeled, then consider any two labeled vertices i, j of $\mathcal{T}$ such that v separates u, i, j in $\mathcal{F}^q$ and $\rho_{ij}^q \neq 0$. We have

$$\prod_{e \in \overline{uv}} \rho_e^2 \;=\; \frac{\rho_{ui}^q \rho_{uj}^q}{\rho_{ij}^q}$$

and again we can require ρ_{ui}^q, ρ_{uj}^q to be nonzero if there are no degree-zero vertices on the path between u and v.   □

6.2.1 Regular q-fibers

Lemma 6.19 suggests that a particularly nice situation occurs when q is such that all unlabeled vertices of $\mathcal{T}$ have degree ≥ 3 in $\mathcal{F}^q$.

Theorem 6.20. *Let $\mathcal{T}$ be a semi-labeled tree. Suppose that $q \in \boldsymbol{M}(\mathcal{T}, 2)$ is such that all unlabeled vertices of $\mathcal{T}$ have degree ≥ 3 in $\mathcal{F}^q$. We have*

$\rho_{iii} = \rho_{iii}^q$ *for* $i \in [m]$ *in the q-fiber and up to sign also the other parameters are identified from q.*

- *For every unlabeled* v, *let* i, j, k *be any three labeled vertices separated from each other by* v *and such that* $\rho_{ij}^q \rho_{ik}^q \rho_{jk}^q \neq 0$, *then*

$$\rho_{vvv}^2 = \frac{\rho_{ijk}^q{}^2}{\rho_{ij}^q \rho_{ik}^q \rho_{jk}^q}.$$

- *For every edge* $u - v$, *where* u, v *are unlabeled, let* i, j, k, l *be any four labeled vertices such that* u *separates* $\{i\}$, $\{j\}$ *and* $\{k, l\}$ *and* v *separates* $\{i, j\}$, $\{k\}$, $\{l\}$; *and* $\rho_{ij}^q, \rho_{kl}^q, \rho_{ik}^q, \rho_{jl}^q$ *are all nonzero. Then*

$$\rho_{uv}^2 = \frac{\rho_{ik}^q \rho_{jl}^q}{\rho_{ij}^q \rho_{kl}^q}.$$

- *For every edge* $u - v$ *such that* u *is labeled but* v *is not, let* j, k *be any two labeled vertices such that* v *separates* u, j, k *and* $\rho_{uj}^q, \rho_{uk}^q, \rho_{jk}^q$ *are all nonzero. Then*

$$\rho_{uv}^2 = \frac{\rho_{uj}^q \rho_{uk}^q}{\rho_{jk}^q}.$$

The above choices of labeled vertices are typically not unique but the parameter values will not depend on these choices.

Proof. Because every unlabeled vertex v of $\mathcal{T}$ has degree ≥ 3 in $\mathcal{F}^q$, we can find three labeled vertices i, j, k separated by v such that ρ_{ij}^q, ρ_{ik}^q, ρ_{jk}^q are all nonzero. Now, the formula for ρ_{vvv}^2 follows directly by applying Theorem 6.10 and the fact that $\bar{\mathsf{t}}_A = \rho_A$ if $|A| \leq 3$ (c.f. Remark 6.3). The formulas for ρ_{uv}^2 and ρ_{iv}^2 were obtained in the proof of Lemma 6.19. $\square$

Corollary 6.21. *Let* $\mathcal{T}$ *be a semi-labeled tree and let* $\boldsymbol{M}(\mathcal{T}, 2)$ *be the general Markov model on* $\mathcal{T}$. *Then* $\dim(\boldsymbol{M}(\mathcal{T}, 2)) = |E(\mathcal{T})| + |V(\mathcal{T})|$ *by which we mean that there exists a dense open subset of* $\boldsymbol{M}(\mathcal{T}, 2)$ *diffeomorphic to a* $(|E(\mathcal{T})| + |V(\mathcal{T})|)$-*dimensional manifold.*

Proof. Let

$$U = \{\omega \in \Omega_{\mathcal{T}} : \rho_e \neq 0, e \in E(\mathcal{T})\}.$$

Every distribution q in the image of U satisfies $\rho_{ij}^q \neq 0$ for all $1 \leq i < j \leq m$. Thus by Theorem 6.20 the parameterization of $\boldsymbol{M}(\mathcal{T}, 2)$ over U is a finite map. Because $\Omega_{\mathcal{T}}$ and U have dimension $|V(\mathcal{T})| + |E(\mathcal{T})|$, it follows that $\dim \boldsymbol{M}(\mathcal{T}, 2) = |V(\mathcal{T})| + |E(\mathcal{T})|$. $\square$

Remark 6.22. Formulas in Theorem 6.20 show that the parameters are identified (up to sign) from 3-way margins of q. This is a special case of a more general result by Chang [1996], which we discuss in Section 6.4.

Theorem 6.23 (The geometry of the q-fiber — the smooth case). *Let $\mathcal{T}$ be a semi-labeled tree and $q \in \boldsymbol{M}(\mathcal{T}, 2)$. If each of the unlabeled vertices of $\mathcal{T}$ has degree ≥ 3 in the q-forest $\mathcal{F}^q$, then the q-fiber is a finite set of points. If each of the unlabeled vertices of $\mathcal{T}$ has degree ≥ 2 in $\mathcal{F}^q$, then the q-fiber is diffeomorphic to a disjoint union of polyhedra. In particular, it is a manifold with corners. Its dimension is $2l_2$ where l_2 is the number of unlabeled degree-two vertices in $\mathcal{F}^q$.*

Proof. If each unlabeled vertex of $\mathcal{T}$ has degree at least three in $\mathcal{F}^q$, then the q-fiber is finite by Theorem 6.20. Assume that some of the unlabeled vertices have degree 2 in $\mathcal{F}^q$. We proceed by describing the ideal I defining the q-fiber in $\Omega_{\mathcal{T}}$. This ideal contains in its set of generators all linear equations of the form $\rho_{iii} - \rho_{iii}^q$ for $i \in [m]$ and all polynomials of the form (6.19), where ρ_{iii}^q and $\bar{\mathfrak{t}}_B^q$ are constants.

Let $u - v$ be an edge, isolated relative to q — we want to show that necessarily $\rho_{uv} = 0$ in the q-fiber. If both u, v are labeled, then this is immediate. Suppose that both u, v are unlabeled. Because both u, v have degree ≥ 2 in $\mathcal{F}^q$, we find leaves i, j, k, l such that ρ_{ij}^q, ρ_{kl}^q are nonzero and $u \in \overline{ij}$, $v \in \overline{kl}$. Now using the fact how correlations are parameterized in $\boldsymbol{M}(\mathcal{T}, 2)$, we easily show that ρ_{ik}^q can be zero only if $\rho_{uv} = 0$ and if it is not zero, then $u - v$ cannot be isolated relative to q. The case when only one of them is labeled can be proved in a similar fashion. This shows that ρ_e for all $e \in E^q$ must lie in I. Adding these generators allows us to remove all the generators in (6.19), where $E(\mathcal{T}^0(B))$ contains an edge in E^q.

The remaining generators depend solely on ρ_{vvv} for $v \in V(\mathcal{T})$ and ρ_e for $e \in E(\mathcal{T}) \setminus E^q$. If v is unlabeled with degree ≥ 3 in $\mathcal{F}^q$, then the value of ρ_{vvv} is fixed (up to sign) in the q-fiber by an argument similar to the one used in Theorem 6.20. If the degree of v is exactly 2, then every generator in (6.19) that depends on ρ_{vvv} already lies in the ideal generated by ρ_e^q for $e \in E^q$. This follows from the fact that for such A the set of edges $E(\mathcal{T}^0(A))$ must contain at least three edges of v but then at least one of these edges, say e, is isolated with respect to q. This implies that I does not depend on ρ_{vvv} if v is an unlabeled degree-two vertex in $\mathcal{F}^q$. Thus, it remains to describe the remaining generators of I that involve only ρ_e for $e \in E(\mathcal{T}) \setminus E^q$. Using Lemma 6.19 we can show that these remaining generators are of the form

$$\prod_{e \in \overline{uv}} \rho_e = c_{uv} \neq 0 \tag{6.21}$$

for all vertices u, v of degree ≥ 3 in $\mathcal{F}^q$ such that all vertices separating them have degree exactly two in $\mathcal{F}^q$. It is now straightforward to show that I defines a smooth set.

We now show that the q-fiber is diffeomorphic to a union of polyhedra. Define $t_v = \sqrt{\frac{1 - \mu_v}{\mu_v}}$ and $z_v = \log t_v$, which gives a change of coordinates from $\mu_v \in (0, 1)$ (or alternatively from $\rho_{vvv} \in \mathbb{R}$) to $z_v \in \mathbb{R}$. Since some of the means are fixed (up to sign) in the q-fiber and some are unconstrained, it follows that some z_v are fixed (up to sign) and some are unconstrained. We also have $\rho_e = 0$ for all $e \in E^q$. Let $s : E(\mathcal{T}) \setminus E^q \to \{-1, 1\}$ be any possible sign assignment for corresponding edge correlations ρ_e for $e \in E(\mathcal{T}) \setminus E^q$ such that $s(e) = \text{sgn}(\rho_e)$. Then s induces an open orthant $\mathbb{R}_s^{|E(\mathcal{T})\setminus E^q|}$ defined by $s(e)\rho_e > 0$ for all $e \in E(\mathcal{T})\setminus E^q$. Consider the space of all ρ_e for $e \in E(\mathcal{T})\setminus E^q$ constrained to one of the orthants $\mathbb{R}_s^{|E(\mathcal{T})\setminus E^q|}$. On this orthant we have a change of coordinates $\nu_e = \log(s_e \rho_e)$. In each of these orthants the constraints (6.21) either become linear or they describe an empty set. By Lemma 6.6 the inequality constraints on ν_e are also linear in z_v.

To show the dimension of the q-fiber is equal to $2l_2$ we show that the dimension of its Zariski closure has dimension $2l_2$. All generators of I as listed above involve disjoint lists of variables. Some indeterminates have fixed values in the q-fiber so the only freedom is in the l_2 indeterminates ρ_{vvv} for all degree-two vertices v in $\mathcal{F}^q$ and an l_2 dimensional set described by equations of the form (6.21), where u, v have degree ≥ 3 and all vertices in $\overline{uv}$ have degree exactly 2. $\square$

Corollary 6.24. *If $\mathcal{T}$ is a binary phylogenetic tree, then a q-fiber is finite if and only if q is such that $\rho_{ij}^q \neq 0$ for all $1 \leq i < j \leq m$. If $\mathcal{T}$ is a phylogenetic star tree, then a q-fiber is finite as long as there are at least two nonzero correlations between the leaves.*

Corollary 6.24 suggests that if $\hat{\rho}_{ij}$ are all far enough from zero, then estimation will be relatively stable. In the next section we describe cases in which everything becomes much more complicated.

6.2.2 Singular q-fibers

By Lemma 6.18 all unlabeled vertices of $\mathcal{T}$ have either degree zero in $\mathcal{F}^q$ or the degree is ≥ 2. The singular case occurs precisely if at least one inner vertex of T has degree zero in $\mathcal{F}^q$. We begin with an example.

Example 6.25. Let $\mathcal{T}$ be the phylogenetic tripod tree as in Figure 6.1 and let $q \in M(\mathcal{T}, 2)$. The degree of h in the q-forest $\mathcal{F}^q$ is zero if and only if $\rho_{ij}^q = 0$ for all $1 \leq i < j \leq 3$. In this situation $E^q = E(\mathcal{T})$ and the q-fiber Ω_q is given as a subset of $\Omega_{\mathcal{T}}$ by equations $\rho_{iii} = \rho_{iii}^q$ for $i = 1, 2, 3$ together with the three additional equations

$$\rho_{1h}\rho_{2h} = \rho_{1h}\rho_{3h} = \rho_{2h}\rho_{3h} = 0.$$

Geometrically, in the subspace of $\Omega_{\mathcal{T}}$ given by $\rho_{iii} = \rho_{iii}^q$ for $i = 1, 2, 3$, this is a union of three planes given by $\{\rho_{1h} = \rho_{2h} = 0\}$, $\{\rho_{1h} = \rho_{3h} = 0\}$ and $\{\rho_{2h} = \rho_{3h} = 0\}$ subject to the additional inequality constraints defining $\Omega_{\mathcal{T}}$ and given by Lemma 6.6. In particular, it is not a regular set since it has self-intersection points.

This example together with Theorem 6.23 provides a fairly good understanding of the local geometry in the case of the tripod tree model. From

(6.2) it follows that if one of the correlations, say ρ_{12}, is zero, then at least one of the remaining correlations has to be zero as well. By Theorem 6.23, if q is such that all the correlations are non-zero, then there are exactly two points in the preimage of q. If exactly two correlations are zero then the q-fiber is a smooth subset of $\Omega_{\mathcal{T}}$. By Example 6.25, if all three correlations are zero, then the q-fiber is a collection of intersecting subvarieties.

Theorem 6.26 (The geometry of the q-fiber — the singular case). *Let $\mathcal{T}$ be a semi-labeled tree. If the q-forest $\mathcal{F}^q$ contains unlabeled degree-zero vertices, then the q-fiber is a singular variety given as a union of intersecting manifolds in $\mathbb{R}^{|V(\mathcal{T})|+|E(\mathcal{T})|}$ restricted to $\Omega_{\mathcal{T}}$. Their common intersection is given by vanishing of all ρ_e for $e \in E^q$.*

Proof. Like in the regular case we have $\rho_{iii} = \rho_{iii}^q$ for $i \in [m]$ and ρ_{vvv}^2 is fixed for all unlabeled vertices v with degree ≥ 3 in $\mathcal{F}^q$. For all the other inner vertices ρ_{vvv} is unconstrained. The constraints on ρ_e for $e \in E(\mathcal{T}) \setminus E^q$ are also obtained like in the proof of Theorem 6.23 and they define a smooth subset. It remains to study the constraints on ρ_e for $e \in E^q$. Let $E_0 \subseteq E^q$ and

$$\Omega_{E_0} = \{\omega \in \Omega_{\mathcal{T}} : \ \rho_e = 0 \text{ for all } e \in E_0\}. \tag{6.22}$$

We say that E_0 is *minimal for* Σ^q if for every point ω in Ω_{E_0} and for every $i, j \in [m]$ such that $\rho_{ij}^q = 0$ we have that $\rho_{ij}(\omega) = 0$ and furthermore that E_0 is minimal with such a property (with respect to inclusion). To illustrate the motivation behind this definition, consider the tripod tree singular case in Example 6.25. If T is rooted in the inner vertex, we have three minimal subsets of E^q: $\{h-1, h-2\}$, $\{h-1, h-3\}$, and $\{h-2, h-3\}$. We now show that the q-fiber satisfies

$$\Omega_q \quad = \quad \bigcup_{E_0 \text{ min.}} \Omega_{E_0} \cap \Omega_q. \tag{6.23}$$

The first inclusion "$\subseteq$" follows from the fact that if $\omega \in \Omega_q$, then $\rho_{ij}(\omega) = \rho_{ij}^q$ for all $i, j \in [m]$. In particular, $\rho_{ij}(\omega) = 0$ whenever $\rho_{ij}^q = 0$. Therefore $\omega \in \Omega_{E_0} \cap \Omega_q$ for some minimal E_0. The second inclusion is obvious. For each minimal E_0, the set $\Omega_{E_0} \cap \Omega_q$ is a smooth set because there are no constraints on ρ_e for $e \in E^q \setminus E_0$. □

6.2.3 Sign patterns on parameters

Let $\mathcal{T}$ be a semi-labeled tree. Let $q \in M(\mathcal{T}, 2)$ be a distribution such that each unlabeled vertex of $\mathcal{T}$ has degree ≥ 3 in the q-forest $\mathcal{F}^q$. By Theorem 6.23, there is a finite set Ω_q of points $\omega \in \Omega_{\mathcal{T}}$ mapping to q, which we call a q-fiber. Theorem 6.20 gives the formulas for the parameters modulo signs which may suggest that $|\Omega_q| = 2^{|V(\mathcal{T})|+|E(\mathcal{T})|}$. However, not all sign choices are possible. We will show that the number of possible choices of signs is in fact equal to $2^{|V(\mathcal{T})|-m}$ where $|V(\mathcal{T})| - m$ is the number of inner vertices of $\mathcal{T}$. We also show how to obtain all the points in Ω_q given one of them.

Let ω be a point in Ω_q and consider again the general case so that Ω_q is not necessarily finite.

Definition 6.27. For every vertex v, we define the operation of local sign switching $\delta_v : \Omega_{\mathcal{T}} \to \Omega_{\mathcal{T}}$ such that $\delta_v(\omega) = \omega'$ where $\rho'_e = -\rho_e$ if one of the ends of e is equal to v and $\rho'_e = \rho_e$ otherwise; $\rho'_{vvv} = -\rho_{vvv}$ and $\rho'_{uuu} = \rho_{uuu}$ for all vertices $u \neq v$.

Note that switching $\rho'_{vvv} = -\rho_{vvv}$ corresponds to label switching $Y_v \mapsto 1 - Y_v$ of the corresponding random variable. To show that δ_v is a well-defined operator on $\Omega_{\mathcal{T}}$ we need to show that if $\omega \in \Omega_{\mathcal{T}}$, then $\omega' = \delta_v(\omega)$ lies in $\Omega_{\mathcal{T}}$. By Lemma 6.6, it remains to check that for every neighbor u of v

$$- \min\{t'_u t'_v, \frac{1}{t'_u t'_v}\} \quad \leq \quad \rho'_{uv} \quad \leq \quad \min\{\frac{t'_u}{t'_v}, \frac{t'_v}{t'_u}\},$$

where $t_u = \sqrt{\frac{1-\mu_u}{\mu_u}}$. However, $\rho'_{uv} = -\rho_{uv}$, $t'_u = t_u$, and $t'_v = \frac{1}{t_v}$ and we get

$$- \min\{\frac{t_u}{t_v}, \frac{t_v}{t_u}\} \quad \leq \quad -\rho_{uv} \quad \leq \quad \min\{t_u t_v, \frac{1}{t_u t_v}\},$$

which holds because $\omega \in \Omega_{\mathcal{T}}$.

Lemma 6.28. *If h is an unlabeled vertex, then the local sign switching δ_h is an automorphism of the q-fiber for every $q \in \boldsymbol{M}(\mathcal{T})$.*

Proof. Let $A \subseteq [m]$, $|A| \geq 2$. By Theorem 6.10

$$\bar{t}_A(\omega') = \prod_{v \in V(\mathcal{T}^0(A))} (\rho'_{vvv})^{\max\{0, \deg(v) - 2\}} \prod_{e \in E(\mathcal{T}^0(A))} \rho'_e,$$

where $\bar{t}$ are $\mathcal{T}^*$-cumulants and $\mathcal{T}^*$ is a binary expansion of $\mathcal{T}$. We have two cases: either h lies in the tree $V(\mathcal{T}^0(A))$ spanned over A or it does not. If not, then $\bar{t}_A(\omega') = \bar{t}_A(\omega)$ because none of the parameters whose sign is switched appears in the formula for $\bar{t}_A$ above. If yes, then without loss we can assume that $\bar{t}_A(\omega') \neq 0$. If v has degree two in $\mathcal{T}^0(A)$, then

$$\bar{t}_A(\omega') = (-1)^{\deg(h)} \bar{t}_A(\omega) = \bar{t}_A(\omega).$$

If the degree of v is ≥ 3, then

$$\bar{t}_A(\omega') = (-1)^{\deg(h)-2}(-1)^{\deg(h)} t_A(\omega) = t_A(\omega).$$

$\square$

The local sign switchings for all unlabeled vertices h form a group $\mathcal{G}$ which is isomorphic to the multiplicative group $Z_2^{|V(\mathcal{T})|-m}$, where Z_2 is the multiplicative group with two elements.

Proposition 6.29. *Let $\mathcal{T}$ be a semi-labeled tree with the labeling set $[m]$. Suppose that every unlabeled vertex of $\mathcal{T}$ has degree ≥ 3 in the q-forest $\mathcal{F}^q$. Then the q-fiber Ω_q is finite and consists of exactly $2^{|V(\mathcal{T})|-m}$ points that form a single orbit of the group $\mathcal{G}$.*

Proof. The fact that Ω_q is finite follows directly from Theorem 6.23. If Ω_q contains a point ω, then acting with $\mathcal{G}$ on ω gives $2^{|V(\mathcal{T})|-m}$ different points in Ω_q. Hence the orbit of ω in Ω_q has exactly $2^{|V(\mathcal{T})|-m}$ elements. It remains to show that there are no other orbits of $\mathcal{G}$ in Ω_q. Let $\omega \in \Omega_q$ and let ω' be a point in $\Omega_{\mathcal{T}}$ such that $(\rho'_e)^2 = \rho_e^{\ 2}$ for all $e \in E(\mathcal{T})$ and $(\rho'_{vvv})^2 = \rho_{vvv}^{\ 2}$ for all unlabeled vertices v of $\mathcal{T}$, which is a necessary condition for ω' to be in Ω_q. Assume that ω' is not in the orbit of ω. Proceed by contradiction. Let $\omega' \in \Omega_q$ and we want to show that $\omega' = \delta(\omega)$ for some $\delta \in \mathcal{G}$. Since ω can be replaced by any other point in its orbit, we can assume that $\mathrm{sgn}(\rho_{vvv}) = \mathrm{sgn}(\rho'_{vvv})$ for all unlabeled $v \in V(\mathcal{T})$. Since $\omega, \omega' \in \Omega_q$, then for every $i, j, k \in [m]$ by the parameterization given in Theorem 6.10 applied for $\bar{\mathsf{t}}_{ij}$ and $\bar{\mathsf{t}}_{ijk}$, respectively, we have that

$$\prod_{e \in \overline{ij}} \mathrm{sgn}(\rho_e) = \prod_{e \in \overline{ij}} \mathrm{sgn}(\rho'_e), \qquad \prod_{e \in E(\mathcal{T}^0(ijk))} \mathrm{sgn}(\rho_e) = \prod_{e \in E(\mathcal{T}^0(ijk))} \mathrm{sgn}(\rho'_e).$$

It follows that $\prod_{e \in \overline{vi}} \mathrm{sgn}(\rho_e) = \prod_{e \in \overline{vi}} \mathrm{sgn}(\rho'_e)$ for each unlabeled vertex v and a labeled vertex i. It immediately implies that $\mathrm{sgn}(\rho_e) = \mathrm{sgn}(\rho'_e)$ for all $e \in E(\mathcal{T})$ and hence $\omega = \omega'$. In this way we have shown that ω' is in the orbit of ω under $\mathcal{G}$. $\square$

Motivated by biological applications, we consider again the positive part $M_{++}(\mathcal{T})$ of the general Markov model (c.f. Definition 5.53). By Lemma 6.15 it is parameterized by all $\omega \in \Omega_T$ such that $\rho_e > 0$ and we denote this parameter space by $\Omega_{\mathcal{T},++}$. We formulate the following result, which can be also inferred from the more general result of Chang [1996], which we introduce in Section 6.4.

Proposition 6.30. *Let $\mathcal{T}$ be a semi-labeled tree and $M_{++}(\mathcal{T})$ the positive part of the general Markov model $M(\mathcal{T})$. For every $q \in M_{++}(\mathcal{T})$ the q-fiber in $\Omega_{\mathcal{T},++}$ consists of exactly one point.*

Proof. This follows directly from the proof of Proposition 6.29 because every local sign switching necessarily changes signs of some edge correlations. $\square$

6.3 Examples, special trees, and submodels

6.3.1 Caterpillar trees and star trees

Caterpillar trees In this section we focus first on the case when $\mathcal{T}$ is a phylogenetic tree, whose underlying tree is a caterpillar tree like in Figure 6.4. As we show, in this case the formulas for $\mathcal{T}$-cumulants are extremely simple. This case is also important because the hidden Markov model can be realized as a submodel of $M(\mathcal{T})$; see Section 6.3.3.

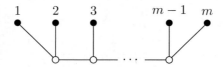

Figure 6.4: *A caterpillar tree with m leaves/legs.*

Directly by definition, the standardized $\mathcal{T}$-cumulants are given for all $|A| \geq 2$ by

$$\bar{t}_A = \sum_{\pi \in \mathbf{\Pi}(\mathcal{T}, A)} \mathfrak{m}(\pi) \prod_{B \in \pi} \rho_B.$$

In general, the Möbius function $\mathfrak{m}$ is not given explicitly. For caterpillar trees we can use the method announced earlier in Example 6.4. To introduce this method, first note that $\mathbf{\Pi}(\mathcal{T})$ contains all trivial splits $\mathcal{S}_0$ obtained by removing from $\mathcal{T}$ the terminal edges. Let $\mathcal{T}'$ be a semi-labeled tree obtained from $\mathcal{T}$ by contracting all terminal edges of $\mathcal{T}$ apart from the first and the last. Hence $\mathcal{T}'$ is a simple chain graph in Figure 5.8.

Denote by $\mathcal{S}'$ the set of splits in $\mathbf{\Pi}(\mathcal{T}')$ so that $\mathbf{\Pi}(\mathcal{T}') = \langle \mathcal{S}' \rangle$. These are the splits of the form $12 \cdots k | (k+1) \cdots m$ for some $1 < k < m$ and so $\mathbf{\Pi}(\mathcal{T}')$ is the lattice of interval partitions $\mathcal{I}$; see Definition 4.15. Let $\mathcal{S}$ be the set of splits of $\mathcal{T}$. Then $\mathcal{S} = \mathcal{S}' \cup \mathcal{S}_0$ and by Proposition 4.38,

$$\bar{t}_A \quad = \quad \sum_{\pi \in \mathcal{I}(A)} \mathfrak{m}(\pi) \prod_{B \in \pi} \rho_B \qquad \text{for } |A| \geq 2.$$

The advantage of this formula is that the lattice of interval partitions $\mathcal{I}(A)$ is isomorphic to the Boolean lattice of $|A| - 1$ elements and hence $\mathfrak{m}(\pi) = (-1)^{|\pi|-1}$.

To compute the standardized $\mathcal{T}$-cumulants for the caterpillar tree we proceed as follows. First we change probabilities $p(x)$ for $x \in \{0,1\}^m$ to moments μ_A for $A \subseteq [m]$ using the formula in (4.23). In `Mathematica`, the following code will do the job for any fixed m (here $m = 4$):

```
m=4;
Do[mu[A] = Sum[p[Union[A,B]], {B,
    Subsets[Complement[Range[m],A]]}], {A, Subsets[Range[m]]}];
```

Note that in the code we index probabilities with subsets of $[m]$ instead of $\{0,1\}^m$. The identification is made in the obvious way so that $x \in \{0,1\}^m$ corresponds to the subset $A = \mathcal{A}(x)$; see the first two columns of Table 6.1. To compute the central standardized moments ρ_A we first compute the central moments μ'_A using Proposition 4.37 and then normalize to get standardized moments. In `Mathematica` this can be done by

```
Do[cmu[A] = Simplify[Sum[(-1)^(Length[A]-Length[B])*mu[B]*
    Product[mu[{i}],{i,Complement[A,B]}], {B, Subsets[A]}]],
    {A, Subsets[Range[m]]}];
Do[rho[A] = cmu[A]/Product[Sqrt[mu[{i}]*(1-mu[{i}])],{i,A}],
    {A, Subsets[Range[m]]}]
```

Finally, to get standardized $\mathcal{T}$-cumulants we use (6.24):

$$\bar{\mathsf{t}}_A = \sum_{\pi \in \mathcal{I}(A)} (-1)^{|\pi|-1} \prod_{B \in \pi} \rho_B \qquad \text{for } |A| \geq 2. \qquad (6.24)$$

The set of all interval partitions of a given set can be found in Mathematica using the following simple function:

```
<< Combinatorica`
intparts=Function[A,Pick[SetPartitions[A],
        Map[OrderedQ, Map[Flatten, SetPartitions[A]]]]];
```

Then, for example, running intparts[1,2,3,4] gives

```
{{{1, 2, 3, 4}}, {{1}, {2, 3, 4}}, {{1, 2}, {3, 4}},
    {{1, 2, 3}, {4}}, {{1}, {2}, {3, 4}}, {{1}, {2, 3}, {4}},
    {{1, 2}, {3}, {4}}, {{1}, {2}, {3}, {4}}}
```

Now the following code changes coordinates from standardized moments to standardized $\mathcal{T}$-cumulants

```
Do[k[A] = Simplify[Sum[(-1)^(Length[pa]-1)*Product[rho[B],
    {B,pa}], {pa, intparts[A]}]], {A, Subsets[Range[m]]}];
```

In particular, we can write the change from probabilities to $\mathcal{T}$-cumulants as a sequence of simple and explicit maps. We do not recommend, however, outputting tree cumulants in terms of probabilities because these are massive polynomials. Expressions for cumulants in terms of moments are much more tractable. Also, it is better to use this change of coordinates on parameterized varieties. We give a numerical example for $m = 4$ in Section 6.3.2.

Star trees We now extend Example 6.14 on the star tree model. Since $\mathcal{T}$ is not a binary phylogenetic tree, Theorem 6.8 does not apply and to obtain a monomial parameterization we apply Theorem 6.10. To this end we need a binary expansion $\mathcal{T}^*$. Since every binary phylogenetic tree is a binary expansion of a star tree, we can take $\mathcal{T}^*$ to be the caterpillar tree and use $\mathcal{T}^*$-cumulants given in (6.24). The new parameters are m edge correlations ρ_{hi} for $i \in [m]$, and $m + 1$ skewnesses ρ_{hhh} and ρ_{iii} for $i \in [m]$. The parameterization is

$$\bar{\mathsf{t}}_A = \rho_{hhh}^{|A|-2} \prod_{i \in A} \rho_{hi} \qquad \text{for } |A| \geq 2,$$

and the skewnesses ρ_{iii} for $i \in [m]$ are free.

By Corollary 6.24 the q-fiber is finite as long as at least two of the correlations $\bar{\mathsf{t}}_{ij} = \rho_{ij}$ are non-zero at q. In this case, by the above parameterization, there exist distinct $i, j, k \in [m]$ such that $\rho_{hi}\rho_{hj}\rho_{hk} \neq 0$ and hence in fact there are three correlations $\rho_{ij}, \rho_{ik}, \rho_{jk}$ that are nonzero. In this case by Theorem 6.19

$$\rho^2_{hhh} \quad \text{is given by} \quad \frac{\rho^2_{ijk}}{\rho_{ij}\rho_{ik}\rho_{jk}},$$

and for each edge correlation ρ_{hl}

$$\rho^2_{hl} \quad \text{is given by} \quad \frac{\rho_{li}\rho_{lj}}{\rho_{ij}}.$$

These formulas identify parameters up to sign. By Proposition 6.29 there are exactly two points in the q-fiber. If

$$(\rho^0_{111}, \ldots, \rho^0_{mmm}, \rho^0_{hhh}, \rho^0_{h1}, \ldots, \rho^0_{hm})$$

is one of them, then the other, obtained by a local sign switching from Definition 6.27 and Lemma 6.28, is

$$(\rho^0_{111}, \ldots, \rho^0_{mmm}, -\rho^0_{hhh}, -\rho^0_{h1}, \ldots, -\rho^0_{hm}).$$

If there is exactly one ρ_{ij} that is nonzero, then we still have $\rho_{hk} = 0$ for all $k \neq i, j$ in the q-fiber. However, the q fiber is not finite because ρ_{hi}, ρ_{hj} can take any value as long as $\rho_{hi}\rho_{hj} = \rho_{ij}$, which defines a smooth set because there are also no constraints on ρ_{hhh}. Finally, if all ρ_{ij} are zero, then the q-fiber is a singular set. It is obtained as the union of sets given by vanishing all but one edge correlations ρ_{hi}.

6.3.2   The quartet tree model

In this section we give a numerical example for the quartet phylogenetic tree model given by the tree in Figure 5.1. The model is parameterized as in (5.13) by the root distribution and conditional probabilities attached to each of the edges. We first set the values of the parameters to $\theta_r(1) = 0.8$, $\theta_{1|r}(1|1) = 0.8$, $\theta_{1|r}(1|0) = 0.3$, $\theta_{2|r}(1|1) = 0.7$, $\theta_{2|r}(1|0) = 0.3$, $\theta_{a|r}(1|1) = 0.8$, $\theta_{a|r}(1|0) = 0.3$, $\theta_{3|a}(1|1) = 0.7$, $\theta_{3|a}(1|0) = 0.3$, $\theta_{4|a}(1|1) = 0.7$, and $\theta_{4|a}(1|0) = 0.3$. Using (5.13), we can then calculate the corresponding probabilities over the observed vertices, which are given in the third column in Table 6.1.

Remark 6.31. We do all our computations over algebraic numbers. This allows us to use the full strength of symbolic computations and check our formulas exactly.

To change from the raw probabilities $p(x)$ to standardized $\mathcal{T}$-cumulants, we use the formulas and code provided in Section 6.3.1. The resulting values are given in Table 6.1. Note that the last two columns in Table 6.1 are exactly the same apart from the last element, where we get $\rho_{1234} = \frac{704}{203\sqrt{12369}}$ and $\bar{t}_{1234} = \frac{256}{203\sqrt{12369}}$. The fact that the other entries are the same is a consequence of Remark 6.3.

Table 6.1 *Moments, standardized central moments, and standardized tree cumulants for a probability assignment in* $M(\mathcal{T}, 2)$, *where* $\mathcal{T}$ *is a quartet tree.*

x	$A = \mathcal{A}(x)$	$p(x)$	μ_A	ρ_A	$\bar{t}_A$
0000	$\emptyset$	$\frac{2221}{50000}$	1	1	1
0001	4	$\frac{1533}{50000}$	$\frac{29}{50}$	0	0
0010	3	$\frac{1533}{50000}$	$\frac{29}{50}$	0	0
0011	34	$\frac{2013}{50000}$	$\frac{37}{100}$	$\frac{4}{29}$	$\frac{4}{29}$
0100	2	$\frac{1729}{50000}$	$\frac{31}{50}$	0	0
0101	24	$\frac{1617}{50000}$	$\frac{931}{2500}$	$\frac{32}{\sqrt{358701}}$	$\frac{32}{\sqrt{358701}}$
0110	23	$\frac{1617}{50000}$	$\frac{931}{2500}$	$\frac{32}{\sqrt{358701}}$	$\frac{32}{\sqrt{358701}}$
0111	234	$\frac{2737}{50000}$	$\frac{1211}{5000}$	$-\frac{256}{609\sqrt{589}}$	$-\frac{256}{609\sqrt{589}}$
1000	1	$\frac{2409}{50000}$	$\frac{7}{10}$	0	0
1001	14	$\frac{2457}{50000}$	$\frac{211}{500}$	$\frac{8}{21\sqrt{29}}$	$\frac{8}{21\sqrt{29}}$
1010	13	$\frac{2457}{50000}$	$\frac{211}{500}$	$\frac{8}{21\sqrt{29}}$	$\frac{8}{21\sqrt{29}}$
1011	134	$\frac{4377}{50000}$	$\frac{11}{40}$	$-\frac{64}{609\sqrt{21}}$	$-\frac{64}{609\sqrt{21}}$
1100	12	$\frac{4141}{50000}$	$\frac{233}{500}$	$\frac{16}{\sqrt{12369}}$	$\frac{16}{\sqrt{12369}}$
1101	124	$\frac{4893}{50000}$	$\frac{7133}{25000}$	$-\frac{32}{7\sqrt{17081}}$	$-\frac{32}{7\sqrt{17081}}$
1110	123	$\frac{4893}{50000}$	$\frac{7133}{25000}$	$\frac{32}{7\sqrt{17081}}$	$\frac{32}{7\sqrt{17081}}$
1111	1234	$\frac{9373}{50000}$	$\frac{9373}{50000}$	$\frac{704}{203\sqrt{12369}}$	$\frac{256}{203\sqrt{12369}}$

Similarly, we compute the values of the moment parameters as given in Definition 6.5. We get

$$\rho_{rrr} = -\frac{3}{2}, \quad \rho_{aaa} = -\frac{4}{\sqrt{21}},$$

$$\rho_{r1} = \frac{2}{\sqrt{21}}, \quad \rho_{r2} = \frac{8}{\sqrt{589}}, \quad \rho_{ra} = \frac{2}{\sqrt{21}}, \quad \rho_{a3} = \frac{2}{\sqrt{29}}, \quad \rho_{a4} = \frac{2}{\sqrt{29}}.$$

It is easy to verify that (6.9) holds in this example. For instance, $\bar{t}_{1234}$ should be equal to the product

$$\rho_{rrr}\rho_{aaa}\rho_{r1}\rho_{r2}\rho_{ra}\rho_{a3}\rho_{a4} = \frac{3}{2}\frac{4}{\sqrt{21}}\frac{2}{\sqrt{21}}\frac{8}{\sqrt{589}}\frac{2}{\sqrt{21}}\frac{2}{\sqrt{29}}\frac{2}{\sqrt{29}} = \frac{256}{203\sqrt{12369}},$$

which agrees with the value of $\bar{t}_{1234}$ in Table 6.1.

Suppose that we have the exact distribution from the model and want to recover parameters. For the distribution in Table 6.1, all correlations are nonzero and hence this can be done easily by Theorem 6.20. We have

$$\rho_{rrr}^2 \quad \text{is given by} \quad \frac{\rho_{123}^2}{\rho_{12}\rho_{13}\rho_{23}} = \frac{9}{4}$$

$$\rho_{aaa}^2 \quad \text{is given by} \quad \frac{\rho_{134}^2}{\rho_{13}\rho_{14}\rho_{34}} = \frac{16}{21}$$

$$\rho_{ra}^2 \quad \text{is given by} \quad \frac{\rho_{13}\rho_{24}}{\rho_{12}\rho_{34}} = \frac{4}{21}$$

$$\rho_{r1}^2 \quad \text{is given by} \quad \frac{\rho_{12}\rho_{13}}{\rho_{23}} = \frac{4}{21}$$

and the formulas for ρ_{r2}^2, ρ_{a3}^2 and ρ_{a4}^2 are obtained in a similar way as for ρ_{r1}^2.
Note that the above formulas require some choices. For example, the formula
for ρ_{rrr}^2 requires picking a triple of leaves separated by r and $\{1, 2, 4\}$ could
be used instead of $\{1, 2, 3\}$. By Theorem 6.20 the obtained values will not
depend on these choices. We have alternatively

$$\rho_{rrr}^2 \quad \text{is given by} \quad \frac{\rho_{124}^2}{\rho_{12}\rho_{14}\rho_{24}} = \frac{9}{4}.$$

By Proposition 6.29 there are exactly four points mapping to the probability
distribution in Table 6.1. One is given by the original set of parameters. The
remaining three, by Lemma 6.28, are obtained from this one by local sign
switchings.

6.3.3 Binary hidden Markov model

The *binary hidden Markov model*, denoted by $\mathrm{HMM}(m, 2)$, is often defined
as follows. Consider the directed caterpillar tree $\mathcal{T}^r$ in Figure 6.5. Then
$\mathrm{HMM}(m, 2)$ is parameterized over this tree like in (5.13) but with assumption
that there exist transition matrices $\alpha, \beta \in \mathbb{R}^{2\times 2}$ such that

$$\begin{aligned}
\theta_{v|u}(y_v|y_u) &= \alpha_{y_u y_v} \quad \text{for all } u \to v \text{ both internal,} \\
\theta_{v|u}(y_v|y_u) &= \beta_{y_u y_v} \quad \text{for all } u \to v \text{ where } v \text{ is a leaf.}
\end{aligned}$$

The matrices α and β are referred to as *transition* and *emission* probabilities.
The root distribution is either assumed to be a free parameter or it is assumed
to be the *stationary distribution* for the Markov chain with the transition
matrix α, that is,

$$\theta_r(1) = \frac{\alpha_{01}}{\alpha_{10} + \alpha_{01}}.$$

In this case we say that the hidden Markov model is *stationary*.

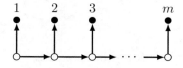

Figure 6.5: *A directed caterpillar tree with m leaves/legs.*

By definition $\mathrm{HMM}(m, 2)$ is contained in the general Markov model defined by the directed caterpillar tree $\mathcal{T}^r$, where the two degree-two vertices
are suppressed. Since rooting does not matter in this model class, we con-
clude that $\mathrm{HMM}(m, 2)$ is contained in $\boldsymbol{M}(\mathcal{T}, 2)$, where $\mathcal{T}$ is the caterpillar
tree in Figure 6.4.

Proposition 6.32. *Let $\mathcal{T}$ be the caterpillar tree in Figure 6.4. In the new parameter space given by ρ_{vvv} for $v \in V(\mathcal{T})$ and ρ_e for $e \in E(\mathcal{T})$, the stationary binary hidden Markov model is given by*

$$\rho_{vvv} = a \qquad \text{for all unlabeled } v \in V(\mathcal{T}),$$

$$\rho_{iii} = b \qquad \text{for all } i \in [m],$$

$$\rho_e = c \qquad \text{for all internal } e \in E(\mathcal{T}),$$

$$\rho_e = d \qquad \text{for all terminal } e \in E(\mathcal{T}),$$

where a, b, c, d are now the new parameters of $\mathrm{HMM}(m, 2)$. *The parameterization in $\mathcal{T}$-cumulants is*

$$\bar{t}_{i_1 \ldots i_l} = a^{l-2} c^{i_l - i_1} d^l$$

for all $\{i_1, \ldots, i_l\} \subseteq [m]$ with $l \geq 2$.

6.4 Higher number of states

In the previous sections we focused on the two-state case, in which we had a simple monomial parameterization given in Theorem 6.10. This enabled us to perform the full analysis of the local geometry of $M(\mathcal{T}, 2)$ with a special focus on the geometry of q-fibers. If $k > 2$, then we currently have no equally nice description of $M(\mathcal{T}, k)$. However, like in the two-state case, generically the model is identified up to permutations of the states of hidden variables. Moreover, the parameters can be identified solely from the marginal distributions over triples. In the remainder of this section we present this result referring to Chang [1996] for proofs.

Recall that the model $M(\mathcal{T}, k)$ is parameterized by the root distribution θ_r and transition matrices $\theta_{v|u}$ for each edge $u \to v$ in $\mathcal{T}$. As always, parameterization is given by a rooted version of $\mathcal{T}$ but the model does not depend on the rooting. The transition matrix represents conditional independence of Y_v given Y_u so that $p_{v|u} = \theta_{v|u}^T$.

Theorem 6.33. *Let $\mathcal{T}$ be a semi-labeled tree. If q is a generic probability distribution from $M(\mathcal{T}, k)$, then the q-fiber is finite and it is identified from marginal distributions over triples of X.*

First, we say what we mean by "generic" in Theorem 6.33. A sufficient condition for the q-fiber to be finite is:

(i) The root distribution θ_r has only positive entries (c.f. (5.8)).

(ii) The matrix $\theta_{v|u}$ is invertible for each $u \to v$ in $\mathcal{T}$ and it is not equal to a permutation matrix.

Second, by saying that a q-fiber is identified from triples of X we mean that we do not actually need the whole distribution q to learn about the preimage of q. It suffices to have only induced marginal distributions q_{ijk} for any triple $1 \le i < j < k \le m$. A refinement of this statement is possible, when all triples are reduced to only some triples.

The basic idea behind Theorem 6.33 is to identify rows of the matrices $\theta_{v|u}$ as eigenvectors of certain matrices that can be formed from joint distributions of triples of observed vertices. To show the idea behind the proof, suppose for simplicity that $\mathcal{T}$ is the tripod tree and $\theta_{2|h}$ has a column j such that all entries of this column are distinct. Consider the probability $\mathbb{P}(X_1 = i, X_2 = j | X_3 = k)$. We have

$$\mathbb{P}(X_1 = i, X_2 = j | X_3 = k) =$$
$$= \sum_l \mathbb{P}(H = l | X_3 = k) \mathbb{P}(X_2 = j | H = l, X_3 = k) \cdot$$
$$\cdot \mathbb{P}(X_1 = i | X_2 = j, X_3 = k, H = l)$$

and using the Markov properties we obtain

$$\mathbb{P}(X_1 = i, X_2 = j | X_3 = k) = \sum_l p_{h|3}(l|k) p_{2|h}(j|l) p_{1|h}(i|l).$$

Denote $J_{13;2}(i, k; j) = \mathbb{P}(X_1 = i, X_2 = j | X_3 = k)$ to obtain

$$J_{13;2}(i, k; j) = p_{h|3}^T \mathrm{diag}(p_{2|h}(j|\cdot)) p_{1|h}^T$$

where $p_{2|h}(j|\cdot)$ denotes the j-th row of $p_{2|h}$ or equivalently the j-th column of $\theta_{2|h}$. Note that $p_{1|3} = p_{1|h} p_{h|3}$ and multiplying both sides of the above equation by $p_{1|3}^{-T} = p_{1|h}^{-T} p_{h|3}^{-T}$ we obtain

$$p_{1|3}^{-T} J_{13;2}(i, k; j) \quad = \quad \theta_{1|h}^{-1} \mathrm{diag}(p_{2|h}(j|\cdot)) \theta_{1|h}, \qquad (6.25)$$

where $\theta_{1|h} = p_{1|h}^T$. Note that the left-hand side of (6.25) is a matrix that depends only on the distribution of the vector (X_1, X_2, X_3). The identity (6.25) is the (left) eigenvalue decomposition, where the j-th column of $\theta_{2|h}$ is the vector of eigenvalues of $p_{1|3}^{-T} J_{13;2}(i, k; j)$, which are distinct by assumption. To recover $\theta_{1|h}$ from this description, note that each eigenvalue corresponds to a one-dimensional (left) eigenspace. Each eigenspace has a unique vector with coordinates summing to 1.

6.5   Bibliographical notes

The moment structures have already been used to understand statistical models with hidden variables by Lazarsfeld and Henry [1968]. Recently this was used to better understand the class of naive Bayes models, that is, general Markov models over a star tree. This was used by Rusakov and Geiger [2005] to approximate a marginal likelihood where the sample size was large, by Geiger et al. [2001] to understand the local geometry of the model class, and by Auvray et al. [2006] to provide the full description of these models in terms of the defining equations and inequalities. More general tree models were studied from this perspective by Settimi and Smith [1998] and Pearl and Tarsi [1986]. Tree cumulants were first used in Zwiernik and Smith [2012] and Zwiernik and Smith [2011]. The local geometry is closely related to the concept of identifiability, which for this model class was studied by Chang [1996] and for other related model classes by Allman et al. [2009].

Chapter 7

The global geometry

[]

Recall that by Propositions 2.33 and 2.34 the image of a semialgebraic set under a polynomial mapping is a semialgebraic set. Therefore, for every semi-labeled tree, the model $M(\mathcal{T})$ is a semialgebraic set and hence it can be described by polynomial equations and inequalities in the probabilities $p(x)$. By Remark 6.2, the map between probabilities and tree cumulants is a polynomial isomorphism and therefore, there exists a polynomial description in terms of tree cumulants.

In this chapter we obtain the full description of these models in terms of implicit polynomial equations and inequalities. Like in the previous chapter, the focus is on the two-state case $M(\mathcal{T}, 2)$. In this case an important part of the information about the global geometry of the general Markov model is contained in its second-order moments. By Theorem 6.10, the correlations of $p \in M(\mathcal{T}, 2)$ satisfy

$$\rho_{ij} \;=\; \prod_{e \in \overline{ij}} \rho_e \qquad \text{for all } 1 \leq i < j \leq m. \tag{7.1}$$

Here we used the fact that second-order standardized tree cumulants coincide with correlations; see Remark 6.3. From this we easily infer the following set of constraints that must be satisfied by any distribution in $M(\mathcal{T}, 2)$.

Lemma 7.1. *Let $\mathcal{T}$ be a phylogenetic tree with labeling set $[m]$. For every $p \in M(\mathcal{T}, 2)$ the corresponding correlations satisfy*

$$\rho_{ij}\rho_{ik}\rho_{jk} \;\geq\; 0 \qquad \text{for all } 1 \leq i < j < k \leq m,$$

or equivalently

$$k_{ij}k_{ik}k_{jk} \;\geq\; 0 \qquad \text{for all } 1 \leq i < j < k \leq m.$$

Later in this chapter we supplement these inequalities with the complete set of constraints describing $M(\mathcal{T}, 2)$ for any phylogenetic tree $\mathcal{T}$.

7.1 Geometry of two-state models

7.1.1 The $2 \times 2 \times 2$ hyperdeterminant

Let A be a 2×2 matrix. Consider a set of linear equations $At = 0$, where $t = (t_0, t_1)$. It is a well-known result of linear algebra that $\det A = 0$ if and only if $At = 0$ has a nonzero solution. The following equivalent formulation will be useful. Define $f(s,t) = s^T At$. Then $\det A = 0$ if and only if there exists a non-trivial solution to the set of equations

$$f(s,t) = \frac{\partial}{\partial s_0} f(s,t) = \frac{\partial}{\partial s_1} f(s,t) = \frac{\partial}{\partial t_0} f(s,t) = \frac{\partial}{\partial t_1} f(s,t) = 0.$$

Let now $A = [a_x] \in \mathbb{C}^{\mathcal{X}}$ for $\mathcal{X} = \{0,\ldots,r_1\} \times \cdots \times \{0,\ldots,r_m\}$ be a tensor. Every such A defines a function

$$f(t) = \sum_{x \in \mathcal{X}} a_x t_{1,x_1} \cdots t_{m,x_m},$$

which is a multilinear function in $m + \sum_i r_r$ indeterminates $t_{i,j}$. Consider the set of $m + \sum_i r_r + 1$ homogeneous multilinear equations of the form

$$f(t) = \frac{\partial f}{\partial t_{i,j}}(t) = 0 \qquad \text{for all } i \in [m], j \in \{0,\ldots,r_i\}.$$

A *hyperdeterminant* of A, denoted by $\mathrm{Det}(A)$, is a generalization of a determinant to higher-order arrays. In particular, $\mathrm{Det}(A)$ is a polynomial in the entries of A such that it vanishes precisely when the above system of multilinear equations has a non-trivial solution. For a formal definition we refer to Gelfand, Kapranov, Zelevinsky [Gelfand et al., 1994, Chapter 14].

Similar to the determinant, there is a formula for the hyperdeterminant. However, this formula can be explicitly written down only in very small cases. In this chapter we are going to use the hyperdeterminant for $2 \times 2 \times 2$ tensors.

Definition 7.2. Let $A = [a_{ijk}]$ be a $2 \times 2 \times 2$ table. The hyperdeterminant of A is given by

$$
\begin{aligned}
\mathrm{Det}\,A \;=\; & (a_{000}^2 a_{111}^2 + a_{001}^2 a_{110}^2 + a_{010}^2 a_{101}^2 + a_{011}^2 a_{100}^2) \\
- \;& 2(a_{000}a_{001}a_{110}a_{111} + a_{000}a_{010}a_{101}a_{111} + a_{000}a_{011}a_{100}a_{111} \\
+ \;& a_{001}a_{010}a_{101}a_{110} + a_{001}a_{011}a_{110}a_{100} + a_{010}a_{011}a_{101}a_{100}) \\
+ \;& 4(a_{000}a_{011}a_{101}a_{110} + a_{001}a_{010}a_{100}a_{111}).
\end{aligned}
$$

If $\sum a_{ijk} = 1$, then treating all entries formally as joint cell probabilities (without positivity constraints) we can simplify this formula expressing it in terms of cumulants

$$\mathrm{Det}\,A = k_{123}^2 + 4k_{12}k_{13}k_{23}. \qquad (7.2)$$

We verify this directly by first expressing cumulants in moments using (4.14), which yields

$$
\begin{aligned}
k_{12} &= \mu_{12} - \mu_1\mu_2 \\
k_{13} &= \mu_{13} - \mu_1\mu_3 \\
k_{23} &= \mu_{23} - \mu_2\mu_3 \\
k_{123} &= \mu_{123} - \mu_1\mu_{23} - \mu_2\mu_{13} - \mu_3\mu_{12} + 2\mu_1\mu_2\mu_3.
\end{aligned}
$$

GEOMETRY OF TWO-STATE MODELS

Then we express moments in the entries of A using (4.23)

$$
\begin{aligned}
\mu_\emptyset &= a_{000} + a_{001} + a_{010} + a_{011} + a_{100} + a_{101} + a_{110} + a_{111} = 1 \\
\mu_1 &= a_{100} + a_{101} + a_{110} + a_{111} \\
\mu_2 &= a_{010} + a_{011} + a_{110} + a_{111} \\
\mu_3 &= a_{001} + a_{011} + a_{101} + a_{111} \\
\mu_{12} &= a_{110} + a_{111} \\
\mu_{13} &= a_{101} + a_{111} \\
\mu_{23} &= a_{011} + a_{111} \\
\mu_{123} &= a_{111}.
\end{aligned}
$$

Now (7.2) can be easily verified.

7.1.2 The group action and $M_+(\mathcal{T}, 2)$

In Section 6.2.3 we described the group that acts on the parameter space by local sign switchings at each inner vertex. The group was generated by local sign switchings at unlabeled vertices of $\mathcal{T}$. In this section we show that the model $M(\mathcal{T}, 2)$ is obtained as the (Euclidean) closure of the orbit of the nonnegative part $M_+(\mathcal{T}, 2)$ of $M(\mathcal{T}, 2)$ (see Definition 5.53) under the action of a group generated by local sign switchings at the leaves of $\mathcal{T}$.

Theorem 7.3. *Let $\mathcal{T}$ be a phylogenetic tree with the labeling set $[m]$. There exists a group G, which is isomorphic to Z_2^{m-1}, such that $M(\mathcal{T}, 2)$ is equal to the G-orbit of the nonnegative part $M_+(\mathcal{T}, 2)$.*

The operation of local sign switching introduced in Definition 6.27 gives an automorphism on any q-fiber by Lemma 6.28. Local sign switching is a crucial tool in the proof of Theorem 7.3.

Lemma 7.4. *Let $q \in M(\mathcal{T}, 2)$, then there exists a point in the q-fiber such that $\rho_e \geq 0$ for all internal edges e.*

Proof. Let ω be a point in the q-fiber with edge correlations ρ_e. We are going to find another point ω' in the q-fiber that satisfies conditions of the lemma. Let v be an internal vertex and u its parent. If $\rho_{uv} < 0$, then act on the parameter vector with the local sign switching δ_v, which switches signs of all ρ_e for e incident with v. To construct ω', after rooting $\mathcal{T}$, we apply this procedure recursively starting from children of the root and moving toward the leaves. After each step, all the edges above v have nonnegative correlations. Also by Lemma 6.28 the new parameter vector lies in the q-fiber. $\square$

Proof of Theorem 7.3. Let q be a distribution in $M(\mathcal{T}, 2)$ parameterized by parameters $\omega = ((\rho_{vvv}), (\rho_e))$. By Lemma 7.4 we can assume that all edge correlations for inner edges are nonnegative. By Theorem 6.10, the correlations of q satisfy (7.1). The set of terminal edges of $\mathcal{T}$ is naturally identified with the labeling set $[m]$. Denote by $\rho_1, \ldots, \rho_m$ the corresponding edge correlations. Consider the sign-switching operators δ_i at all leaves $i \in [m]$. The set $\{\delta_1, \ldots, \delta_m\}$ generates a multiplicative group $\widehat{G}$ isomorphic to Z_2^m.

The $\widehat{G}$-orbit of ω contains ω' such that the image of ω' lies in $\boldsymbol{M}_+(\mathcal{T}, 2)$. Indeed, just act on ω with $g \in \widehat{G}$ given as a product of all δ_i such that $\rho_i \leq 0$. Since q was arbitrary, this shows that $\boldsymbol{M}(\mathcal{T}, 2)$ can be described as a G-orbit of $\boldsymbol{M}_+(\mathcal{T}, 2)$. However, if we define $\bar{g} = g \prod_{i=1}^m \delta_i$, then the image of $\bar{g} \cdot \omega$ also lies in $\boldsymbol{M}_+(\mathcal{T})$. This shows that $\delta_1 \cdots \delta_m$ maps $\boldsymbol{M}_+(\mathcal{T}, 2)$ to itself and therefore, in fact, $\boldsymbol{M}(\mathcal{T}, 2)$ can be described as a G-orbit of $\boldsymbol{M}_+(\mathcal{T}, 2)$. $\quad \square$

By Theorem 7.3, given a distribution q in $\boldsymbol{M}(\mathcal{T}, 2)$, we can always switch labels of the observed variables to obtain another distribution q' with nonnegative correlations between the leaves. For q', without loss of generality, we can assume that all edge correlations in the parameter vector mapping to it are nonnegative and thus $q' \in \boldsymbol{M}_+(\mathcal{T}, 2)$. By Proposition 6.30, if $q' \in \boldsymbol{M}_{++}(\mathcal{T}, 2)$, all parameters are identified uniquely. Moreover, to obtain the corresponding parameter vector for q, we just need to adjust signs of the edge correlations for terminal edges. We state this slightly informally as follows.

Theorem 7.5. *To understand the model $\boldsymbol{M}(\mathcal{T}, 2)$ it is enough to understand $\boldsymbol{M}_+(\mathcal{T}, 2)$.*

The following result extends [Klaere and Liebscher, 2012, Theorem 6, Proposition 11] and is a rephrased version of [Zwiernik and Smith, 2012, Corollary 5.5].

Theorem 7.6. *If $q \in \boldsymbol{M}_{++}(\mathcal{T}, 2)$, then the original parameters satisfy the following formulas:*

1. Let i, j, k be any three leaves separated in $\mathcal{T}$ by the root r. Then

$$\theta_r(1) = \frac{1}{2}\left(1 - \frac{k_{ijk}}{\sqrt{\mathrm{Det}(p_{ijk})}}\right).$$

2. If $u \to i$ is a terminal edge and i, j, k are separated by u in $\mathcal{T}$, then

$$\theta_{i|u}(1|0) = \mu_i + \frac{k_{ijk} - \sqrt{\mathrm{Det}(p_{ijk})}}{2k_{jk}}$$

$$\theta_{i|u}(1|1) = \mu_i + \frac{k_{ijk} + \sqrt{\mathrm{Det}(p_{ijk})}}{2k_{jk}}. \tag{7.3}$$

3. If $u \to v$ is an inner edge and i, j, k, l are leaves such that u separates $i, j, \{k, l\}$ and v separates $\{i, j\}, k, l$ in $\mathcal{T}$, then

$$\theta_{v|u}(1|0) = \frac{1}{2} + \frac{k_{il}k_{ijk} - k_{ij}k_{ikl} - k_{il}\sqrt{\mathrm{Det}(p_{ijk})}}{2k_{ij}\sqrt{\mathrm{Det}(p_{ikl})}}$$

$$\theta_{v|u}(1|1) = \frac{1}{2} + \frac{k_{il}k_{ijk} - k_{ij}k_{ikl} + k_{il}\sqrt{\mathrm{Det}(p_{ijk})}}{2k_{ij}\sqrt{\mathrm{Det}(p_{ikl})}}. \tag{7.4}$$

Proof. By Theorem 6.8 (and Remark 6.3), if leaves i, j, k are separated by v, then

$$\rho_{ijk} = \rho_{vvv} \prod_{e \in \overline{ijk}} \rho_e,$$

where $\overline{ijk}$ denotes the set of edges in the induced tree $\mathcal{T}^0(\{i, j, k\})$ spanned over $\{i, j, k\}$ (c.f. Definition 5.51). Moreover, $\rho_{ij} = \prod_{e \in \overline{ij}} \rho_e$ and hence

$$\rho_{ij}\rho_{ik}\rho_{jk} = \prod_{e \in \overline{ijk}} \rho_e^2.$$

Denote $\sigma_i = \sqrt{\mu_i(1 - \mu_i)}$; then $\rho_{ijk} = \frac{k_{ijk}}{\sigma_i \sigma_j \sigma_k}$ and $\rho_{ij} = \frac{k_{ij}}{\sigma_i \sigma_j}$. Since all edge correlations are strictly positive and $\text{Det}(p_{ijk}) = k_{ijk}^2 + 4k_{ij}k_{ik}k_{jk}$, we have

$$\frac{1}{\sigma_i \sigma_j \sigma_k} \sqrt{\text{Det}(p_{ijk})} = \sqrt{\rho_{vvv}^2 + 4} \prod_{e \in \overline{ijk}} \rho_e.$$

Because $\rho_{vvv} = \frac{1 - 2\mu_v}{\sqrt{\mu_v(1 - \mu_v)}}$,

$$\frac{1}{2}\left(1 - \frac{k_{ijk}}{\sqrt{\text{Det}(p_{ijk})}}\right) = \frac{1}{2}\left(1 - \frac{\rho_{vvv}}{\sqrt{4 + \rho_{vvv}^2}}\right) = \mu_v.$$

Now the first formula follows by taking $v = r$ and using the fact that $\theta_r(1) = \mu_r$. For any edge $u \to v$ we have

$$\mu_v = (1 - \mu_u)\theta_{v|u}(1|0) + \mu_u\theta_{v|u}(1|1) = \theta_{v|u}(1|0) + \mu_u\left(\theta_{v|u}(1|1) - \theta_{v|u}(1|0)\right).$$

Moreover, because $k_{uv} = p_{uv}(1, 1) - p_u(1)p_v(1)$,

$$\theta_{v|u}(1|1) - \theta_{v|u}(1|0) = \frac{p_{uv}(1, 1)}{p_u(1)} - \frac{p_{uv}(0, 1)}{p_u(0)} = \frac{k_{uv}}{\mu_u(1 - \mu_u)},$$

and hence

$$\theta_{v|u}(1|0) = \mu_v - \frac{k_{uv}}{1 - \mu_u}, \qquad \theta_{v|u}(1|1) = \mu_v + k_{uv}\mu_u. \qquad (7.5)$$

Now to prove formulas in (7.3) and (7.4) we proceed like in the first case. We first use the parameterization in Theorem 6.8, then we show that everything reduces to $\mu_v - \frac{k_{uv}}{1 - \mu_u}$ or $\mu_v + k_{uv}\mu_u$ when we use (7.5). We follow this procedure to show the formula for $\theta_{i|u}(1|0)$. We have

$$\mu_i + \frac{k_{ijk} - \sqrt{\text{Det}(p_{ijk})}}{2k_{jk}} = \mu_i + \frac{\sigma_i \sigma_j \sigma_k}{\sigma_j \sigma_k} \frac{\rho_{ijk} - \sqrt{\rho_{ijk}^2 + 4\rho_{ij}\rho_{ik}\rho_{jk}}}{2\rho_{jk}} =$$

$$= \mu_i + \frac{\sigma_i \sigma_j \sigma_k}{\sigma_j \sigma_k} \frac{\rho_{vvv} \prod_{e \in \overline{ijk}} \rho_e - \sqrt{4 + \rho_{vvv}^2} \prod_{e \in \overline{ijk}} \rho_e}{2 \prod_{e \in \overline{jk}} \rho_e} =$$

$$= \mu_i + \frac{\sigma_i}{2}\left(\rho_{vvv} - \sqrt{4 + \rho_{vvv}^2}\right) \prod_{e \in \overline{iv}} \rho_e = \mu_i - \sigma_i \frac{\mu_v}{\sqrt{\mu_v(1 - \mu_v)}} \rho_{ik} =$$

$$= \mu_i - \frac{k_{vi}}{1 - \mu_v},$$

which is equal to $\theta_{i|v}(1|0)$ by (7.5). The other formulas are proved in a similar fashion. □

7.1.3 Model stratification

Every general Markov model $\boldsymbol{M}(\mathcal{T}, 2)$ on a semi-labeled tree $\mathcal{T}$ admits a natural stratification. More precisely, we can stratify $\boldsymbol{M}_+(\mathcal{T}, 2)$ into pieces that are isomorphic to $\boldsymbol{M}_{++}(\mathcal{F}, 2)$, where $\mathcal{F}$ are semi-labeled subforests of $\mathcal{T}$ (c.f. Definition 5.53). In this section we briefly discuss this construction.

Let $q \in \boldsymbol{M}_+(\mathcal{T}, 2)$ and let ρ_{ij} for $1 \leq i < j \leq m$ be the corresponding correlations. Define the equivalence relation on the labeling set $[m]$ by $i \sim j$ if and only $\rho_{ij} \neq 0$. To show that this is indeed equivalence relation, we first note that $i \sim i$, since $\rho_{ii} = 1$, and if $i \sim j$, then $j \sim i$. It remains to show that the relation is transitive. This follows from the parameterization of $\boldsymbol{M}(\mathcal{T}, 2)$ given in Theorem 6.10 that either $\rho_{ij}, \rho_{ik}, \rho_{jk}$ are all positive or at least two of them vanish. Equivalently, if $\rho_{ij} \neq 0$ and $\rho_{ik} \neq 0$, then $\rho_{jk} \neq 0$ and hence this relation is transitive.

Every equivalence relation on $[m]$ defines a set partition. Recall from Section 5.3.3 that $\boldsymbol{M}(\mathcal{F}, 2)$ for a semi-labeled forest $\mathcal{F}$ is defined as a Cartesian product of the latent graphical tree models $\boldsymbol{M}(\mathcal{T}_A, 2)$ where $\mathcal{T}_A$ are the tree components of $\mathcal{F}$. We have the following result.

Theorem 7.7. *Let $\mathcal{T}$ be a semi-labeled tree. For every partition $\pi \in \boldsymbol{\Pi}(\mathcal{T})$, denote by $\boldsymbol{M}_\pi$ the set of all distributions $q \in \boldsymbol{M}_+(\mathcal{T}, 2)$ that induce π, that is i, j lie in the same block of π if and only if $\rho_{ij} \neq 0$. Then*

$$\boldsymbol{M}_\pi = \boldsymbol{M}_{++}(\mathcal{F}^\pi, 2), \qquad (7.6)$$

where $\mathcal{F}^\pi$ is defined by (5.5). Moreover,

$$\boldsymbol{M}_{++}(\mathcal{T}, 2) = \boldsymbol{M}_+(\mathcal{T}, 2) \setminus \bigcup_{\pi < [m]} \boldsymbol{M}_+(\mathcal{F}^\pi, 2). \qquad (7.7)$$

Proof. Note that for every edge $u - v$ we have $\det(\theta_{u|v}) = 0$ if and only if $\rho_{uv} = 0$ and $\det(\theta_{u|v}) > 0$ if and only if $\rho_{uv} > 0$. By (7.1) we have $\rho_{ij} = \prod_{e \in \overline{ij}} \rho_e$ for every $i, j \in [m]$ and so a distribution in $\boldsymbol{M}_+(\mathcal{T}, 2)$ lies in $\boldsymbol{M}_{++}(\mathcal{T}, 2)$ if and only if $\rho_{ij} > 0$ for all labeled vertices i, j. This also implies (7.6) for $\pi = \hat{1}$. For general π, note that the parameterization of the underlying Markov process on $\mathcal{F}^\pi$ given by (5.9) implies that all subvectors of X corresponding to connected components of $\mathcal{F}^\pi$ are independent. This in turn implies that $\rho_{ij} = 0$ in $\boldsymbol{M}_+(\mathcal{F}^\pi)$ if i, j lie in two different blocks of π. By definition of $\boldsymbol{M}_{++}(\mathcal{F}^\pi, 2)$, there are no other marginal independencies and so (7.6) holds for every $\pi \in \boldsymbol{\Pi}(\mathcal{T})$. The same reasoning implies (7.7). □

7.1.4 Phylogenetic invariants

The monomial parameterization of $\boldsymbol{M}(\mathcal{T}, 2)$ given in Theorem 6.10 shows that the Zariski closure $\boldsymbol{V}_\mathcal{T}$ of this model, in the complex space of standardized tree cumulants, forms a toric variety; see Definition 2.20. By Theorem

2.22, the ideal defining $\boldsymbol{V}_{\mathcal{T}}$ is generated by binomials. By Proposition 2.38, the same ideal defines the real part of $\boldsymbol{V}_{\mathcal{T}}$.

Theorem 7.8. *Let $\mathcal{T}$ be a binary phylogenetic tree. Denote by t_A the $\mathcal{T}^*$-cumulants, where $\mathcal{T}^*$ is a binary expansion of $\mathcal{T}$; see Definition 5.46. The Zariski closure $\boldsymbol{V}_{\mathcal{T}}$ of $\boldsymbol{M}(\mathcal{T}, 2)$ is defined by all binomials*

$$\mathsf{t}_{I \cup J}\mathsf{t}_{I' \cup J'} - \mathsf{t}_{I \cup J'}\mathsf{t}_{I' \cup J} \qquad \textit{for all nonempty } I, I' \subseteq A, J, J' \subseteq B,$$

where $A|B$ is a non-trivial tree split of $[m]$ induced by $\mathcal{T}$.

Proof. First note that the equations are formulated in terms of tree cumulants instead of standardized tree cumulants. A tree cumulant t_A is equal to the corresponding central tree cumulant t'_A as long as $|A| \geq 2$; see Lemma 4.35. It follows that the standardized tree cumulants $\bar{\mathsf{t}}_A$ satisfy

$$\bar{\mathsf{t}}_A = \frac{\mathsf{t}_A}{\prod_{i \in A} \sqrt{\sigma_i}},$$

where $\sigma_i = \mathrm{var}(X_i)$. In particular, $\bar{\mathsf{t}}_{I \cup J}\bar{\mathsf{t}}_{I' \cup J'} - \bar{\mathsf{t}}_{I \cup J'}\bar{\mathsf{t}}_{I' \cup J}$ vanishes if and only if $\mathsf{t}_{I \cup J}\mathsf{t}_{I' \cup J'} - \mathsf{t}_{I \cup J'}\mathsf{t}_{I' \cup J}$ does.

Denote by $\boldsymbol{V}$ the toric variety defined by the equations in Theorem 7.8. We first show that $\boldsymbol{V}_{\mathcal{T}} \subseteq \boldsymbol{V}$. Let $u - v$ be an inner edge defining a non-trivial tree split $A|B$ such that leaves in A are closer to u than to v. By Theorem 6.10 we can rewrite

$$\bar{\mathsf{t}}_{I \cup J} = \bar{\mathsf{t}}_{I \cup \{u\}}\bar{\mathsf{t}}_{J \cup \{v\}}(\rho_{uuu}^{\deg(u;\mathcal{T}^0(I \cup J))-2}\rho_{vvv}^{\deg(v;\mathcal{T}^0(I \cup J))-2}\rho_{uv}).$$

Moreover, $\deg(u; \mathcal{T}^0(I \cup J)) = \deg(u; \mathcal{T}^0(I \cup \{v\}))$ and $\deg(v; \mathcal{T}^0(I \cup J)) = \deg(v; \mathcal{T}^0(J \cup \{u\}))$. This already implies that all equations of Theorem 7.8 must hold and therefore $\boldsymbol{V}_{\mathcal{T}} \subseteq \boldsymbol{V}$.

Now we show that $\boldsymbol{V} \subseteq \boldsymbol{V}_{\mathcal{T}}$. We constrain ourselves to the subset $\boldsymbol{U}$ of $\boldsymbol{V}$ whose points have nonzero coordinates, which form a Zariski-open subset of $\boldsymbol{V}$. Because the toric variety $\boldsymbol{V}$ is irreducible, the Zariski closure of $\boldsymbol{U}$ is equal to $\boldsymbol{V}$ and hence it is enough to show that $\boldsymbol{U}$ admits the same parameterization as $\boldsymbol{V}_{\mathcal{T}}$. For any internal edge $e = u - v$, let e, e_1, e_2 be the edges incident with u and e, e_3, e_4 be the edges incident with v. Consider any four non-empty subsets I_1, I_2, J_1, J_2 of $[m]$ such that u separates I_1, I_2 and $J_1 \cup J_2$ and v separates $I_1 \cup I_2, J_1$ and J_2. Moreover, we assume that e_1 is closer to I_1, e_2 is closer to I_2, e_3 is closer to J_1, and e_4 is closer to J_2. We first check that in $\boldsymbol{U}$ the ratio

$$\frac{\bar{\mathsf{t}}_{I_1 \cup J_1}\bar{\mathsf{t}}_{I_2 \cup J_2}}{\bar{\mathsf{t}}_{I_1 \cup I_2}\bar{\mathsf{t}}_{J_1 \cup J_2}} = \frac{\mathsf{t}_{I_1 \cup J_1}\mathsf{t}_{I_2 \cup J_2}}{\mathsf{t}_{I_1 \cup I_2}\mathsf{t}_{J_1 \cup J_2}}$$

does not depend on the choice of I_1, I_2, J_1, J_2. That is, for any other choice of four nonempty sets I'_1, I'_2, J'_1, J'_2 satisfying the conditions above, this ratio will be the same. Indeed, for a tree split induced by e_1 equations of Theorem 7.8 give

$$L = \mathsf{t}_{I_1 \cup J_1}\mathsf{t}_{I_2 \cup J_2}\mathsf{t}_{I'_1 \cup I'_1}\mathsf{t}_{J'_1 \cup J'_2} = \mathsf{t}_{I_1 \cup I'_2}\mathsf{t}_{I'_1 \cup J_1}\mathsf{t}_{I_2 \cup J'_2}\mathsf{t}_{J'_1 \cup J_2}.$$

Now for a tree split induced by e_2 we further obtain

$$L = t_{I_1 \cup I_2} t_{J_1 \cup J_2} t_{I_1' \cup J_1'} t_{I_2' \cup J_2'},$$

which implies that

$$\frac{t_{I_1 \cup J_1} t_{I_2 \cup J_2}}{t_{I_1 \cup I_2} t_{J_1 \cup J_2}} = \frac{t_{I_1' \cup J_1'} t_{I_2' \cup J_2'}}{t_{I_1' \cup I_2'} t_{J_1' \cup J_2'}}$$

as claimed. This allows us to denote

$$\rho_e := \sqrt{\frac{\bar{t}_{I_1 \cup J_1} \bar{t}_{I_2 \cup J_2}}{\bar{t}_{I_1 \cup I_2} \bar{t}_{J_1 \cup J_2}}}.$$

Similarly, for any terminal edge $e = v - i$ we define

$$\rho_e := \sqrt{\frac{\bar{t}_{\{i\} \cup I} \bar{t}_{\{i\} \cup J}}{\bar{t}_{I \cup J}}},$$

which again in U does not depend on I, J as long as they are nonempty subsets of $[m]$ such that v separates i, I, and J. Finally, for each inner vertex v, if I, J, K are any three nonempty subsets of $[m]$ separated by v, then we define

$$\rho_{vvv} := \sqrt{\frac{\bar{t}_{I \cup J \cup K}^2}{\bar{t}_{I \cup J} \bar{t}_{I \cup K} \bar{t}_{J \cup K}}}.$$

Now we show that with these definitions the parameterization in (6.15) holds. We first prove it for $A = \{i, j\}$ for some $i, j \in [m]$, in which case $\bar{t}_{ij} = \rho_{ij}$. For every vertex h on the path between i and j, pick a leaf k such that h separates i, j, and k. We label these leaves $k_1, \ldots, k_l$. Then

$$\rho_{ij} = \sqrt{\frac{\rho_{ij} \rho_{ik_1}}{\rho_{jk_1}}} \sqrt{\frac{\rho_{ik_2} \rho_{jk_1}}{\rho_{ik_1} \rho_{jk_2}}} \cdots \sqrt{\frac{\rho_{ik_l} \rho_{jk_{l-1}}}{\rho_{ik_{l-1}} \rho_{jk_l}}} \sqrt{\frac{\rho_{ij} \rho_{jk_l}}{\rho_{ik_l}}} = \prod_{e \in \overline{ij}} \rho_e.$$

Now our argument proceeds by induction. We prove it holds for $\bar{t}_{[m]}$ assuming it holds for all subsets of cardinality $\leq m - 1$. Take any non-trivial tree split $A|B$ and let $i \in A$, $j \in B$. Then in U

$$\bar{t}_{[m]} = \frac{\bar{t}_{\{i\} \cup B} \bar{t}_{A \cup \{j\}}}{\bar{t}_{ij}}.$$

The rest follows by induction.   □

A more popular approach to listing phylogenetic invariants for $M(\mathcal{T}, 2)$ is by using tensor flattenings; see Definition 2.55.

Definition 7.9 (Edge flattenings). Let $\mathcal{T}$ be a phylogenetic tree and $q \in M(\mathcal{T}, 2)$. An edge flattening of q is any (matrix) flattening of q denoted $q_{A;B}$, where $A|B$ is a tree split.

The following theorem was proved by Allman and Rhodes [2008].

Theorem 7.10. *Let $\mathcal{T}$ be a binary phylogenetic tree. The ideal describing the Zariski closure of the model $\boldsymbol{M}(\mathcal{T}, 2)$ is generated by all 3×3 minors of all edge flattenings of $p = [p(x)]$ together with the trivial phylogenetic invariant $\sum_{x \in \mathcal{X}} p(x) - 1$.*

Remark 7.11. If $\mathcal{T}$ is a tripod tree, then there are no non-trivial polynomials in the corresponding ideal because each edge flattening is a 2×4 matrix and hence has no 3×3 minors.

One direction of the proof of Theorem 7.10 is simple and it follows by a purely probabilistic argument. Each edge flattening $p_{A;B}$ can be considered as a joint probability table of two vector-valued random variables X_A and X_B. By the global Markov properties on $\mathcal{T}$ (c.f. Proposition 5.37) we know that $X_A \perp\!\!\!\perp X_B | X_u$, where X_u is a binary hidden variable corresponding to one of the vertices of the edge inducing the tree split $A|B$. This means that $p_{A;B}$ has rank at most two and hence all 3×3 minors must vanish. Note that this reasoning easily extends to the case when all random variables in the system have k states. In this case all $(k + 1) \times (k + 1)$ minors must vanish; see also Section 7.2.4.

Instead of proving the other direction, we show that Theorem 7.10 is equivalent to Theorem 7.8. The argument that we use also gives more insight into the nature of tree cumulants. First consider a probability distribution over $\{0, \ldots, r\} \times \{0, \ldots, s\}$

$$p = \begin{bmatrix} p_{00} & p_{01} & p_{02} & \cdots & p_{0s} \\ p_{10} & p_{11} & p_{12} & \cdots & p_{1s} \\ \vdots & & & & \vdots \\ p_{r0} & p_{r1} & p_{r2} & \cdots & p_{rs} \end{bmatrix}.$$

We now add all rows to the first row and all columns to the first column obtaining the matrix of moments (c.f. Section 4.3.2)

$$\begin{bmatrix} 1 & p_{+1} & p_{+2} & \cdots & p_{+s} \\ p_{1+} & p_{11} & p_{12} & \cdots & p_{1s} \\ \vdots & & & & \vdots \\ p_{r+} & p_{r1} & p_{r2} & \cdots & p_{rs} \end{bmatrix}.$$

This matrix has of course the same rank as p. We further perform elementary operations to get rid of the first row by subtracting from the i-th column the first column multiplied by p_{+i}. Denoting $y_{ij} = p_{ij} - p_{i+}p_{+j}$ and $y_{i0} = p_{i+}$, $y_{0j} = p_{+j}$ for $i, j > 0$ we conclude that

$$\text{rank}(p) = \text{rank} \begin{bmatrix} 1 & 0 & 0 & \cdots & 0 \\ y_{10} & y_{11} & y_{12} & \cdots & y_{1s} \\ \vdots & & & & \vdots \\ y_{r0} & y_{r1} & y_{r2} & \cdots & y_{rs} \end{bmatrix} = 1 + \text{rank} \begin{bmatrix} y_{11} & y_{12} & \cdots & y_{1s} \\ \vdots & & & \vdots \\ y_{r1} & y_{r2} & \cdots & y_{rs} \end{bmatrix}.$$

The only thing that allowed us for this reduction was that the sum of all elements of p is 1. More generally, let p be a binary probability tensor. Suppose that a flattening $p_{A;B}$ of a probability tensor p has rank ≤ 2. Like in Section 4.3.2, define

$$M = \begin{bmatrix} 1 & 1 \\ 1 & 0 \end{bmatrix}$$

and let M_A be the Kronecker product of $|A|$ times M, then

$$\mu_{A;B} = M_A\, p_{A;B}\, M_B^T$$

is the flattening of the moment tensor, which has the same rank as $p_{A;B}$. Since the left-top entry of $\mu_{A;B}$ is 1, then it is clear that using elementary row and column operations we can make all non-diagonal entries in the first row and the first column to be zero. This shows that for flattenings of distributions, after a change of coordinates, rank constraints can be reduced by one. The magic of tree cumulants is that we can do it consistently for a collection of flattenings.

Denote by $t_{A;B}$ the corresponding flattening of the tree cumulants tensor t. By $\widetilde{t}_{A;B}$ denote the flattening $t_{A;B}$ with the first row and the first column removed. Here, the first row corresponds to elements t_J for $J \subseteq B$ and the first column corresponds to t_I for $I \subseteq A$.

Proposition 7.12. *Let $\mathcal{T}$ be a binary phylogenetic tree and let p be a probability distribution in $\mathbf{M}(\mathcal{T}, 2)$. If $A|B \in \mathbf{\Pi}(\mathcal{T})$, then $\mathrm{rank}(p_{A;B}) \leq 2$ if and only if $\mathrm{rank}(\widetilde{t}_{A;B}) \leq 1$.*

Proof. First note that the flattening of the moment tensor $\mu_{A;B} = M_A p_{A;B} M_B^T$ has the same rank as $p_{A;B}$. Let $I \subseteq A$, $J \subseteq B$. Then for each $\pi \in \mathbf{\Pi}(\mathcal{T}(I \cup J))$ there is at most one block containing elements from both I and J. Otherwise, removing e would increase the number of blocks in π by more than one, which is not possible. Denote this block by $(I'J')$ where $I' \subseteq I$, $J' \subseteq J$. Note that, by construction, either both I', J' are empty sets if $\pi \leq A|B$ in $\mathbf{\Pi}(\mathcal{T}(I \cup J))$ or both $I', J' \neq \emptyset$ otherwise. Equation (6.5) gives the formula for moments in terms of tree cumulants. It can be rewritten as

$$\mu_{IJ} = \sum_{\pi \in \mathbf{\Pi}(\mathcal{T}(IJ))} \left(t_{I'J'} \prod_{I \supseteq C \in \pi} t_C \prod_{J \supseteq C \in \pi} t_C \right). \tag{7.8}$$

Note that $t_{I'J'}$ is the (I', J')-th entry of a block diagonal matrix with the first 1×1 block 1 and the second block $\widetilde{t}$. Denoting this matrix by $1 \oplus \widetilde{t}$ we can further write

$$\mu_{IJ} = \sum_{I' \subseteq I} \sum_{J' \subseteq J} u_{II'} (1 \oplus \widetilde{t})_{I'J'} v_{J'J}$$

where

$$u_{II'} = \sum_{\pi \in \mathbf{\Pi}(\mathcal{T}(I \setminus I'))} \prod_{C \in \pi} t_C \quad \text{and} \quad v_{J'J} = \sum_{\pi \in \mathbf{\Pi}(\mathcal{T}(J \setminus J'))} \prod_{C \in \pi} t_C.$$

Setting $u_{II'} = 0$ for $I' \not\subseteq I$, $v_{J'J} = 0$ for $J' \not\subseteq J$, we can write these coefficients as elements of a lower triangular matrix U and an upper triangular matrix V. By construction $u_{II} = 1$ for all $I \subseteq A$ and $v_{JJ} = 1$ for all $J \subseteq B$ and hence $\det U = \det V = 1$. Therefore, $\mu_{A;B}$ has the same rank as $1 \oplus \widetilde{t}$ and one less than $\widetilde{t}$. $\qquad\qquad\square$

Proposition 7.12 shows that the condition that each edge flattening has rank ≤ 2 translates directly into all edge flattenings of $\widetilde{t}$ having rank ≤ 1, which is equivalent to vanishing all quadratic equations of Theorem 7.8.

7.2 Full semialgebraic description

7.2.1 The tripod tree model

Let $\mathcal{T}$ be the tripod tree in Figure 6.1 and $\boldsymbol{M}(\mathcal{T}, 2)$ be the corresponding general Markov model. By Proposition 5.45, to describe the set of all probability distributions in this model it is enough to restrict to the subspace of the parameter space for which all variables in the system are non-degenerate. In this case we can work with standardized moments. Simple parameterization of $\boldsymbol{M}(\mathcal{T}, 2)$ together with the constraints on $\Omega_{\mathcal{T}}$ given in Lemma 6.6 can now be used to obtain the full description of this model; see Zwiernik and Smith [2011].

Consider the conditional covariances

$$k_{12|3} = \mathrm{cov}(X_1, X_2|X_3), \quad k_{12|3} = \mathrm{cov}(X_1, X_2|X_3), \quad k_{12|3} = \mathrm{cov}(X_1, X_2|X_3).$$

They are themselves two-state random variables and we write $k_{12|3}(x)$ for the value of $k_{12|3}$ when $X_3 = x$. Let p be the $2 \times 2 \times 2$ probability tensor, whose entries will be denoted by p_{ijk}. For $x = 0, 1$ we have

$$
\begin{array}{rclcl}
k_{12|3}(x) & = & p_{11x}p_{++x} - p_{1+x}p_{+1x} & = & p_{11x}p_{00x} - p_{10x}p_{01x} \\
k_{13|2}(x) & = & p_{1x1}p_{+x+} - p_{1x+}p_{+x1} & = & p_{1x1}p_{0x0} - p_{1x0}p_{0x1} \qquad (7.9) \\
k_{23|1}(x) & = & p_{x11}p_{x++} - p_{x1+}p_{x+1} & = & p_{x11}p_{x00} - p_{x10}p_{x01},
\end{array}
$$

where $+$ denotes summing over all values in a given index.

Lemma 7.13. *The following formulas express conditional covariances in terms of unconditional cumulants:*

$$
\begin{aligned}
k_{12|3}(1) &= k_3^2 k_{12} + k_3 k_{123} - k_{13} k_{23} \\
k_{12|3}(0) &= (1 - k_3)^2 k_{12} - (1 - k_3) k_{123} - k_{13} k_{23} \\
k_{13|2}(1) &= k_2^2 k_{13} + k_2 k_{123} - k_{12} k_{23} \\
k_{13|2}(0) &= (1 - k_2)^2 k_{13} - (1 - k_2) k_{123} - k_{12} k_{23} \\
k_{23|1}(1) &= k_1^2 k_{23} + k_1 k_{123} - k_{12} k_{13} \\
k_{23|1}(0) &= (1 - k_1)^2 k_{23} - (1 - k_1) k_{123} - k_{12} k_{13}.
\end{aligned}
\qquad (7.10)
$$

We have the following proposition.

Proposition 7.14. *Let $\mathcal{T}$ be the tripod tree and $M(\mathcal{T}, 2)$ be the corresponding general Markov model. Let p be a $2 \times 2 \times 2$ probability table for three binary random variables (X_1, X_2, X_3) with cumulants (central moments) $k_{12}, k_{13}, k_{23}, k_{123}$, and means μ_1, μ_2, μ_3. Then $p \in M_+(\mathcal{T}, 2)$ if and only if*

$$
\begin{aligned}
k_{12} &\geq 0, \quad k_{13} \geq 0, \quad k_{23} \geq 0, \\
k_{12|3}(0) &\geq 0, \quad k_{12|3}(1) \geq 0, \\
k_{13|2}(0) &\geq 0, \quad k_{13|2}(1) \geq 0, \\
k_{23|1}(0) &\geq 0, \quad k_{23|1}(1) \geq 0.
\end{aligned}
\tag{7.11}
$$

Proof. To show that the constraints are necessary, we write them in terms of the parameters

$$
\begin{aligned}
k_{12} &= p_{11+} - p_{1++}p_{+1+} = \theta_r(0)\theta_r(1)\det(\theta_{1|r})\det(\theta_{2|r}) \geq 0, \\
k_{13} &= p_{1+1} - p_{1++}p_{++1} = \theta_r(0)\theta_r(1)\det(\theta_{1|r})\det(\theta_{3|r}) \geq 0, \\
k_{23} &= p_{+11} - p_{+1+}p_{++1} = \theta_r(0)\theta_r(1)\det(\theta_{2|r})\det(\theta_{3|r}) \geq 0
\end{aligned}
$$

because $\det\theta_{i|r} \geq 0$ in $M_+(\mathcal{T}, 2)$. Moreover

$$
\begin{aligned}
k_{12|3}(0) &= \theta_r(0)\theta_r(1)\det(\theta_{1|r})\det(\theta_{2|r})\theta_{3|r}(1|1)\theta_{3|r}(1|0) \geq 0, \\
k_{12|3}(1) &= \theta_r(0)\theta_r(1)\det(\theta_{1|r})\det(\theta_{2|r})\theta_{3|r}(0|1)\theta_{3|r}(0|0) \geq 0 \\
k_{13|2}(0) &= \theta_r(0)\theta_r(1)\det(\theta_{1|r})\det(\theta_{3|r})\theta_{2|r}(1|1)\theta_{2|r}(1|0) \geq 0, \\
k_{13|2}(1) &= \theta_r(0)\theta_r(1)\det(\theta_{1|r})\det(\theta_{3|r})\theta_{2|r}(0|1)\theta_{2|r}(0|0) \geq 0 \\
k_{23|1}(0) &= \theta_r(0)\theta_r(1)\det(\theta_{2|r})\det(\theta_{3|r})\theta_{1|r}(1|1)\theta_{3|r}(1|0) \geq 0, \\
k_{23|1}(1) &= \theta_r(0)\theta_r(1)\det(\theta_{2|r})\det(\theta_{3|r})\theta_{1|r}(0|1)\theta_{3|r}(0|0) \geq 0.
\end{aligned}
$$

Now we show that the constraints are also sufficient. If $k_{12}, k_{13}, k_{23} \geq 0$, then necessarily

$$
\mathrm{Det}(p) = k_{123}^2 + 4k_{12}k_{13}k_{23} \geq 0.
$$

We consider two cases. First consider the case when p is such that $\mathrm{Det}(p) = 0$. This implies that $k_{123} = 0$ and $k_{12}k_{13}k_{23} = 0$ and so one covariance vanishes. Suppose that $k_{12} = 0$. Now, since $k_{12|3}(1) \geq 0$, by (7.10), $-k_{13}k_{23} \geq 0$. This implies that at least one more covariance vanishes, say $k_{12} = k_{13} = 0$. We can now set $\rho_{1h} = 0$, $\rho_{hhh} = \rho_{222}$ (or equivalently $\mu_h = \mu_2$) and $\rho_{2h} = 1$, $\rho_{3h} = \rho_{23}$. With this choice of parameters, parameterization in (6.2) holds. It remains to check that the given parameter vector lies in $\Omega_{\mathcal{T}, ++}$. The only non-trivial check is for ρ_{2h} and ρ_{3h}. Constraints in Lemma 6.6 are satisfied because $\rho_{hhh} = \rho_{222}$. In the second case, when $\mathrm{Det}(p) > 0$, we first show that necessarily all covariances are strictly positive. Indeed, suppose that $k_{12} = 0$. Since $\mathrm{Det}(p) > 0$, $k_{123} \neq 0$ and in particular $k_3 \in (0, 1)$. If $k_{123} > 0$, then by (7.10) we obtain contradiction with $k_{12|3}(0) \geq 0$ and if $k_{123} < 0$, then we obtain contradiction with $k_{12|3}(1) \geq 0$. This shows that if $\mathrm{Det}(p) > 0$ and p

satisfies (7.11), then $k_{12}, k_{13}, k_{23} > 0$. In this case we can directly compute all parameters in the original parameterization of $M_+(\mathcal{T}, 2)$ using Theorem 7.6. It remains to show that the following inequalities are satisfied for every $i = 1, 2, 3$ and $j < k \in \{1, 2, 3\} \setminus \{i\}$

$$0 \ \leq \ \frac{1}{2}(1 - \frac{k_{123}}{\sqrt{\mathrm{Det}(p)}}) \ \leq \ 1$$

$$0 \ \leq \ \mu_i + \frac{k_{123} - \sqrt{\mathrm{Det}(p)}}{2k_{jk}} \ \leq \ 1$$

$$0 \ \leq \ \mu_i + \frac{k_{123} + \sqrt{\mathrm{Det}(p)}}{2k_{jk}} \ \leq \ 1.$$

The first equation is automatically satisfied when $k_{12}k_{13}k_{23} > 0$ and, because $k_{123} < \sqrt{\mathrm{Det}(p)}$, the other two can be rewritten as

$$-\mu_i \ \leq \ \frac{k_{123} - \sqrt{\mathrm{Det}(p)}}{2k_{jk}}$$
$$\frac{k_{123} + \sqrt{\mathrm{Det}(p)}}{2k_{jk}} \ \leq \ 1 - \mu_i. \tag{7.12}$$

Now note that $\frac{k_{123} - \sqrt{\mathrm{Det}(p)}}{2k_{jk}}$ is the unique negative root of quadratic equation

$$k_{jk}x^2 - k_{123}x - k_{ij}k_{ik} = 0$$

and $\frac{k_{123} + \sqrt{\mathrm{Det}(p)}}{2k_{jk}}$ is its unique positive root. Since this quadratic polynomial is negative at zero, (7.12) translates into

$$k_{jk|i}(1) \ = \ k_{jk}(-\mu_i)^2 - k_{123}(-\mu_i) - k_{ij}k_{ik} \ \geq \ 0$$
$$k_{jk|i}(0) \ = \ k_{jk}(1 - \mu_i)^2 - k_{123}(1 - \mu_i) - k_{ij}k_{ik} \ \geq \ 0.$$

□

The inequality formulation in Proposition 7.14 is far from unique and many other formulations have been proposed; see Allman et al. [2014], Auvray et al. [2006], Pearl and Tarsi [1986], Settimi and Smith [1998], Zwiernik and Smith [2011]. Here we used the formulation of [Klaere and Liebscher, 2012, Theorem 7], which, we believe, is the most elegant one. This formulation also links to the main results of Section 4.4. The approach of Allman et al. [2014] is the most useful because it can be extended to general trees with any number of states. We present this result in Section 7.2.4.

Consider again the constraints in Proposition 7.14. Note that by $k_{12|3}(0), k_{12|3}(1) \geq 0$ it follows that

$$(1 - k_3)k_{12|3}(1) + k_3 k_{12|3}(0) = k_3(1 - k_3)k_{12} - k_{13}k_{23} \geq 0.$$

By symmetry, this implies that $\frac{1}{\mu_i(1-\mu_i)}k_{ij}k_{ik} \leq k_{jk}$, or after proper normalization,

$$\rho_{ij}\rho_{ik} \leq \rho_{jk} \qquad \text{for all } 1 \leq j < k \leq 3,$$

which gives a set of simple necessary inequalities. Comparing this with inequalities in Example 2.44 shows that $\boldsymbol{M}_+(\mathcal{T}, 2)$ constrained to correlations is a subset of the space of phylogenetic oranges on a tripod tree. This is of course true for any tree, which follows directly from how both sets are parameterized. We will generalize this result for any phylogenetic tree and exploit links to phylogenetic oranges later in Section 7.2.3.

7.2.2   Semialgebraic description of $\boldsymbol{M}(\mathcal{T}, 2)$

If $\mathcal{T}$ is a tripod tree, then, as the previous section shows, the semialgebraic description of $\boldsymbol{M}_+(\mathcal{T}, 2)$ is relatively simple, and the model has full dimension in the ambient probability simplex. The situation gets much more complicated for the quartet tree. In this section, we provide the complete semialgebraic description of $\boldsymbol{M}(\mathcal{T}, 2)$ for any phylogenetic tree $\mathcal{T}$.

First note that, by Proposition 5.52, for any three leaves i, j, k of $\mathcal{T}$ the corresponding marginal model over this triplet is a general Markov model on a tripod tree. Therefore, for all triples i, j, k, inequalities like in Proposition 7.14 must hold. It turns out, however, that the model has additional constraints that come from quartets.

Theorem 7.15. *Let $\mathcal{T}$ be a phylogenetic tree with the labeling set $[m]$. Suppose p is a joint probability distribution on m binary variables. Then $p \in \boldsymbol{M}_{++}(\mathcal{T}, 2)$ if and only if the following conditions hold:*

(C1) *p satisfies the set of equations given in Theorem 7.8.*

(C2) *For all $1 \leq i < j < k \leq m$, the corresponding marginal distribution p_{ijk} satisfies constraints in Proposition 7.14.*

(C3) *For any distinct $i, j, k, l \in [m]$ such that there exists $u - v \in E(\mathcal{T})$ with a property that u separates i, j and $\{k, l\}$ and v separates $\{i, j\}$, k, and l we have*

$$\begin{aligned}
\frac{k_{ikl} - \sqrt{\text{Det}(p_{ikl})}}{2k_{il}} &\leq \frac{k_{ijk} - \sqrt{\text{Det}(p_{ijk})}}{2k_{ij}} \\
\frac{k_{ijk} + \sqrt{\text{Det}(p_{ijk})}}{2k_{ij}} &\leq \frac{k_{ikl} + \sqrt{\text{Det}(p_{ikl})}}{2k_{il}}.
\end{aligned} \tag{7.13}$$

Proof. It is clear that (C1) and (C2) are necessary. By Theorem 7.6, expressions in (7.4) need to be between 0 and 1, which gives (C3). Indeed, the first expression in (7.4) gives

$$-\frac{1}{2} \leq \frac{k_{il}k_{ijk} - k_{ij}k_{ikl} - k_{il}\sqrt{\text{Det}(p_{ijk})}}{2k_{ij}\sqrt{\text{Det}(p_{ikl})}} \leq \frac{1}{2}$$

which is equivalent to

$$\frac{k_{ikl} - \sqrt{\text{Det}(p_{ikl})}}{2k_{il}} \leq \frac{k_{ijk} - \sqrt{\text{Det}(p_{ijk})}}{2k_{ij}} \leq \frac{k_{ikl} + \sqrt{\text{Det}(p_{ikl})}}{2k_{il}}.$$

If all correlations are positive, then the last inequality is void and thus can be disregarded and we get the first inequality in (7.13). The second inequality is obtained in the same way from the second expression in (7.4).

Now we show that (C1)–(C3) are sufficient. If p satisfies constraints in (C1), then there exist, possibly in $\mathbb{C}$, parameters $\theta_r(1)$ and $\theta_{v|u}(1|1)$, $\theta_{v|u}(1|0)$ such that p is parameterized as in (5.13). Constraints (C2) and (C3) assure that these parameters lie in $\Theta_{\mathcal{T}}$. □

Remark 7.16. Similar to the proof of Proposition 7.14, the constraints in (C3) can be expressed in terms of relative positioning of roots of two quadratic polynomials

$$k_{ij}x^2 - k_{ijk}x - k_{ik}k_{jk} = 0 \quad \text{and} \quad k_{il}x^2 - k_{ikl}x - k_{ik}k_{kl} = 0,$$

which we denote by x_L, x_R and y_L, y_R, respectively. Together with (C2) we necessarily have

$$-\mu_i \leq y_L \leq x_L \leq 0 \leq x_R \leq y_R \leq 1 - \mu_i.$$

7.2.3 Constraints on the second-order moments

Equation (7.1) suggests a link between $\boldsymbol{M}(\mathcal{T}, 2)$ and phylogenetic oranges parameterized by (5.2). This further links this model to the space of tree metrics.

The following proposition gives a set of simple constraints on probability distribution in tree models. This may be particularly useful in practice since it involves only computing pairwise margins of the data and it enables us to check if a data point may come from a phylogenetic tree model.

Proposition 7.17. *Let* $p \in \Delta_\mathcal{X}$ *be a probability distribution. If* $p \in \boldsymbol{M}_{++}(\mathcal{T}, 2)$ *for some tree* $\mathcal{T}$ *with the labeling set* $[m]$, *then*

$$0 < \min\left\{\frac{k_{ik}k_{jl}}{k_{ij}k_{kl}}, \frac{k_{il}k_{jk}}{k_{ij}k_{kl}}\right\} \leq 1 \qquad (7.14)$$

for all (not necessarily distinct) $i, j, k, l \in [m]$.

We now present a very important generalization of this result that links the concept of phylogenetic oranges and general Markov models $\boldsymbol{M}(\mathcal{T}, k)$ for $k \geq 2$. For any two leaves $i, j \in [m]$, the corresponding marginal distribution $p_{ij} \in \mathbb{R}^{k \times k}$ comes from a latent tree model of the following simple tree

$$\overset{i}{\bullet} \leftarrow \overset{v_k}{\circ} \leftarrow \cdots \leftarrow \overset{v_1}{\circ} \leftarrow \overset{r(ij)}{\circ} \rightarrow \overset{u_1}{\circ} \rightarrow \cdots \rightarrow \overset{u_l}{\circ} \rightarrow \overset{i}{\bullet}$$

The standard parameterization implies that

$$p_{ij} = \theta_{i|v_k}^T \cdots \theta_{v_1|r(ij)}^T \operatorname{diag}(\theta_{r(ij)}) \theta_{u_1|r(ij)} \cdots \theta_{j|u_l}.$$

Here $\theta_{v|u}$ for every edge $u \to v$ is the stochastic matrix of the conditional distribution of Y_v given Y_u. Note that $\theta_{r(ij)}$, unless $r(ij)$ is the root of $\mathcal{T}$, is not a parameter of the original model $M(\mathcal{T}, k)$ but only a function of the parameter vector. We can write the marginal distributions of X_i and X_j as

$$\operatorname{diag}(p_i) = \theta_{i|v_k}^T \cdots \theta_{v_1|r}^T \operatorname{diag}(\theta_{r(ij)}), \quad \operatorname{diag}(p_j) = \operatorname{diag}(\theta_{r(ij)}) \theta_{u_1|r} \cdots \theta_{j|u_l}.$$

We now easily check that defining for all leaves i, j

$$u_{ij} := \frac{\det p_{ij}}{\sqrt{\det(\operatorname{diag}(p_i)) \det(\operatorname{diag}(p_j))}}$$

we obtain

$$|u_{ij}| = \prod_{u \to v \in \overline{ij}} \sqrt{|\det \theta_{v|u}|}. \tag{7.15}$$

For every stochastic matrix $\theta_{v|u}$ we have $\det \theta_{v|u} \in [-1, 1]$ and this determinant is equal to ± 1 only if $\theta_{v|u}$ is a permutation matrix. This implies that $|u_{ij}| \in [0, 1]$ and $|u_{ij}| = 1$ only if X_i and X_j are functionally related. Moreover, it shows that the space of all $(|u_{ij}|)$ for a fixed semi-labeled tree $\mathcal{T}$ is equal to the space $\mathcal{E}(\mathcal{T})$ defined in Section 5.1.4. In particular, Proposition 5.27 gives a set of simple equations that need to hold for $M(\mathcal{T}, k)$ for every k. The following proposition was first formulated by Steel [1994]. The proof follows immediately from the above considerations.

Proposition 7.18. *Suppose that* $p \in M(\mathcal{T}, k)$ *for* $k \geq 2$; *then, whenever there is a* $\mathcal{T}$-*split* A/B *of* $[m]$ *such that* $i, j \in A$ *and* $k, l \in B$, *we have*

$$\det p_{ik} \det p_{jl} - \det p_{il} \det p_{jk} = 0.$$

In the next section we provide a more complete set of inequalities for $M(\mathcal{T}, k)$.

We conclude this section with two remarks.

Remark 7.19. For every three leaves $i, j, k \in [m]$ we have $u_{ij} u_{ik} u_{jk} \geq 0$. It will follow from Proposition 8.7, that the space of all u is equal to the set of all correlations in the Gaussian latent tree model.

Remark 7.20. If $\det \theta_{v|u} > 0$ for every $u \to v$, which is a common assumption in phylogenetics, $u_{ij} \in (0, 1]$ and so $-\log u_{ij} > 0$ is well defined. By (7.15) the collection of all $d_{ij} := -\log u_{ij}$ forms a tree metric.

7.2.4 Description of general $M(\mathcal{T})$

The problem in the case of general state spaces is that we do not have such a nice parameterization as in the two-state case. In this section we present a more direct approach proposed by Allman et al. [2014]. In this approach $k \geq 2$ and the model $\boldsymbol{M}(\mathcal{T}, k)$ are as given in Definition 5.13 where we also assume that (M1) and (M2) hold; see Section 5.2.1. The model, where the parameters satisfy (M1), (M2), will be denoted by $\boldsymbol{M}_0(\mathcal{T}, k)$. We describe results on the tripod tree and the quartet tree and explain how we can obtain the complete set of constraints for any binary phylogenetic tree.

In the beginning, $\mathcal{T}$ is assumed to be a tripod tree as in Figure 6.1. In this case $\mathcal{X} = \{0, \ldots, k-1\}^3$ and $\pi \in \mathbb{R}^k$, $M_i \in \mathbb{R}^{k \times k}$ denote the root distribution and the transition matrices for $i = 1, 2, 3$. Using the tensor notation of Section 4.3.1 the model is given by all tensors in $\mathbb{R}^{\mathcal{X}}$ given by

$$p \quad = \quad (M_1^T, M_2^T, M_3^T) \cdot \operatorname{diag}(\pi).$$

A *principal minor* of a matrix is the determinant of a submatrix chosen with the same row and column indices, and the *leading* principal minor is one of these where the chosen indices are $\{1, \ldots, k\}$ for any k.

Theorem 7.21 (Sylvester's Theorem). *Let A be an $n \times n$ real symmetric matrix. Then*

1. *A is positive semidefinite if and only if, all leading principal minors of A are nonnegative, and*

2. *A is positive definite, if and only if, all leading principal minors of A are strictly positive.*

For a $k \times k \times k$ tensor and a vector $\boldsymbol{v} \in \mathbb{R}^k$ we use the shortcut notation $\boldsymbol{v}^T *_i p$ for $(\boldsymbol{v}^T, \boldsymbol{I}, \boldsymbol{I}) \cdot p$, $(\boldsymbol{I}, \boldsymbol{v}^T, \boldsymbol{I}) \cdot p$ and $(\boldsymbol{I}, \boldsymbol{I}, \boldsymbol{v}^T) \cdot p$ for $i = 1, 2, 3$, respectively. The following result was first formulated by Allman et al. [2013].

Theorem 7.22. *Let $\mathcal{T}$ be the tripod tree model and p a $k \times k \times k$ probability distribution. Then p lies in $\boldsymbol{M}_0(\mathcal{T}, k)$ if and only if for all $i = 1, 2, 3$*

(i) *$(e_j^T *_i p)\operatorname{adj}(x^T *_i p)(e_l^T *_i p) - (e_l^T *_i p)\operatorname{adj}(x^T *_i p)(e_j^T *_i p) = \boldsymbol{0}$ for all $j, k = 1, \ldots, k$. Here adj denotes the classical adjoint.*

(ii) *$f_i(p; x)$ is not identically zero as a polynomial in x.*

(iii) *$\det(\boldsymbol{1}^T *_i p) \neq 0$.*

(iv) *All leading principal minors of*

$$\det(\boldsymbol{1}^T *_3 p) \cdot (\boldsymbol{1}^T *_1 p)^T \cdot \operatorname{adj}(\boldsymbol{1}^T *_3 p) \cdot (\boldsymbol{1}^T *_2 p)$$

are positive, and all principal minors of the following matrices are nonnegative:

$$\det(\boldsymbol{1}^T *_3 p) \cdot (e_i^T *_1 p)^T \cdot \operatorname{adj}(\boldsymbol{1}^T *_3 p) \cdot (\boldsymbol{1}^T *_2 p) \quad \text{for } i = 1, \ldots, k,$$
$$\det(\boldsymbol{1}^T *_3 p) \cdot (\boldsymbol{1}^T *_1 p)^T \cdot \operatorname{adj}(\boldsymbol{1}^T *_3 p) \cdot (e_i^T *_2 p) \quad \text{for } i = 1, \ldots, k,$$
$$\det(\boldsymbol{1}^T *_1 p) \cdot (\boldsymbol{1}^T *_2 p)^T \cdot \operatorname{adj}(\boldsymbol{1}^T *_1 p) \cdot (e_i^T *_3 p) \quad \text{for } i = 1, \ldots, k,$$

where e_i are unit vectors in $\mathbb{R}^k$.

Similar to the two-state case, the main difficulty in obtaining the full description for a general tree is to understand the constraints on the quartet tree.

Proposition 7.23. *Let $\mathcal{T}$ be the quartet tree $12|34$, and p a $k \times k \times k \times k$ probability distribution. Then p lies in $\boldsymbol{M}_0(\mathcal{T}, k)$ if and only if*

(i) $\mathbf{1} *_i p$ *lies in* $\boldsymbol{M}_0(\mathcal{T}(\{1,2,3,4\} \setminus \{i\}, k)$ *for every* $i = 1, 2, 3, 4$.

(ii) *All* $(k+1) \times (k+1)$ *minors of the matrix flattening* $p_{12;34}$ *vanish.*

(iii) *The following* $k^2 \times k^2$ *is positive definite*

$$\det(p_{+\cdot\cdot+}) \det(p_{\cdot+\cdot+}) p_{13;24}\big((p *_2 (\mathrm{adj}(p^T_{+\cdot\cdot+}) p^T_{\cdot+\cdot+})) *_3 (\mathrm{adj}(p_{\cdot+\cdot+}) p_{\cdot++\cdot})\big).$$

Here for compactness we denote by $+$ summation over all indices in a given dimension of p.

Using the above result and following the ideas of Allman et al. [2014], the complete set of constraints can be given for any tree and any k. This set, however, will be complicated. It seems necessary to provide a list of simple necessary conditions that can be easily tested and that closely approximate $\boldsymbol{M}(\mathcal{T}, k)$.

7.3 Examples, special trees, and submodels

7.3.1 The quartet tree model

We extend the numerical example given in Section 6.3.2 for the probability distribution p provided in Table 6.1. By construction p lies in $\boldsymbol{M}_{++}(\mathcal{T}, 2)$, where $\mathcal{T}$ is a quartet tree, and we first verify that all the constraints in Theorem 7.15 are satisfied. Consider again the Mathematica code given in Section 6.3.1. To verify constraints in Theorem 7.15 it is more convenient to work with tree cumulants that are not standardized. The quickest way to do it is to replace rho[B] with cmu[B] in the definition of tree cumulants. So putting everything together we have

```
<< Combinatorica`
m=4;
Do[mu[A] = Sum[p[Union[A,B]], {B,
    Subsets[Complement[Range[m],A]]}], {A, Subsets[Range[m]]}];
Do[cmu[A] = Simplify[Sum[(-1)^(Length[A]-Length[B])*mu[B]*
    Product[mu[{i}],{i,Complement[A,B]}], {B, Subsets[A]}]],
    {A, Subsets[Range[m]]}];
intparts=Function[A,Pick[SetPartitions[A],
        Map[OrderedQ, Map[Flatten, SetPartitions[A]]]]];
Do[k[A] = Simplify[Sum[(-1)^(Length[pa]-1)*Product[cmu[B],
    {B,pa}], {pa, intparts[A]}]], {A, Subsets[Range[m]]}];
```

In this way, for the given distribution p, we obtain

$$t_1 = \frac{7}{10} \qquad t_2 = \frac{31}{50} \qquad t_3 = \frac{29}{50} \qquad t_4 = \frac{29}{50}$$

$$t_{12} = \frac{4}{125} \qquad t_{13} = \frac{2}{125} \qquad t_{14} = \frac{2}{125} \qquad t_{23} = \frac{8}{625}$$

$$t_{24} = \frac{8}{625} \qquad t_{34} = \frac{21}{625} \qquad t_{123} = -\frac{12}{3125} \qquad t_{124} = -\frac{12}{3125}$$

$$t_{134} = -\frac{8}{3125} \qquad t_{234} = -\frac{32}{15625} \qquad t_{1234} = \frac{48}{78125}$$

where the tree cumulants up to order three are equal to the corresponding cumulants and so we use interchangeably letter k and t. Recall also that k_1, k_2, k_3, k_4 are equal to the means $\mu_1, \mu_2, \mu_3, \mu_4$. To verify (C1), check among others that

$$t_{13}t_{24} - t_{14}t_{23} = \frac{2}{125} \cdot \frac{8}{625} - \frac{2}{125} \cdot \frac{8}{625} = 0,$$

$$t_{123}t_{134} - t_{1234}t_{13} = (-\frac{12}{3125}) \cdot (-\frac{8}{3125}) - \frac{48}{78125} \cdot \frac{2}{125} = 0$$

and other invariants can be checked in the same way. We now verify that (C2) holds for the triple $\{1, 2, 3\}$. This follows because all covariances k_{ij} are positive and the conditional covariances satisfy

$$k_{12|3}(1) = \frac{651}{78125} \qquad\qquad k_{12|3}(0) = \frac{551}{78125}$$

$$k_{13|2}(1) = \frac{21}{6250} \qquad\qquad k_{13|2}(0) = \frac{21}{6250}$$

$$k_{23|1}(1) = \frac{48}{15625} \qquad\qquad k_{23|1}(0) = \frac{28}{15625}$$

and hence they are also positive. To verify (C3), compute

$$\sqrt{\text{Det}(p_{123})} = \sqrt{\text{Det}(p_{134})} = \frac{4}{625}.$$

Now

$$\frac{k_{123} - \sqrt{\text{Det}(p_{123})}}{2k_{12}} = -\frac{4}{25} \qquad \frac{k_{123} + \sqrt{\text{Det}(p_{123})}}{2k_{12}} = \frac{1}{25}$$

$$\frac{k_{134} - \sqrt{\text{Det}(p_{134})}}{2k_{14}} = -\frac{7}{25} \qquad \frac{k_{134} + \sqrt{\text{Det}(p_{134})}}{2k_{14}} = \frac{3}{25}$$

and hence (7.13) holds. Note that the fact that

$$-k_3 = -\frac{29}{50} \leq \frac{k_{123} - \sqrt{\text{Det}(p_{123})}}{2k_{12}} = -\frac{4}{25}$$

$$\frac{k_{123} + \sqrt{\text{Det}(p_{123})}}{2k_{12}} = \frac{1}{25} \leq 1 - k_3 = \frac{21}{50}$$

follows already from (C2); see (7.12).

Suppose that we do not know the original parameters that mapped to p. We can easily identify those using Theorem 7.6. We have

$$\theta_r(1) = \frac{1}{2}\left(1 - \frac{k_{ijk}}{\sqrt{\operatorname{Det}(p_{ijk})}}\right) = \frac{8}{10},$$

which is equal to the true value of $\theta_r(1)$ given in Section 6.3.2. To find the values for $\theta_{1|r}(1|1)$ and $\theta_{1|r}(1|0)$ we compute

$$k_1 + \frac{k_{123} - \sqrt{\operatorname{Det}(p_{123})}}{2k_{23}} = \frac{3}{10}$$

$$k_1 + \frac{k_{123} + \sqrt{\operatorname{Det}(p_{123})}}{2k_{23}} = \frac{8}{10}.$$

The value of $\theta_{a|r}(1|i)$ is verified in the same way.

Table 7.1 *Moments, central moments, and tree cumulants for a probability assignment that does not lie in* $\mathbf{M}(\mathcal{T}, 2)$.

x	$A = \mathcal{A}(x)$	$p(x)$	μ_A	μ'_A	t_A
0000	$\emptyset$	$\frac{601}{10000}$	1	1	1
0001	4	$\frac{1533}{50000}$	$\frac{137}{250}$	0	$\frac{137}{250}$
0010	3	$\frac{1533}{50000}$	$\frac{137}{250}$	0	$\frac{137}{250}$
0011	34	$\frac{1229}{50000}$	$\frac{169}{500}$	$\frac{589}{15625}$	$\frac{589}{15625}$
0100	2	$\frac{413}{10000}$	$\frac{31}{50}$	0	$\frac{31}{50}$
0101	24	$\frac{1617}{50000}$	$\frac{907}{2500}$	$\frac{72}{3125}$	$\frac{72}{3125}$
0110	23	$\frac{1617}{50000}$	$\frac{907}{2500}$	$\frac{72}{3125}$	$\frac{72}{3125}$
0111	234	$\frac{2401}{50000}$	$\frac{1163}{5000}$	$-\frac{864}{390625}$	$-\frac{864}{390625}$
1000	1	$\frac{549}{10000}$	$\frac{7}{10}$	0	$\frac{7}{10}$
1001	14	$\frac{2457}{50000}$	$\frac{1031}{2500}$	$\frac{18}{625}$	$\frac{18}{625}$
1010	13	$\frac{2457}{50000}$	$\frac{1031}{2500}$	$\frac{18}{625}$	$\frac{18}{625}$
1011	134	$\frac{4041}{50000}$	$\frac{1327}{5000}$	$-\frac{216}{78125}$	$-\frac{216}{78125}$
1100	12	$\frac{857}{10000}$	$\frac{233}{500}$	$\frac{4}{125}$	$\frac{4}{125}$
1101	124	$\frac{4893}{50000}$	$\frac{7061}{25000}$	$-\frac{108}{15625}$	$-\frac{108}{15625}$
1110	123	$\frac{4893}{50000}$	$\frac{7061}{25000}$	$-\frac{108}{15625}$	$-\frac{108}{15625}$
1111	1234	$\frac{9229}{50000}$	$\frac{9229}{50000}$	$\frac{3652}{1953125}$	$\frac{1296}{1953125}$

Consider now another probability distribution given in Table 7.1. We check as before that all constraints in (C1) hold, so for example

$$t_{13}t_{24} - t_{14}t_{23} = \frac{18}{625} \cdot \frac{72}{3125} - \frac{18}{625} \cdot \frac{72}{3125} = 0,$$

$$t_{123}t_{134} - t_{1234}t_{13} = \left(-\frac{108}{15625}\right) \cdot \left(-\frac{216}{78125}\right) - \frac{1296}{1953125} \cdot \frac{18}{625} = 0.$$

It turns out that conditions (C2) also hold for any triple of leaves but still p does not lie in $M_+(\mathcal{T}, 2)$ because (C3) fails to hold. Now, as expected, Theorem 7.6 gives valid values for all parameters apart from the ones associated to the inner edge. Equations in (7.4) become

$$\frac{1}{2} + \frac{k_{14}k_{123} - k_{12}k_{134} - k_{14}\sqrt{\text{Det}(p_{123})}}{2k_{12}\sqrt{\text{Det}(p_{134})}} = -\frac{1}{10}$$

$$\frac{1}{2} + \frac{k_{14}k_{123} - k_{12}k_{134} + k_{14}\sqrt{\text{Det}(p_{123})}}{2k_{12}\sqrt{\text{Det}(p_{134})}} = \frac{8}{10}.$$

In particular, $\theta_{a|r}(1|0)$ is equal to $-\frac{1}{10}$ and hence this parameter cannot represent a conditional probability.

7.3.2   Two-state Neyman model

In phylogenetics, general Markov models are used to identify a tree underlying the observed data. As we will see later, the tree topology can be uniquely identified from the pairwise marginal distributions, that is, marginal distributions over all pairs of observed variables. This motivates the analysis of a special symmetric model of $M(\mathcal{T}, 2)$, which we now introduce.

Let $\mathcal{T}$ be a phylogenetic tree. Suppose that we constrain the parameters in Θ_T so that $\theta_{v|u}(1|1) = 1 - \theta_{v|u}(1|0)$ for each $u \to v$. Denote $a_v = \theta_{v|u}(1|1)$. Then each transition matrix is of the form

$$\begin{bmatrix} a_v & 1 - a_v \\ 1 - a_v & a_v \end{bmatrix} \qquad \text{for some } a_v \in [\tfrac{1}{2}, 1].$$

In particular, it is doubly stochastic, that is, both the row and the column sums are 1. Often in this context it is also assumed that the root distribution is $[1/2, 1/2]$. So constrained model is called the *symmetric two-state general Markov model* and denoted by $JN(\mathcal{T})$ (for Jerzy Neyman, who first introduced this model; see Neyman [1971]).

The root distribution $(\frac{1}{2}, \frac{1}{2})$ is the stationary distribution for any doubly stochastic matrix because

$$\begin{bmatrix} \frac{1}{2}, \frac{1}{2} \end{bmatrix} \begin{bmatrix} a_v & 1 - a_v \\ 1 - a_v & a_v \end{bmatrix} = \begin{bmatrix} \frac{1}{2}, \frac{1}{2} \end{bmatrix}.$$

This implies that, if $\theta_r(1) = \mu_r = 1/2$, then $\mu_v = 1/2$ for all $v \in V(\mathcal{T})$. In the moment parameters this means that $\rho_{vvv} = 0$ for all $v \in V(\mathcal{T})$ and, by Lemma 6.6, the edge correlations ρ_e can freely take values in $[-1, 1]$. By Theorem 6.10, the model $JN(\mathcal{T})$ lies in the subspace of the space of all $\mathcal{T}$-cumulants given by two sets of constraints:

$$\mu_i = \frac{1}{2} \quad \text{for } i \in [m] \qquad \text{and} \qquad \bar{t}_A = 0 \quad \text{for all } |A| \geq 3.$$

In this space the coordinates are given by correlations $\bar{t}_{ij} = \rho_{ij}$ that are parameterized via $\rho_{ij} = \prod_{e \in \overline{ij}} \rho_e$. Applying Lemma 6.6 we see that the parameter space for this model is a subset of $\Omega_{\mathcal{T}}$ given by $\rho_{vvv} = 0$ for all inner vertices and $0 \le \rho_{uv} \le 1$. The model is exactly the space of phylogenetic oranges for $\mathcal{T}$ (c.f. (5.2)), which we denoted by $\mathcal{E}(\mathcal{T})$. If $\mathcal{T}$ is the tripod tree, the model is described in Example 2.48 and depicted in Figure 2.3.

7.3.3 Submodularity on trees

Let $\mathcal{T} = (T, \phi)$ be a semi-labeled tree. In this section we show that all distributions in $\boldsymbol{M}_+(\mathcal{T}, 2)$ are log-supermodular; see Definition 4.49. Establishing this fact is relatively simple and amounts to showing two basic results: first, that all distributions in $\boldsymbol{N}_+(T, 2)$ (in the fully observed Markov process on T) are log-supermodular, and then that log-supermodularity is preserved under taking margins. This result was first stated by Steel and Faller [2009].

Proposition 7.24. *Let $\mathcal{T}$ be a semi-labeled tree. If $p \in \boldsymbol{M}_+(\mathcal{T}, 2)$, then p is log-supermodular, that is, $p_x p_y \le p_{x \vee y} p_{x \wedge y}$ for all $x, y \in \{0, 1\}^m$, where $x \wedge y$ and $x \vee y$ denote the coordinatewise minimum and maximum, respectively.*

Proof. By definition, p is obtained as a marginal distribution of q over the labeled vertices of $\mathcal{T}$, where q is parameterized as in (5.7) by

$$q_x = \theta_r(x_r) \prod_{u \to v} \theta_{v|u}(x_v | x_u) \qquad \text{for all } x \in \{0, 1\}^{V(\mathcal{T})}$$

where for every $u \to v$ the determinant of the matrix $\theta_{v|u}$ is nonnegative. By Theorem 4.56 it is enough to show that q is log-supermodular. To compare $L = q_x q_y$ and $R = q_{x \vee y} q_{x \wedge y}$ we proceed term by term. First note that in L we have $\theta_r(x_r) \theta_r(y_r)$, which is equal to $\theta_r(x_r \vee y_r) \theta_r(x_r \wedge y_r)$ in R. Now for every $u \to v$ in L we have $\theta_{v|u}(x_v | x_u) \theta_{v|u}(y_v | y_u)$ and $\theta_{v|u}(x_v \vee y_v | x_u \vee y_u) \theta_{v|u}(x_v \wedge y_v | x_u \wedge y_u)$ in R. If either $x_u = y_u$ or $x_v = y_v$, then we have an equality. Similarly, if either $x_u = x_v$ or $y_u = y_v$, then we have an equality. Now we have two remaining cases to consider, when $x_u = 1, x_v = 0, y_u = 0, y_v = 1$ and $x_u = 0, x_v = 1, y_u = 1, y_v = 0$. However, in this case

$$\theta_{v|u}(0|1) \theta_{v|u}(1|0) \le \theta_{v|u}(0|0) \theta_{v|u}(1|1)$$

follows from the fact that the determinant of the matrix $\theta_{v|u}$ is nonnegative. In particular, q is log-supermodular. $\square$

A general log-supermodular probability distribution over $\{0, 1\}^m$ does not come from a tree model, because it needs to satisfy some additional equations and potentially also some further inequalities. However, if $\mathcal{T}$ is the tripod tree, then $\boldsymbol{M}_+(\mathcal{T}, 2)$ are precisely the log-supermodular distributions over $\{0, 1\}^3$. Recall from Example 4.51 that these are the probability distributions satisfying

$$
\begin{array}{lll}
p_{000}p_{111} \ge p_{001}p_{110} & p_{000}p_{111} \ge p_{010}p_{101} & p_{000}p_{111} \ge p_{100}p_{011} \\
p_{001}p_{111} \ge p_{011}p_{101} & p_{010}p_{111} \ge p_{011}p_{110} & p_{100}p_{111} \ge p_{101}p_{110} \quad (7.16) \\
p_{000}p_{011} \ge p_{001}p_{010} & p_{000}p_{101} \ge p_{001}p_{100} & p_{000}p_{110} \ge p_{010}p_{100}.
\end{array}
$$

Theorem 7.25. *Let $\mathcal{T}$ be a tripod tree. A probability distribution p lies in $M_+(\mathcal{T}, 2)$ if and only if it satisfies (7.16).*

Proof. Proposition 7.24 shows that (7.16) are necessary. To show that they are sufficient it is now enough to show that they imply constraints of Proposition 7.14. First we note that, by (7.9), the last six constraints above are exactly the last six constraints in (7.11). Hence it is enough to show that $k_{12} \geq 0$, $k_{13} \geq 0$, $k_{23} \geq 0$ is implied by (7.16). By symmetry it is enough to show the first. We have

$$k_{12} = k_{12|3}(0) + k_{12|3}(1) + (p_{000}p_{111} + p_{110}p_{001} - p_{100}p_{011} - p_{010}p_{101}). \quad (7.17)$$

We show that the expression in parentheses is nonnegative for $p \in M_+(\mathcal{T}, 2)$. We write this expression as $R = f_{00} + f_{11} - f_{01} - f_{10}$, where

$$f_{00} = p_{000}p_{111}, \quad f_{01} = p_{010}p_{101}, \quad f_{10} = p_{100}p_{011}, \quad f_{11} = p_{110}p_{001}.$$

Note that, by (7.16), we have $f_{00} \geq \max\{f_{01}, f_{10}\}$. This implies $R \geq 0$ if either $p_{ijk} = 0$ for some i, j, k, or if $f_{11} \geq \min\{f_{01}, f_{10}\}$. Thus, we assume that $f_{ij} > 0$ and $f_{11} < \min\{f_{01}, f_{10}\}$. The supermodular inequalities $p_{010}p_{100} \leq p_{000}p_{110}$ and $p_{101}p_{011} \leq p_{111}p_{001}$ imply

$$f_{01}f_{10} = p_{010}p_{101}p_{100}p_{011} \leq p_{000}p_{111}p_{110}p_{001} = f_{00}f_{11}.$$

Hence $[f_{ij}]$ is supermodular itself. As a consequence, we have

$$\frac{f_{10}}{f_{00}} - 1 \leq \frac{f_{11}}{f_{01}} - 1 \leq \left(\frac{f_{11}}{f_{01}} - 1\right)\frac{f_{01}}{f_{00}},$$

where the second inequality holds since $f_{11} < f_{01} \leq f_{00}$.
After multiplying both sides by f_{00} we obtain

$$f_{10} - f_{00} \leq f_{11} - f_{01}$$

or equivalently $R \geq 0$. It follows that $k_{12} \geq 0$ and, by symmetry, that $k_{ij} \geq 0$ for all $1 \leq i < j \leq 3$. □

It is instructive to look at a 3-dimensional picture of our 7-dimensional model $M(\mathcal{T}, 2)$ of the tripod tree. We consider the slice given by

$$\begin{bmatrix} p_{000} & p_{001} \\ p_{010} & p_{011} \end{bmatrix} = \begin{bmatrix} x & y \\ z & w \end{bmatrix} \quad \text{and} \quad \begin{bmatrix} p_{100} & p_{101} \\ p_{110} & p_{111} \end{bmatrix} = \begin{bmatrix} w & z \\ y & x \end{bmatrix}.$$

Under this specialization, the hyperdeterminant factors as

$$\text{Det}(p) = (x+y+z+w)(x+y-z-w)(x-y+z-w)(x-y-z+w). \quad (7.18)$$

Consider the tetrahedron $\{(x, y, z, w) \in \mathbb{R}^4_{\geq 0} : x+y+z+w = 1/2\}$. Fixing the signs of the last three factors in (7.18) divides the tetrahedron into four bipyramids and four smaller tetrahedra. Inside our slice, the four cells occupy the bipyramids. Each cell is precisely the toric cube in Figure 2.3, its convex hull is the bipyramid, and it contains six of the nine edges. The whole slice is depicted in Figure 2.5. Any two of the cells meet in a line segment such as $\{x+y-z-w = x-y+z-w = 0,\ x-y-z+w \geq 0\}$. The algebraic boundary of each cell consists of the same three quadrics $\{xy = zw\}$, $\{xz = yw\}$ and $\{xw = yz\}$. Neither the three planes in (7.18) nor the four facet planes of the tetrahedron are in the algebraic boundary.

The fact that the slice described above is isomorphic to the simplest toric cube depicted in Figure 2.5 is not a coincidence. Indeed, this slice corresponds to the two-state Neyman model on the tripod tree, and we observed in Section 7.3.2 that this corresponds to the space of phylogenetic oranges, that form toric cubes.

7.3.4   The mixture model

In this section we study the model $\boldsymbol{M}(\mathcal{T})$, where $\mathcal{T}$ is a star tree; the inner vertex represents a binary random variable that is not observed but the leaves have arbitrary number of states. In machine learning these type of models are often referred to as *naive Bayes models*. We denote these models by $\boldsymbol{M}_{\mathcal{X}}$, where $\mathcal{X} = \prod_{i=1}^{m} \mathcal{X}_i$ and $\mathcal{X}_i = \{0, \ldots, k_i\}$. The observed variables in the system are $X_1, \ldots, X_m$ and the hidden binary variable is denoted by H.

The parameters of the model are conditional probability matrices $M_i \in \mathbb{R}^{2 \times (k_i+1)}$ representing the conditional distribution of X_i given the binary latent variable H. The first row of M_i is denoted by $a_i = [a_{ij}]$ and the second row by $b_i = [b_{ij}]$. The joint distributions $p = [p_x] \in \boldsymbol{M}_{\mathcal{X}}$ are parameterized by

$$p_x = (1 - \pi) \prod_{i=1}^{m} a_{ix_i} + \pi \prod_{i=1}^{m} b_{ix_i} \quad \text{for all } x = (x_1, \ldots, x_m) \in \mathcal{X}, \qquad (7.19)$$

where $\pi \in [0, 1]$ and $a_i, b_i \in \Delta_{\mathcal{X}_i}$ for all $i = 1, \ldots, m$. By our identification $\Delta_{\mathcal{X}} \simeq \mathbb{RP}^{\mathcal{X}}_{\geq 0}$ (see Lemma 2.37) this model corresponds exactly to the mixture model discussed in Section 2.4.

We say that a probability distribution $p \in \Delta_{\mathcal{X}}$ has flattening rank ≤ 2 if for any split $A|B$ of $\{1, \ldots, m\}$ the resulting matrix flattening $p_{A;B}$ (c.f. Section 4.3.1) has rank ≤ 2. In this section we discuss a general class of models where log-supermodularity together with flattening conditions gives the complete description.

Definition 7.26. The nonnegative part $\boldsymbol{M}_{\mathcal{X}}^{+}$ of $\boldsymbol{M}_{\mathcal{X}}$ is the image in $\Delta_{\mathcal{X}}$ of the parameterization in (7.19) constrained to parameters such that all 2×2 minors of all M_i are nonnegative.

Note that if $\mathcal{X} = \{0, 1\}^m$, then the above definition corresponds to the regular definition of the nonnegative part of the model $\boldsymbol{M}(\mathcal{T}, 2)$, where $\mathcal{T}$ is a star graph. It turns out that the description of all probability distributions in $\boldsymbol{M}_{\mathcal{X}}^{+}$ is extremely simple.

Theorem 7.27. *A probability distribution $p \in \Delta_{\mathcal{X}}$ lies in $\boldsymbol{M}_{\mathcal{X}}^{+}$ if and only if p is log-supermodular and it has flattening rank ≤ 2.*

This result follows from Theorem 1.1 in Allman et al. [2015], which gives a full semialgebraic description of all tensors in $\mathbb{R}^{\mathcal{X}}$ of *nonnegative rank* ≤ 2, that is, all tensors u that can be written as a sum of two nonnegative tensors of rank 1

$$u = a_1 \otimes \cdots \otimes a_m + b_1 \otimes \cdots \otimes b_m, \qquad a_i, b_i \in \mathbb{R}_{\geq 0}^{\mathcal{X}_i}.$$

The reason why these two descriptions are equivalent is that the set of all tensors in $\mathbb{R}^{\mathcal{X}}$ of nonnegative rank ≤ 2 is a cone over $\boldsymbol{M}_{\mathcal{X}}^{+}$, which follows from the following lemma.

Lemma 7.28. *Let $u \in \mathbb{R}^{\mathcal{X}}$. Then u has nonnegative rank ≤ 2 if and only if $u/\sum_{x \in \mathcal{X}} u_x$ lies in $\boldsymbol{M}_{\mathcal{X}}^{+}$.*

Proof. Let $A = (\mathbf{1}^T a_1) \cdots (\mathbf{1}^T a_m)$ and $B = (\mathbf{1}^T b_1) \cdots (\mathbf{1}^T b_m)$. The sum of elements of u is

$$\sum_{x \in \mathcal{X}} u_x = (\mathbf{1}^T, \ldots, \mathbf{1}^T) \cdot u = A + B.$$

Therefore

$$\frac{u}{\sum_x u_x} = \frac{A}{A+B} \left(\frac{a_1}{\mathbf{1}^T a_1} \right) \otimes \cdots \otimes \left(\frac{a_m}{\mathbf{1}^T a_m} \right) +$$
$$\frac{B}{A+B} \left(\frac{b_1}{\mathbf{1}^T b_1} \right) \otimes \cdots \otimes \left(\frac{b_m}{\mathbf{1}^T b_m} \right),$$

where by construction $a_i/(\mathbf{1}^T a_i)$ and $b_i/(\mathbf{1}^T b_i)$ lie in $\Delta_{\mathcal{X}_i}$. We now compare this with (7.19) to conclude that $u/\sum_{x \in \mathcal{X}} u_x$ lies in $\boldsymbol{M}_{\mathcal{X}}^{+}$. The opposite direction is straightforward. $\square$

Similar to general Markov models, to get the full description of $\boldsymbol{M}_{\mathcal{X}}$, it is enough to describe its nonnegative part and the nonnegative part forms a fundamental domain of the action of a group described as follows. Let $S_{\mathcal{X}_i}$ be the symmetric group on $\mathcal{X}_i$ and denote $S_{\mathcal{X}} = \prod_{i=1}^{m} S_{\mathcal{X}_i}$. The group $S_{\mathcal{X}}$ acts on the state space $\mathcal{X}$ by permutation $x \mapsto \sigma(x)$. This action induces the action on the parameter space, so that $S_{\mathcal{X}_i}$ permutes columns of M_i, and on the model $\boldsymbol{M}_{\mathcal{X}}$ so that $p(x)$ is mapped by $\sigma \in S_{\mathcal{X}}$ to $p(\sigma(x))$.

Proposition 7.29. *The model $\boldsymbol{M}_{\mathcal{X}}$ is a $S_{\mathcal{X}}$-orbit of its nonnegative part $\boldsymbol{M}_{\mathcal{X}}^{+}$. Moreover, $\boldsymbol{M}_{\mathcal{X}}$ consists of $(k_1 + 1)! \cdots (k_m + 1)!/2$ cells that are copies of $\boldsymbol{M}_{\mathcal{X}}^{+}$.*

Proof. By Theorem 7.27, it suffices to check that the only non-trivial $\sigma = (\sigma_i) \in S_{\mathcal{X}}$ such that it maps log-supermodular distributions to log-supermodular distributions is such that each σ_i maps j to $k_i - j$ for $j \in \mathcal{X}_i$. $\square$

We now describe the algebraic boundary of this model, which follows from [Allman et al., 2015, Theorem 1.2].

Theorem 7.30. *The algebraic boundary of $\boldsymbol{M}_{\mathcal{X}}$ has $\sum_{i=1}^{m}(k_i+1)$ irreducible components, given by slices having rank ≤ 1. The algebraic boundary of $\boldsymbol{M}_{\mathcal{X}}^{+}$ has the same irreducible components plus $\sum_{i=1}^{m}\binom{k_i+1}{2}$ additional components given by linearly dependent double slices.*

A double slice is *linearly dependent* if its two slices are identical up to a multiplicative scalar. In this situation, one X_i is *marginally* independent of the other variables in the system. In the second component count of Theorem 7.30 we exclude the special case $2 \times 2 \times 2$ because the "further components" fail to be hypersurfaces.

7.4 Inequalities and estimation

In this section we show how our understanding of the geometry of the model $\boldsymbol{M}(\mathcal{T})$ gives an insight into potential estimation problems. For simplicity we focus on the case where all variables in the system are binary.

Let $\boldsymbol{X}$ be a random vector with values in $\mathcal{X} = \{0,1\}^{m}$. Suppose that a random sample of size n was observed. We summarize the data in the tensor of sample proportions $\hat{p}$. Recall from Section 3.3.1 that the multinomial log-likelihood is a function $\ell : \Delta_{\mathcal{X}} \to \mathbb{R}$ defined for a fixed tensor $\hat{p} \in \Delta_{\mathcal{X}}$ by

$$\ell(p;\hat{p}) \quad = \quad n\langle \log p, \hat{p} \rangle \quad = \quad n\sum_{x \in \mathcal{X}} \hat{p}_x \log p_x.$$

In view of Section 3.3.3, the likelihood for the model $\boldsymbol{M}(\mathcal{T})$ can be defined as the multinomial likelihood above constrained to tensors in $\boldsymbol{M}(\mathcal{T}) \subseteq \Delta_{\mathcal{X}}$. Therefore, we are interested in the following optimization problem

$$\text{for given } \hat{p} \quad \text{maximize} \quad \ell(p;\hat{p}) \quad \text{s.t. } p \in \boldsymbol{M}(\mathcal{T},2). \qquad (7.20)$$

By Proposition 3.35 the maximum likelihood estimator in the parameter space of the model is given by any point of the q-fiber, where q is the maximizer of (7.20).

In some applications, especially in phylogenetics, we replace $\boldsymbol{M}(\mathcal{T},2)$ with $\boldsymbol{M}_{+}(\mathcal{T},2)$. However, by Theorem 7.3, $\boldsymbol{M}(\mathcal{T},2)$ can be obtained as a $(Z_2)^{m}$-orbit of $\boldsymbol{M}_{+}(\mathcal{T},2)$ and the multinomial likelihood function is invariant with respect to the action of $(Z_2)^{m}$, namely $\ell(g\cdot\boldsymbol{p}; g\cdot\hat{p}) = \ell(p;\hat{p})$ for all $g \in (Z_2)^{m}$. This also implies that $\ell(g\cdot\boldsymbol{p}; \hat{p}) = \ell(p; g\cdot\hat{\boldsymbol{p}})$. So, even if our primary interest is in $\boldsymbol{M}(\mathcal{T},2)$, instead of optimizing the log-likelihood over $\boldsymbol{M}(\mathcal{T},2)$ we can run several optimization queries over $\boldsymbol{M}_{+}(\mathcal{T},2)$, one for each element of $(Z_2)^{m}$.

In the ideal situation, the likelihood function has a unique maximum. However, for the model $\boldsymbol{M}(\mathcal{T})$ it will never happen. Even if we find the global maximum of (7.20), the corresponding q-fiber, and hence the set of all points in the parameter space giving the same likelihood value, always has more than one element. A straightforward way around this is to take a quotient with respect to the sign switching group given in Definition 6.27. This can be done by assuming that the determinant of each transition matrix in the model is nonnegative. By Proposition 6.30, at least generically, the parameterization of the model becomes one-to-one.

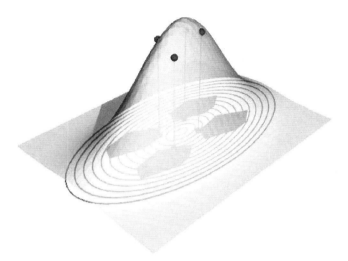

Figure 7.1 *The multinomial likelihood and a submodel of the saturated model given by four disjoint regions.*

Much more serious problems are related directly with the complicated structure of the maximization problem (7.20). The multinomial likelihood function constrained to $M(\mathcal{T}, 2)$ will typically have many local optima. One reason is that the model consists of several regions that touch only along lower dimensional faces; see Figure 7.1. Another reason is that each of these regions is not convex itself. This, in turn, can make estimation schemes unstable, which was observed in many applied analyzes of these models (see, e.g., Chor et al. [2000]).

Probably the biggest problem with latent tree models is that the maxima of the constrained likelihood function will often lie on the boundary of the parameter space. Moreover, these boundary points always correspond to some degenerate cases where the usual interpretation of the hidden process breaks down. This problem becomes especially serious if either the model is misspecified or correlations between the observed variables are weak, which was observed by Wang and Zhang [2006].

To illustrate all these issues, we focus on the simplest case of the tripod tree model, whose complete description was given in Proposition 7.14 and Theorem 7.25. The model has full dimension in the ambient probability simplex and Figure 2.5 shows a slice of this 7-dimensional object. This slice is a bit misleading because it suggests that the model fills a large proportion of the ambient probability simplex. To get a better idea about this model

Figure 7.2 *The space of all possible correlations $\rho_{12}, \rho_{13}, \rho_{23}$ for the tripod tree model.*

we draw several slices. Each slice assumes $\mu_1 = \mu_2 = \mu_3 = 1/2$ and is depicted in the space given by correlations for a fixed value of the third-order standardized moment $\rho_{123} \in \{0, 0.005, 0.02\}$; see Figure 7.2. Simple Monte Carlo simulations show that the volume of the model accounts for only 8% of the ambient probability simplex. This means that typically the sample proportions end up outside of the model even if the true data generating distribution lies in the tripod tree model.

Let $\hat{p}$ be the sample proportions for some observed data on the tripod tree model. We have three possible scenarios:

(i) $\hat{p} \in \boldsymbol{M}(\mathcal{T}, 2)$ and then $\ell(p; \hat{p})$ is unimodal.

(ii) $\hat{p} \notin \boldsymbol{M}(\mathcal{T}, 2)$ and $\ell(p; \hat{p})$ is multimodal but there exists only one global maximum.

(iii) $\hat{p} \notin \boldsymbol{M}(\mathcal{T}, 2)$ and $\ell(p; \hat{p})$ has multiple global maxima.

Although the situation in (iii), generically never happens, it raises an interesting question related to the model identifiability. For every data point satisfying (iii) we are not able to identify the parameters using the maximum likelihood estimation even if we take into account the label switching problem.

From the numerical point of view, the situation in (ii) and (iii) may describe equally bad scenarios since in both cases the algorithms become unstable. If the true data-generating distribution lies only approximately in $\boldsymbol{M}(\mathcal{T}, 2)$, then this happens even for arbitrary large sample sizes.

We illustrate this with a simple simulation. Suppose that a sample of size 10,000 has been observed

$$\begin{bmatrix} u_{000} & u_{001} & u_{100} & u_{101} \\ u_{010} & u_{011} & u_{110} & u_{111} \end{bmatrix} = \begin{bmatrix} 2069 & 16 & 2242 & 331 \\ 2678 & 863 & 442 & 1359 \end{bmatrix}. \quad (7.21)$$

By direct computations we check that the sample proportions obtained from this table satisfy only some of the defining constraints and hence $\hat{p}$ does not lie in the tripod tree model. The corresponding maximum likelihood estimator will lie on the boundary of the parameter space. The most popular way of estimating this model class is the EM algorithm presented in Section 3.3.4. The algorithm starts from an arbitrary point in the parameter space and moves around it at each step strictly increasing the likelihood function. We perform the following simulation. At each iteration, sample uniformly from $\Theta_T = [0,1]^7$ the starting parameters for the EM algorithm (see Algorithm 3.41) and record the corresponding EM estimate. For 100 iterations the procedure found six different isolated maxima given in Table 7.2.

Table 7.2: *Results of the EM algorithm for data in (7.21).*

	$\theta_1^{(r)}$	$\theta_{1\|0}^{(1)}$	$\theta_{1\|1}^{(1)}$	$\theta_{1\|0}^{(2)}$	$\theta_{1\|1}^{(2)}$	$\theta_{1\|0}^{(3)}$	$\theta_{1\|1}^{(3)}$
1	0.466	0.337	0.552	1.000	0.000	0.416	0.074
2	0.534	0.552	0.337	0.000	1.000	0.074	0.416
3	0.257	0.361	0.658	0.420	0.865	0.000	1.000
4	0.743	0.658	0.361	0.865	0.420	1.000	0.000
5	0.437	0.000	1.000	0.629	0.412	0.156	0.386
6	0.563	1.000	0.000	0.412	0.629	0.386	0.156

Up to label switching on the inner vertex, these are three distinct maximizers of the log-likelihood function $\ell(\theta; \hat{p})$ corresponding to rows $2, 4, 6$. The value of the log-likelihood function is equal to $-18{,}281$ and $-18{,}387$ and $-18{,}881$, respectively so the first point seems to be the global maximizer. All points correspond to somewhat degenerate tripod tree models where one of the observed variables is functionally related to the hidden variable. For example, the first point lies on the submodel given by $X_1 \perp\!\!\!\perp X_3 | X_2$. In conjunction with Corollary 6.24 we also can theoretically construct data for which the likelihood function $\ell(\theta; \hat{p})$ is maximized over an infinite number of points. This, for example, holds for any data such that the constrained multinomial likelihood is maximized over a point such that $p_{0ij} = \lambda p_{1ij}$ for some λ and each $i, j = 0, 1$. In this case $\rho_{12} = \rho_{13} = 0$ and the MLEs form a set of a positive dimension by Corollary 6.24.

Finally, in situation (ii), even if we are able to identify the global maximum, often we will encounter a further complication, where another distant parameter gives a similar likelihood value. *In this case the maximum likelihood estimate is meaningless.*

7.5   Bibliographical notes

Various earlier papers listed some necessary constraints on general Markov models in the binary case; see Pearl and Tarsi [1986] and Steel and Faller [2009]. Under some special symmetric assumptions and in the case when $k = 4$, Matsen [2009] gave a set of some necessary inequalities using the Fourier transformation of the raw probabilities. The semialgebraic description we obtain here also has an elegant mathematical structure. The basic idea was given by Cavender [1997] who linked the correlation system on tree models to tree metrics. The hyperdeterminant was first applied to the analysis of the semialgebraic structure of tensors of positive rank in de Silva and Lim [2008]. A more detailed discussion of the cumulant representation of this polynomial function is given in Sturmfels and Zwiernik [2012]. Log-supermodularity constraints are interesting for its own sake. In statistics they are typically referred to as MTP_2 (multivariate positivity of order two); see, for example, Bartolucci and Forcina [2000] and references therein. The two-state Neyman model is studied from an algebraic point of view by Tuffley and Steel [1997]. The discussion of the naive Bayes model is taken from Allman et al. [2015]. A different perspective on the study of mixture models is given in Montúfar [2013]. The tree metric structure given in Remark 7.20 was independently discovered by Chang and Hartigan [1991], Steel [1994] and then further studied by Lake [1994], Lockhart et al. [1994].

In this book we focus on the general Markov model because we are interested in applications outside of phylogenetics, where specific submodels considered in biology may not make sense. In particular, we omitted a whole series of beautiful results on the geometry of group-based models. For more on this topic, see Draisma and Kuttler [2009], Eriksson et al. [2005], Evans and Speed [1993], Matsen [2009], Sturmfels and Sullivant [2005]. In the end we note that very little is known about models $M(\mathcal{T})$ where the state spaces differ across vertices. Some basic results are provided by Zhang [2003/04].

Chapter 8

Gaussian latent tree models

[]

In this chapter we extent our analysis of discrete models on trees to latent Gaussian tree models. It turns out that this model has all the basic features of other models introduced earlier.

8.1 Gaussian models

Denote by PD_m the set of symmetric $m \times m$ matrices that are positive definite. We say that a random vector Y with values y in $\mathbb{R}^m$ has *multivariate Gaussian distribution* with the mean parameter $\mu \in \mathbb{R}^m$ and the *covariance matrix* $\Sigma \in \mathrm{PD}_m$, which we denote by $Y \sim N(\mu, \Sigma)$, if it has density with respect to the Lebesgue measure on $\mathbb{R}^m$ of the form:

$$f_{\mu,\Sigma}(y) = \frac{1}{(2\pi)^{m/2}}(\det \Sigma)^{-1/2} \exp(-\frac{1}{2}(y - \mu)^T \Sigma^{-1}(y - \mu)).$$

The inverse of the covariance matrix Σ is denoted by K and called the *concentration matrix*. In this section, for simplicity, we assume that $\mu = \mathbf{0}$. In this case a Gaussian model can be uniquely identified with a subset of PD_m. By the following proposition, conditional independence statements for Gaussian models translate directly to minor constraints on the covariance matrix.

Proposition 8.1 (Proposition 3.1.13, Drton et al. [2009]). *The conditional independence statement $Y_A \perp\!\!\!\perp Y_B | Y_C$ holds for a multivariate Gaussian random vector $Y \sim N(\mu, \Sigma)$ if and only if the submatrix $\Sigma_{A \cup C, B \cup C}$ has rank at most $|C|$.*

Given n independent observations $Y^1, \ldots, Y^n$ of a Gaussian vector $Y \sim N(\mathbf{0}, \Sigma)$, we construct the *sample covariance matrix* $S = \frac{1}{n}\sum_{i=1}^{n} Y^i Y^{iT}$, which by construction is always positive semidefinite and it is positive definite with probability 1 as long as $n \geq m$, which we always assume here. The log-likelihood function in the Gaussian case is a function on PD_m defined in terms of concentration matrices by

$$\ell(K; S) = \frac{n}{2}\log \det K - \frac{n}{2}\mathrm{Tr}(SK). \tag{8.1}$$

Proposition 8.2. *The Gaussian likelihood defined in (8.1) is strictly concave over the whole cone PD_m. Its unique maximum, if it exists, is given by $K^* = S^{-1}$ and thus $\Sigma^* = S$.*

Proof. Because $\text{tr}(SK)$ is a linear function of K, the first part of the above result follows by showing that $\log \det K$ is strictly concave over PD_m. This result is well known (see, e.g., [Boyd and Vandenberghe, 2004, Section 3.1.5]). The second statement follows directly by evaluating the derivatives at zero. $\qquad\square$

8.2 Gaussian tree models and Chow–Liu algorithm

Let $Y = (Y_v)_{v \in V}$ be a random vector whose components are indexed by the vertices of an undirected tree $T = (V, E)$. Denote by $\mathbf{N}(T)$ the *Gaussian tree model* for Y induced by T. The model $\mathbf{N}(T)$ is the collection of all multivariate Gaussian distributions on $\mathbb{R}^{|V|}$ under which Y_i and Y_j are conditionally independent given a subvector Y_C whenever the set $C \subset V \setminus \{i, j\}$ contains a vertex on $\overline{ij}$. Suppose that $k \in \overline{ij}$. Then $Y_i \perp\!\!\!\perp Y_j | Y_k$ and by Proposition 8.1 $\sigma_{ij}\sigma_{kk} = \sigma_{ik}\sigma_{jk}$. Equivalently

$$\rho_{ij} \quad = \quad \rho_{ik}\rho_{jk},$$

where $\rho_{ij} := \sigma_{ij}/\sqrt{\sigma_{ii}\sigma_{jj}}$ is the *correlation* between X_i and X_j. Using this argument recursively implies that a normal distribution belongs to $\mathbf{N}(T)$ if and only if

$$\rho_{ij} \quad = \quad \prod_{e \in \overline{ij}} \rho_e, \qquad (8.2)$$

where $\rho_e := \rho_{kl}$ for $e = k - l \in E$.

The model $\mathbf{N}(T)$ is equivalently given by linear restrictions on K:

$$K_{ij} = 0 \quad \text{whenever } i \neq j \text{ and there is no edge between } i \text{ and } j \text{ in } T.$$

Denote by $\mathbf{N}^{-1}(T)$ the space of all concentration matrices corresponding to covariance matrices in $\mathbf{N}(T)$. Let E_{ij} be an $|V| \times |V|$ matrix with 1 on the (i, j)-th entry and zeros otherwise. Every concentration matrix in $\mathbf{N}^{-1}(T)$ is of the form

$$K_\theta \quad = \quad \sum_{i \in V} \theta_{ii} E_{ii} + \sum_{i-j \in E} \theta_{ij}(E_{ij} + E_{ji}),$$

where $\theta = (\theta_{ii}, \theta_{ij})$ are real parameters such that K_θ is positive definite. We are going to use this parametric representation in the analysis of the likelihood function.

The fact that the set of concentration matrices in $\mathbf{N}(T)$ forms a linear subset of $\text{PD}_{|V|}$ implies that the likelihood function

$$\ell(\theta) \quad = \quad \log \det K_\theta - \text{tr}(SK_\theta) \qquad (8.3)$$

is a strictly concave function. This follows from Proposition 8.2 and the fact that a strictly concave function constrained to a linear subspace remains strictly concave.

Proposition 8.3. *The maximum likelihood estimate over the model* $\mathbf{N}(T)$ *is given by the unique matrix* $\widehat{K} := K_{\hat{\theta}}$ *such that*

$$(\widehat{K}^{-1})_{ii} = S_{ii}, \quad (\widehat{K}^{-1})_{ij} = S_{ij} \quad for\ all \quad i \in V,\ i-j \in E.$$

Proof. Denote $\partial_{ij} := \frac{\partial}{\partial \theta_{ij}}$, then

$$\partial_{ii}\ell(\theta) = \frac{n}{2}\mathrm{tr}(K_\theta^{-1}E_{ii}) - \frac{n}{2}\mathrm{tr}(SE_{ii}), \tag{8.4}$$

$$\partial_{ij}\ell(\theta) = \frac{n}{2}\mathrm{tr}(K_\theta^{-1}(E_{ij}+E_{ji})) - \frac{n}{2}\mathrm{tr}(S(E_{ij}+E_{ji})), \tag{8.5}$$

which vanishes only if $(K_\theta^{-1})_{ii} = S_{ii}$ and $(K_\theta^{-1})_{ij} = S_{ij}$. $\qquad\square$

Corollary 8.4. *If* $\widehat{K}$ *is a critical point of the likelihood function over* $\mathbf{N}(T)$, *then* $\mathrm{tr}(S\widehat{K}) = |V|$.

Proof. If the partial derivatives in (8.4) vanish, then in particular

$$\sum_{i \in V} \hat{\theta}_{ii}\partial_{ii}\ell(\theta) + \sum_{i-j \in E} \hat{\theta}_{ij}\partial_{ij}\ell(\theta) = 0.$$

Using linearity of the trace, we conclude that $\mathrm{tr}(SK_{\hat{\theta}}) = \mathrm{tr}(K_{\hat{\theta}}^{-1}K_{\hat{\theta}})$ and the latter is equal to $|V|$. $\qquad\square$

The main aim of the remainder of this section is to show that the Chow–Liu algorithm presented in Section 5.2.3 for discrete variables has a straightforward generalization to Gaussian tree models. Suppose that we observe sample covariance matrix $S \in \mathrm{PD}_{|V|}$ and we want to find the tree T with vertices in V that yields the largest value of the maximum likelihood function over $\mathbf{N}(T)$. In the corresponding maximum of (8.3), by Corollary 8.4, the trace term does not depend on T and hence to find the best tree it suffices to pick the one that yields the largest value of $\log \det \widehat{K}$.

Lemma 8.5. *For every* $\mathbf{N}(T)$ *we have*

$$\log \det \widehat{K} = -\sum_{i \in V} \log S_{ii} - \sum_{i-j \in E} \log(1 - \hat{\rho}_{ij}^2),$$

where $\hat{\rho}_{ij} := S_{ij}/\sqrt{S_{ii}S_{jj}}$ *are the sample correlations.*

Proof. By [Lauritzen, 1996, Proposition 5.9],

$$\det \widehat{K} = n^{|V|}\frac{\prod_{i \in V}(nS_{ii})^{\deg(i)-1}}{\prod_{i-j \in E}\det(nS_{ij,ij})} = \frac{\prod_{i=1}^m S_{ii}^{\deg(i)-1}}{\prod_{i-j \in E}\det(S_{ij,ij})}.$$

Because $\det(S_{ij,ij}) = S_{ii}S_{jj}(1 - \hat{\rho}_{ij}^2)$, we have

$$\det \widehat{K} = \frac{1}{\prod_{i \in V} S_{ii}}\frac{1}{\prod_{i-j \in E}(1 - \hat{\rho}_{ij}^2)}$$

and the lemma follows by taking the logarithm on both sides. $\qquad\square$

Since the first term in the expression for $\log \det \widehat{K}$ in Lemma 8.5 does not depend on T, the tree that yields maximal value of $\log \det \widehat{K}$ is given as the minimum-cost spanning tree (see Section 5.2.3) for a complete graph on m vertices with weights given by $\log(1 - \hat{\rho}_{ij}^2)$. To obtain such a tree, we order all pairs (i, j) according to the increasing value of $\log(1 - \hat{\rho}_{ij}^2)$ and subsequently join vertices by edges as long as no cycles are introduced. Alternatively, we could order pairs (i, j) according to the decreasing value of $\hat{\rho}_{ij}^2$.

8.3   Gaussian latent tree models

In latent Gaussian tree models, only the components of Y corresponding to the leaves of T are observed. For a phylogenetic tree $\mathcal{T} = (T, \phi)$, we are thus interested in the *Gaussian latent tree model* $\boldsymbol{M}(\mathcal{T})$, which is the marginal model for $X := (Y_v)_{v \in \phi([m])}$ induced from $\boldsymbol{N}(T)$. The correlations in this model are parameterized by (8.2) with parameters ω, given by all edge correlations. This already describes all distributions in the model because the variances of observed variables are not constrained and in the estimation process they can be always fixed to their sample values.

Define a multiplicative group $Z_2^m = \{-1, 1\}^m$ that acts on the sample space $\mathbb{R}^m$ by reflections across axes. A typical element of Z_2^m is $\epsilon = (\epsilon_1, \ldots, \epsilon_m)$, where $\epsilon_i \in \{-1, 1\}$ and it acts on $\mathbb{R}^m$ by $x \mapsto \epsilon \cdot x = (\epsilon_i x_i)$. The action of Z_2^m on the sample space induces the action on the observed correlations

$$R \mapsto \epsilon \cdot R = R' = [\rho'_{ij}], \qquad \text{where } \rho'_{ij} = \epsilon_i \epsilon_j \rho_{ij}.$$

It also induces the action on the parameter space. We have $\epsilon \cdot \omega = \omega'$, where ω' satisfies $R(\epsilon \cdot \omega) = R(\omega')$. More precisely, denote by e_i the terminal edge connected to the leave i. Then $\epsilon \cdot \omega =: \omega'$ is defined so that $\omega'_e = \omega_e$ for all inner edges e, and $\omega'_{e_i} = \epsilon_i \omega_{e_i}$ for the corresponding terminal edge. We conclude the following result.

Proposition 8.6. *For every $\mathcal{T}$ the Gaussian model $\boldsymbol{M}(\mathcal{T})$ is invariant under the action of Z_2^m, that is, $Z_2^m \cdot \boldsymbol{M}(\mathcal{T}) = \boldsymbol{M}(\mathcal{T})$. Moreover, the element $(-1, \ldots, -1)$ acts trivially on $\boldsymbol{M}(\mathcal{T})$.*

We distinguish the nonnegative part $\boldsymbol{M}_+(\mathcal{T})$ of $\boldsymbol{M}(\mathcal{T})$ with correlation matrices parameterized by (8.2) where $\rho_e \geq 0$. Both from the geometric and from the inferential point of view it is enough to understand the model $\boldsymbol{M}_+(\mathcal{T})$, which follows from the proposition below.

Proposition 8.7. *Let $R = [\rho_{ij}] \in \mathrm{PD}_m$ be a correlation matrix. Then $R \in \boldsymbol{M}(\mathcal{T})$ if and only if*

(a) $R' = [|\rho_{ij}|] \in \boldsymbol{M}_+(\mathcal{T})$

(b) $\rho_{ij}\rho_{ik}\rho_{jk} \geq 0$ *for any* $1 \leq i < j < k \leq m$.

Proof. For the "if" part, if $R \in \boldsymbol{M}(\mathcal{T})$, then each ρ_{ij} has representation (8.2). Thus $|\rho_{ij}| = \prod_{e \in \overline{ij}} |\rho_e|$ and hence R' lies in $\boldsymbol{M}_+(\mathcal{T})$. To show that

(b) holds, note that the tree spanned over three leaves i, j, k has necessary a unique vertex v that lies on the intersection of paths $\overline{ij}, \overline{ik}$ and $\overline{jk}$. Moreover,

$$\rho_{ij}\rho_{ik}\rho_{jk} = \prod_{e \in \overline{ij}} \rho_e \prod_{e \in \overline{ik}} \rho_e \prod_{e \in \overline{jk}} \rho_e = \prod_{e \in \overline{iv}} \rho_e^2 \prod_{e \in \overline{jv}} \rho_e^2 \prod_{e \in \overline{kv}} \rho_e^2 \geq 0.$$

For the "only if" part we use the action of Z_2^m. Let $\epsilon \in Z_2^m$ be such that for $i = 1, \ldots, m$, $\epsilon_i = -1$ if $\rho_{1i} < 0$ and $\epsilon_i = 1$ otherwise. Then $R = \epsilon \cdot R'$ because: $\epsilon_1\epsilon_i|\rho_{1i}| = \rho_{1i}$ for all $i = 2, \ldots, m$ and $\epsilon_i\epsilon_j|\rho_{ij}| = \rho_{ij}$ for $i, j > 1$. This last equality follows from the fact that by (b) the sign of $\rho_{1i}\rho_{1j}$ is equal to the sign of ρ_{ij}. Now, since $R' \in \boldsymbol{M}_+(\mathcal{T})$ and $R = \epsilon \cdot R'$, Proposition 8.6 implies that $R \in \boldsymbol{M}(\mathcal{T})$.   □

The complete description of $\boldsymbol{M}_+(\mathcal{T})$ follows from Proposition 5.27, which gives the description of the space of phylogenetic oranges, and the following result, which establishes the link between $\boldsymbol{M}_+(\mathcal{T})$ and phylogenetic oranges.

Theorem 8.8. *Let $\mathcal{T}$ be a phylogenetic tree. The set of all possible correlations in $\boldsymbol{M}_+(\mathcal{T})$ is equal to the model $\boldsymbol{JN}(\mathcal{T})$ of Section 7.3.2, which in turn is equal to the space of phylogenetic oranges; see Section 5.1.4.*

The boundary points of $\boldsymbol{M}(\mathcal{T})$ are given by some of the inequalities in Proposition 5.27 becoming equations. The analysis of the boundary is, however, more complex because of complicated relations between these inequalities (many become equalities simultaneously). An efficient boundary description of $\boldsymbol{M}_+(\mathcal{T})$ is obtained in Section 5.1.4 in terms of the *Tuffley poset*. For example, if $\mathcal{T}$ is a star graph with m leaves, then the maximal dimensional pieces of the boundary are represented by semi-labeled forests $\mathcal{T}_i$ with $m - 1$ leaves labeled by $[m] \setminus \{i\}$ and the inner vertex labeled by $\{i\}$. These models correspond to submodels where one of the edge correlations is equal to 1. The underlying model is a fully observed graphical model, whose maximum likelihood estimators σ_{ij}^* are well understood. For example, on $\boldsymbol{M}(\mathcal{T}_1)$ we have

$$\sigma_{ij}^* = \hat{\sigma}_{1i}\hat{\sigma}_{1j} \quad \text{and} \quad \sigma_{1i}^* = \hat{\sigma}_{1i} \quad \text{for } i, j > 1, \tag{8.6}$$

where $\hat{\sigma}_{ij}$ are the elements of the *sample covariance matrix*. This shows that for star graph trees, maximizing the log-likelihood function is easy on the boundary and potentially hard in the relative interior of the model.

8.4   The tripod tree

In this section we provide a thorough analysis of the simplest non-trivial Gaussian latent tree model given by the tripod tree. We denote the three observed variables by X_1, X_2, X_3 and the hidden variable by X_0. Denote by σ_{ij} for $0 \leq i \leq j \leq 3$ the covariance between X_i and X_j and by ρ_{ij} the corresponding correlation. We denote the corresponding tripod tree model by M.

8.4.1   Shape of the model

By (8.2), $\rho_{ij} = \rho_{0i}\rho_{0j}$ for all distributions in M. Because $\rho_{0i} \in [-1, 1]$ we will be interested in a subset of the cube $[-1, 1]^3$ given as the image of

$$f : [-1, 1]^3 =: \Theta \mapsto C := [-1, 1]^3,$$
$$(\rho_{01}, \rho_{02}, \rho_{03}) \mapsto (\rho_{01}\rho_{02}, \rho_{01}\rho_{03}, \rho_{02}\rho_{03})$$

which we denote by U and often identify with the model M.

Although both the domain and the codomain of f are given by the cube $[-1, 1]^3$, it is important to distinguish them, which explains our notation Θ, C. The coordinates on Θ are $\rho_{01}, \rho_{02}, \rho_{03}$ and the coordinates on C are $\rho_{12}, \rho_{13}, \rho_{23}$. We identify the cube C with the space of symmetric 3×3 matrices with 1's on the diagonal

$$(\rho_{12}, \rho_{13}, \rho_{23}) \equiv \begin{bmatrix} 1 & \rho_{12} & \rho_{13} \\ \rho_{12} & 1 & \rho_{23} \\ \rho_{13} & \rho_{23} & 1 \end{bmatrix}$$

and we often switch between these two representations. Such a matrix is a correlation matrix if and only if

$$1 - \rho_{12}^2 - \rho_{13}^2 - \rho_{23}^2 + 2\rho_{12}\rho_{13}\rho_{23} > 0.$$

The proof of the following result is straightforward.

Proposition 8.9. *A matrix in the set U is positive definite if and only if it lies in the image of the cube Θ with all faces of dimension ≤ 1 removed.*

The symmetry group of the 3-cube, denoted by G, is the semidirect product of the symmetric group S_3, and the abelian group $Z_2^3 = \{-1, 1\}^3$. The action of Z_2^3 was given in the previous section and the action of S_3 is by permuting indices. The group G plays an important role in our understanding of the model M.

Proposition 8.10. *The set of correlation matrices in the model M is given by all correlation matrices satisfying*

$$(\rho_{23} - \rho_{12}\rho_{13})(\rho_{13} - \rho_{12}\rho_{23})(\rho_{12} - \rho_{13}\rho_{23}) \geq 0. \tag{8.7}$$

Proof. Denote by W the subset of C described by (8.7). We are going to show that $W = U$. This statement will imply our claim by constraining both sets to PD$_3$. The fact that $U \subseteq W$ follows by replacing ρ_{ij} with $\rho_{0i}\rho_{0j}$. For the opposite inclusion first expand (8.7) to obtain

$$\rho_{12}\rho_{13}\rho_{23}(1 + \rho_{12}^2 + \rho_{13}^2 + \rho_{23}^2) - (\rho_{12}^2\rho_{13}^2 + \rho_{12}^2\rho_{23}^2 + \rho_{13}^2\rho_{23}^2 + \rho_{12}^2\rho_{13}^2\rho_{23}^2) \geq 0.$$

Since the second term is a sum of squares, we conclude that in W necessarily $\rho_{12}\rho_{13}\rho_{23} \geq 0$. Now define

$$W^+ = \{(\rho_{12}, \rho_{13}, \rho_{23}) \in W : 0 \leq \rho_{12} \leq \rho_{13} \leq \rho_{23}\}.$$

We claim that $G \cdot W^+ = W$. The first inclusion $G \cdot W^+ \subseteq W$ follows from $W^+ \subseteq W$ and the fact that the inequality (8.7) is G-invariant. For the opposite inclusion, let $(\rho_{12}, \rho_{13}, \rho_{23}) \in W$ and define

$$\epsilon = (\epsilon_1, \epsilon_2, \epsilon_3) := (1, \operatorname{sgn}(\rho_{12}), \operatorname{sgn}(\rho_{13})),$$

where $\operatorname{sgn}(\rho_{ij}) = -1$ if $\rho_{ij} < 0$ and it is 1 otherwise. We have

$$\epsilon \cdot (\rho_{12}, \rho_{13}, \rho_{23}) = (\operatorname{sgn}(\rho_{12})\rho_{12}, \operatorname{sgn}(\rho_{13})\rho_{13}, \operatorname{sgn}(\rho_{12})\operatorname{sgn}(\rho_{13})\rho_{23}).$$

Since $\rho_{12}\rho_{13}\rho_{23} \geq 0$ in W, we have $\operatorname{sgn}(\rho_{12})\operatorname{sgn}(\rho_{13}) = \operatorname{sgn}(\rho_{23})$ and hence $\epsilon \cdot (\rho_{12}, \rho_{13}, \rho_{23})$ has only nonnegative entries. We can now easily find a permutation $\pi \in G$ such that $\pi \cdot (\epsilon \cdot (\rho_{12}, \rho_{13}, \rho_{23}))$ lies in W^+. This proves that $G \cdot W^+ = W$. Now, to show $W \subseteq U$ it is enough to show that all points in W^+ lie in U. All these points satisfy $\rho_{23} - \rho_{12}\rho_{13} \geq 0$ and $\rho_{13} - \rho_{12}\rho_{23} \geq 0$. Thus, by (8.7) also $\rho_{12} - \rho_{13}\rho_{23} \geq 0$. Now the claim follows by defining

$$\rho_{01} := \sqrt{\frac{\rho_{12}\rho_{13}}{\rho_{23}}}, \quad \rho_{02} := \sqrt{\frac{\rho_{12}\rho_{23}}{\rho_{13}}}, \quad \rho_{03} := \sqrt{\frac{\rho_{13}\rho_{23}}{\rho_{12}}} \tag{8.8}$$

and attaching appropriate signs.   □

The set defined by (8.7) is depicted in Figure 2.5 and it is strictly contained in the closure of PD_3.

It is convenient to think about the set U as a union of four isomorphic blobs. The *positive* blob U^+ is the image of $[0,1]^3 \to [0,1]^3$ under the map f, where now this map is one-to-one. Elementary analysis shows that the volume of U^+ is $1/4$ and therefore the volume of U takes $1/8$ of the volume of the ambient cube C, and $2/\pi^2 \approx 0.202$ of the volume of the set of all 3×3 correlation matrices. In particular, a "typical" Gaussian model on three variables will *not* lie in U even though the model is full dimensional, which we discuss in more detail in Section 8.4.2.

The usual way of providing constraints for M is by providing the description of U^+ and taking advantage of the action of Z_2^3. Our approach of dealing with the whole model enables the following compact reformulation in terms of concentration matrices.

Proposition 8.11. *The set of all concentration matrices in the model M is given by all symmetric 3×3 positive definite matrices K satisfying $k_{12}k_{13}k_{23} \leq 0$.*

Proof. Multiplying both sides of (8.7) by $(\sigma_{11}\sigma_{22}\sigma_{33})^2$ we obtain that the set of all covariance matrices in M is given by

$$(\sigma_{11}\sigma_{23} - \sigma_{12}\sigma_{13})(\sigma_{22}\sigma_{13} - \sigma_{12}\sigma_{23})(\sigma_{33}\sigma_{12} - \sigma_{13}\sigma_{23}) \geq 0. \tag{8.9}$$

Let $\Sigma = [\sigma_{ij}]$. Direct computations show that

$$k_{12} = \frac{\sigma_{33}\sigma_{12} - \sigma_{13}\sigma_{23}}{\det(\Sigma)}, \quad k_{13} = \frac{\sigma_{22}\sigma_{13} - \sigma_{12}\sigma_{23}}{\det(\Sigma)}, \quad k_{23} = \frac{\sigma_{11}\sigma_{23} - \sigma_{12}\sigma_{13}}{\det(\Sigma)}.$$

Thus dividing both sides of (8.9) by $\det(\Sigma)^3$, which is strictly positive, gives $k_{12}k_{13}k_{23} \leq 0$. $\square$

Proposition 8.11 can be rephrased by saying that the concentration matrices in the model M are all positive definite matrices that lie in one of the four orthants that satisfy $k_{12}k_{13}k_{23} \leq 0$.

8.4.2   The maximum likelihood

By Proposition 8.11, the model M is given as a union of four orthants in the space of concentration matrices. Over each of these orthants, maximizing the likelihood function (8.1) is a convex optimization problem, which becomes trivial if S^{-1}, the inverse of the sample covariance matrix, lies in a given orthant. This suggests the following procedure of finding the global maximum of $\ell(K; S)$ over M: First, check if the global unconstrained maximum given by the sample covariance matrix S lies in M. If yes, then S is the maximum likelihood estimate for M and the parameters can be recovered using (8.8). Otherwise, compute the maximum likelihood estimators for each of the three components of the boundary. Then the maximum over M is one of the three reported maxima.

Each piece of the boundary corresponds to a simple conditional independence model. For example, $k_{12} = 0$ correspond to the conditional independence statement $X_1 \perp\!\!\!\perp X_2 | X_3$. These are special undirected graphical models represented by chain graphs of the form $\bullet - \bullet - \bullet$. To maximize the likelihood over the boundary of M we use the fact that the three conditional independence models are exactly all possible trees over three vertices $\{1, 2, 3\}$. Thus, to find the maximum over the union of these three models we can use the Chow–Liu algorithm. To this end, we compute three sample correlations $\hat{\rho}_{12}$, $\hat{\rho}_{13}$, $\hat{\rho}_{23}$. Then we take a complete graph on $\{1, 2, 3\}$ and we give weight $\hat{\rho}_{ij}^2$ to the edge joining vertices i and j. Now the boundary with the maximum likelihood estimate is the one that corresponds to the *maximal* weight spanning tree. For example, if $\hat{\rho}_{12}^2 \leq \min\{\hat{\rho}_{13}^2, \hat{\rho}_{23}^2\}$, the maximal spanning tree is $\overset{1}{\bullet} - \overset{3}{\bullet} - \overset{2}{\bullet}$ and the maximum likelihood estimate of the correlation matrix is

$$\rho_{12}^* = \hat{\rho}_{13}\hat{\rho}_{23}, \quad \rho_{13}^* = \hat{\rho}_{13}, \quad \rho_{23}^* = \hat{\rho}_{23}. \tag{8.10}$$

For any permutation (i, j, k) of $\{1, 2, 3\}$, define the subset U_{ijk} of PD_3

$$U_{ijk} := \{(\rho_{12}, \rho_{13}, \rho_{23}) \in \mathrm{PD}_3 : \rho_{ij}^2 \leq \rho_{ik}^2 \leq \rho_{jk}^2\},$$

so for example, U_{321} is given by $\rho_{23}^2 \leq \rho_{13}^2 \leq \rho_{12}^2$. Each of these sets is a *fundamental domain* for the action of S_3 on PD_3. The example in (8.10) gives a straightforward proof of the following result.

Lemma 8.12. *If $(\hat{\rho}_{12}, \hat{\rho}_{13}, \hat{\rho}_{23})$ lies in U_{ijk}, then the maximum likelihood estimate also lies in U_{ijk}.*

Define further sets V_ϵ for $\epsilon = (\epsilon_{12}, \epsilon_{13}, \epsilon_{23}) \in \{-1, 1\}^3$ that contain all points such that the sign of ρ_{ij} is ϵ_{ij}. Every subset V_ϵ is a fundamental domain of the action of Z_2^3 on PD$_3$. Moreover, each set of the form $U_{ijk} \cap V_\epsilon$ is a fundamental domain of the G action.

Proposition 8.13. *Suppose that $(\hat{\rho}_{12}, \hat{\rho}_{13}, \hat{\rho}_{23}) \in U_{ijk} \cap V_\epsilon$. If $\epsilon_{12}\epsilon_{13}\epsilon_{23} = 1$, then the maximum likelihood estimate also lies in $U_{ijk} \cap V_\epsilon$. If $\epsilon_{12}\epsilon_{13}\epsilon_{23} = -1$, then the maximum likelihood estimate lies in $U_{ijk} \cap V_{\epsilon'}$, where $\epsilon'_{ij} = -\epsilon_{ij}$, $\epsilon'_{ik} = \epsilon_{ik}$, and $\epsilon'_{jk} = \epsilon_{jk}$.*

Proof. The fact that the maximum likelihood estimate lies always in U_{ijk} follows from Lemma 8.12. If $\hat{\rho}_{ij}^2 \leq \hat{\rho}_{ik}^2 \leq \hat{\rho}_{jk}^2$, then $\rho_{ik}^* = \hat{\rho}_{ik}$, $\rho_{jk}^* = \hat{\rho}_{jk}$ and $\rho_{ij}^* = \hat{\rho}_{ik}\hat{\rho}_{jk}$. If $\epsilon_{12}\epsilon_{13}\epsilon_{23} = 1$, then the sign of ρ_{ij}^* must be the same as the sign of $\hat{\rho}_{ij}$ and hence the maximum likelihood estimate must lie in V_ϵ. Otherwise, the sign of ρ_{ij}^* is the opposite of the sign of $\hat{\rho}_{ij}$ and the maximum likelihood estimate lies in $V_{\epsilon'}$. $\qquad\square$

Corollary 8.14. *Suppose that $(\hat{\rho}_{12}, \hat{\rho}_{13}, \hat{\rho}_{23}) \in U_{ijk}$. Then the global maximum of the likelihood function over the model M can be found by maximizing the likelihood function over a single orthant (in the space of concentration matrices) given by $\operatorname{sgn}(k_{ik}) = \operatorname{sgn}(\hat{\rho}_{ik}) =: \epsilon_{ik}$, $\operatorname{sgn}(k_{jk}) = \operatorname{sgn}(\hat{\rho}_{jk}) =: \epsilon_{jk}$, and $\operatorname{sgn}(k_{ij}) = -\epsilon_{ik}\epsilon_{jk}$.*

Proof. Suppose that $(\hat{\rho}_{12}, \hat{\rho}_{13}, \hat{\rho}_{23}) \in U_{123}$. By Lemma 8.12, $(\rho_{12}^*, \rho_{13}^*, \rho_{23}^*)$ also lies in U_{123}. In this case, because $\rho_{12}^{*\,2} \leq \min\{\rho_{13}^{*\,2}, \rho_{23}^{*\,2}\}$,

$$\operatorname{sgn}(k_{23}^*) = \operatorname{sgn}(\rho_{12}^*\rho_{13}^* - \rho_{23}^*) = -\operatorname{sgn}(\hat{\rho}_{23})$$
$$\operatorname{sgn}(k_{13}^*) = \operatorname{sgn}(\rho_{12}^*\rho_{23}^* - \rho_{13}^*) = -\operatorname{sgn}(\hat{\rho}_{13}).$$

Because the maximum likelihood estimate lies in the model, it necessarily satisfies $k_{12}^* k_{13}^* k_{23}^* \leq 0$, which already determines the sign of k_{12}^*. For any other U_{ijk} the proof is the same up to symmetry. $\qquad\square$

8.4.3   Boundary problem

The relative volume of the model in the ambient cube C is $2/\pi^2 \approx 0.202$, which suggests that the sample covariance may typically not fall into the model and in consequence the maximum likelihood estimate will lie on the boundary of the parameter space. The problem with this is that often in applications the hidden variable has some interpretation and for the boundary points this interpretation breaks down.

We present simulation results to illustrate how big this problem is. If Σ is the covariance matrix of the true data generating distribution, then the statistic nS has *Wishart distribution* $\mathcal{W}_3(n, \Sigma)$. As our simulations show, the probability that S lies in the model is typically small for reasonable sample sizes even if the true data generating distribution lies in the model. So suppose that the true correlation matrix comes from parameters $\rho_{01} = 0.1$, $\rho_{02} = 0.8$, $\rho_{03} = 0.9$. The corresponding probabilities are

n	10	20	50	100	1000
$\mathbb{P}(S \in M)$	0.24	0.26	0.28	0.33	0.68

For more centered parameter values, the probabilities are higher but still small for reasonable sample sizes. For example, if $\rho_{01} = 0.3$, $\rho_{02} = 0.6$, $\rho_{03} = 0.7$ we obtain the following estimates.

n	10	20	50	100	1000
$\mathbb{P}(S \in M)$	0.36	0.47	0.64	0.78	1

The points in the model that correspond to high values of the parameters also lie closer to the boundary of the cone of positive definite matrices. Therefore, on average the corresponding Wishart distribution should cover the model better. So, for example, if $\rho_{01} = 0.8$, $\rho_{02} = 0.8$, $\rho_{03} = 0.9$, we obtain the following estimates.

n	10	20	50	100	1000
$\mathbb{P}(S \in M)$	0.65	0.84	0.96	0.99	1

This suggests that the model M should be used with great care unless correlations between variables are relatively high.

8.4.4 Different aspects of non-uniqueness

We conclude this chapter by discussing briefly three types of non-uniqueness that can arise in the maximum likelihood estimation for the model M. Proposition 8.13 enables us, for example, to easily understand how different data sets may give the same maximum likelihood estimates. For example, every data set that gives $\hat{\rho}_{13} = 0.5$, $\hat{\rho}_{23} = 0.6$ will give the same maximum likelihood estimator for every value of $\hat{\rho}_{12}$ within the region $(\frac{3}{10} - \frac{2\sqrt{3}}{5}, \frac{1}{2})$, where $\frac{3}{10} - \frac{2\sqrt{3}}{5} \approx -0.39$. The maximum likelihood estimator is given by $\rho_{12}^* = 0.3$, $\rho_{13}^* = 0.5$, and $\rho_{23}^* = 0.6$.

Another type of non-uniqueness is related to the problem of identifiability of this model. For a generic point in the model, we can easily recover the corresponding parameters up to sign using (8.8). However, if any of the correlations vanishes, these formulas cannot be used. For example, if $\rho_{12} = 0$, then necessarily either $\rho_{01} = 0$ or $\rho_{02} = 0$. If $\rho_{01} = 0$, then also $\rho_{13} = 0$ and hence this point corresponds to a distribution in which X_1 is marginally independent from X_2 and X_3. Now there are infinitely many combinations of parameters ρ_{02}, ρ_{03} that give $\rho_{02}\rho_{03} = \rho_{12}$. Thus, if any of the sample correlations is exactly zero, then we are guaranteed that the likelihood is going to be maximized over a whole ridge in the parameter space.

Finally, suppose that $\hat{\rho}_{12} = \hat{\rho}_{13} < \hat{\rho}_{23}$. In that case, there are exactly two points maximizing the likelihood. They are given by

$$(\hat{\rho}_{12}\hat{\rho}_{23}, \hat{\rho}_{12}, \hat{\rho}_{23}) \quad \text{and} \quad (\hat{\rho}_{12}, \hat{\rho}_{12}\hat{\rho}_{23}, \hat{\rho}_{23}).$$

For example, if $\hat{\rho}_{12} = \hat{\rho}_{13} = 0.5$, $\hat{\rho}_{23} = 0.8$, then $(0.4, 0.5, 0.8)$ and $(0.5, 0.4, 0.8)$ give the same likelihood value. Similarly, if all sample correlations are equal, we have three maximizers of the likelihood function. Like in the previous paragraph, all these cases happen with zero probability. However, if the data lie close to these extreme cases, this will be reflected in the geometry of the likelihood function.

8.5   Bibliographical notes

Gaussian distributions and Gaussian graphical models are discussed in great detail by Lauritzen [1996]. The Chow–Liu algorithm for Gaussian data is discussed for example by [Tan et al., 2010, Section II.B]. Linear concentration models and their maximum likelihood inference are discussed by Anderson [1970], Sturmfels and Uhler [2010]. The earliest examples of Gaussian latent tree models are related to factor analysis with a single hidden factor; see Spearman [1928], Thurstone [1934]. The inference for this model class is generally very hard. The study of the underlying geometry was central to developing estimation procedures at the early stage of the model development. We refer to Bekker and de Leeuw [1987] for a very detailed account of the history of this geometric approach. A more recent account is given in Drton et al. [2007]. The full geometric description for the star tree is given by Bekker and de Leeuw [1987]. More general trees were treated in some special cases by Pearl and Xu [1987]. Defining equations in the general case were provided by Sullivant [2008]. The set of implied equations contains so-called tetrad equations; see Scheines et al. [1998]. Finally, the complete description is given by Aston et al. [2015]. Latent Gaussian tree models were introduced in the phylogenetic setting by Felsenstein [1973, 1981].

Bibliography

Alan Agresti. *An Introduction to Categorical Data Analysis*. John Wiley & Sons, New York, 1996.

Alan Agresti. *Categorical Data Analysis*. Wiley Series in Probability and Statistics. Wiley-Interscience [John Wiley & Sons], New York, second edition, 2002. ISBN 0-471-36093-7.

Rudolf Ahlswede and David E. Daykin. An inequality for the weights of two families of sets, their unions and intersections. *Z. Wahrsch. Verw. Gebiete*, 43(3):183–185, 1978. ISSN 0178-8051.

Martin Aigner. *Combinatorial Theory*. Classics in Mathematics. Springer-Verlag, Berlin, 1997. ISBN 3-540-61787-6. doi: 10.1007/978-3-642-59101-3. URL http://dx.doi.org/10.1007/978-3-642-59101-3. Reprint of the 1979 original.

Elizabeth S. Allman and John A. Rhodes. Phylogenetic invariants for the general Markov model of sequence mutation. *Math. Biosci.*, 186(2):113–144, 2003. ISSN 0025-5564.

Elizabeth S. Allman and John A. Rhodes. Phylogenetic invariants. In *Reconstructing Evolution: New Mathematical and Computational Advances*, pages 108–146. Oxford Univ. Press, Oxford, 2007.

Elizabeth S. Allman and John A. Rhodes. Phylogenetic ideals and varieties for the general Markov model. *Adv. in Appl. Math.*, 40(2):127–148, 2008. ISSN 0196-8858.

Elizabeth S. Allman, Catherine Matias, and John A. Rhodes. Identifiability of parameters in latent structure models with many observed variables. *Ann. Statist*, 37(6A):3099–3132, 2009.

Elizabeth S. Allman, Peter D. Jarvis, John A. Rhodes, and Jeremy G. Sumner. Tensor rank, invariants, inequalities, and applications. *SIAM J. Matrix Anal. Appl.*, 34(3):1014–1045, 2013. ISSN 0895-4798. doi: 10.1137/120899066.

Elizabeth S. Allman, John A. Rhodes, and Amelia Taylor. A semialgebraic description of the general Markov model on phylogenetic trees. *SIAM Journal on Discrete Mathematics*, 28(2):736–755, 2014.

Elizabeth S. Allman, John A. Rhodes, Bernd Sturmfels, and Piotr Zwiernik. Tensors of nonnegative rank two. *Linear Algebra and its Applications*, 473 (0):37 – 53, 2015. ISSN 0024-3795. doi: http://dx.doi.org/10.1016/j.laa.

2013.10.046. URL http://www.sciencedirect.com/science/article/pii/S0024379513006812. Special issue on Statistics.

Shun-ichi Amari. *Differential-Geometrical Methods in Statistics*, volume 28 of *Lecture Notes in Statistics*. Springer-Verlag, New York, 1985. ISBN 3-540-96056-2. doi: 10.1007/978-1-4612-5056-2.

Ian Anderson. *Combinatorics of Finite Sets*. Oxford Science Publications. The Clarendon Press, Oxford University Press, New York, 1987. ISBN 0-19-853367-5.

Theodore W. Anderson. Estimation of covariance matrices which are linear combinations or whose inverses are linear combinations of given matrices. In *Essays in Probability and Statistics*, pages 1–24. Univ. of North Carolina Press, Chapel Hill, N.C., 1970.

Steen Arne Andersson, David Madigan, and Michael D. Perlman. A characterization of Markov equivalence classes for acyclic digraphs. *Ann. Statist.*, 25(2):505–541, 1997. ISSN 0090-5364. doi: 10.1214/aos/1031833662.

Steen Arne Andersson, David Madigan, and Michael D. Perlman. Alternative Markov properties for chain graphs. *Scand. J. Statist.*, 28(1):33–85, 2001. ISSN 0303-6898. doi: 10.1111/1467-9469.00224.

John Aston, Nathaniel Shiers, Jim Q. Smith, and Piotr Zwiernik. The geometry of latent tree Gaussian models with applications. In preparation, 2015.

Michael F. Atiyah. Resolution of singularities and division of distributions. *Communications on Pure and Applied Mathematics*, 23(2):145–150, 1970.

Vincent Auvray, Pierre Geurts, and Louis Wehenkel. A semi-algebraic description of discrete naive Bayes models with two hidden classes. In *Proc. Ninth International Symposium on Artificial Intelligence and Mathematics*, Fort Lauderdale, Florida, January 2006.

Francis Bach. Learning with submodular functions: A convex optimization perspective. *arXiv preprint arXiv:1111.6453*, 2011.

Ole E. Barndorff–Nielsen. *Information and Exponential Families in Statistical Theory*. John Wiley & Sons Ltd., Chichester, 1978. ISBN 0-471-99545-2. Wiley Series in Probability and Mathematical Statistics.

Ole E. Barndorff–Nielsen and David R. Cox. *Asymptotic Techniques for Use in Statistics*. Monographs on Statistics and Applied Probability. Chapman & Hall, London, 1989. ISBN 0-412-31400-2.

Daniel Barry and JA Hartigan. Statistical analysis of hominoid molecular evolution. *Statistical Science*, pages 191–207, 1987.

David Bartholomew, Martin Knott, and Irini Moustaki. *Latent Variable Models and Factor Analysis*. Wiley Series in Probability and Statistics. John Wiley & Sons Ltd., Chichester, third edition, 2011. ISBN 978-0-470-97192-5. doi: 10.1002/9781119970583.

Francesco Bartolucci and Antonio Forcina. A likelihood ratio test for MTP$_2$

within binary variables. *Ann. Statist.*, 28(4):1206–1218, 2000. ISSN 0090-5364. doi: 10.1214/aos/1015956713.

Saugata Basu, Richard Pollack, and Marie-Françoise Roy. *Algorithms in Real Algebraic Geometry*, volume 10 of *Algorithms and Computation in Mathematics*. Springer-Verlag, Berlin, second edition, 2006. ISBN 978-3-540-33098-1; 3-540-33098-4.

Saugata Basu, Andrei Gabrielov, and Nicolai Vorobjov. Semi-monotone sets. *arXiv preprint arXiv:1004.5047*, 2010.

Saugata Basu, Andrei Gabrielov, and Nicolai Vorobjov. Monotone functions and maps. *Revista de la Real Academia de Ciencias Exactas, Fisicas y Naturales. Serie A. Matematicas*, pages 1–29, 2012.

Paul A Bekker and Jan de Leeuw. The rank of reduced dispersion matrices. *Psychometrika*, 52(1):125–135, 1987.

Louis J. Billera, Susan P. Holmes, and Karen Vogtmann. Geometry of the space of phylogenetic trees. *Adv. in Appl. Math.*, 27(4):733–767, 2001. ISSN 0196-8858. doi: 10.1006/aama.2001.0759. URL http://dx.doi.org/10.1006/aama.2001.0759.

Louis J. Billera, Susan P. Holmes, and Karen Vogtmann. Erratum to: "Geometry of the space of phylogenetic trees" [Adv. in Appl. Math. **27** (2001), no. 4, 733–767; MR1867931 (2002k:05229)] by L. J. Billera, S. P. Holmes and K. Vogtmann. *Adv. in Appl. Math.*, 29(1):136, 2002. ISSN 0196-8858. doi: 10.1016/S0196-8858(02)00016-7. URL http://dx.doi.org/10.1016/S0196-8858(02)00016-7.

Jacek Bochnak, Michel Coste, and Marie-Françoise Roy. *Real Algebraic Geometry*. Springer, 1998.

Stephen P. Boyd and Lieven Vandenberghe. *Convex Optimization*. Cambridge university press, 2004.

David R. Brillinger. The calculation of cumulants via conditioning. *Annals of the Institute of Statistical Mathematics*, 21(1):215–218, 1969.

Lawrence D. Brown. *Fundamentals of Statistical Exponential Families with Applications in Statistical Decision Theory*. Institute of Mathematical Statistics Lecture Notes—Monograph Series, 9. Institute of Mathematical Statistics, Hayward, CA, 1986. ISBN 0-940600-10-2.

Peter Buneman. The recovery of trees from measures of dissimilarity. In F. Hodson et al., editor, *Mathematics in the Archaeological and Historical Sciences*, pages 387–395. Edinburgh University Press, 1971.

Peter Buneman. A note on the metric properties of trees. *J. Combinatorial Theory Ser. B*, 17:48–50, 1974.

Marta Casanellas and Jesús Fernández–Sánchez. Relevant phylogenetic invariants of evolutionary models. *Journal de mathématiques pures et appliquées*, 96(3):207–229, 2011.

Marta Casanellas and Jesús Fernández-Sánchez. Performance of a new invariants method on homogeneous and nonhomogeneous quartet trees. *Molecular Biology and Evolution*, 24(1):288, 2007.

George Casella and Roger L. Berger. *Statistical inference*. The Wadsworth & Brooks/Cole Statistics/Probability Series. Wadsworth & Brooks/Cole Advanced Books & Software, Pacific Grove, CA, 1990. ISBN 0-534-11958-1.

Fabrizio Catanese, Serkan Hoşten, Amit Khetan, and Bernd Sturmfels. The maximum likelihood degree. *American Journal of Mathematics*, 128(3): 671–697, 2006.

James A. Cavender. Taxonomy with confidence. *Math. Biosci.*, 40(3-4): 271–280, 1978. ISSN 0025-5564.

James A. Cavender. Erratum: "Taxonomy with confidence" [Math. Biosci. **40** (1978), no. 3-4, 271–280; MR **58** #20548]. *Math. Biosci.*, 44(3-4): 309, 1979. ISSN 0025-5564. doi: 10.1016/0025-5564(79)90091-9. URL http://dx.doi.org/10.1016/0025-5564(79)90091-9.

James A. Cavender. Mechanized derivation of linear invariants. *Molecular Biology and Evolution*, 6(3):301, 1989.

James A. Cavender. Letter to the editor. *Molecular Phylogenetics and Evolution*, 8(3):443–444, 1997. ISSN 1055-7903. doi: DOI:10.1006/mpev. 1997.0451. URL http://www.sciencedirect.com/science/article/B6WNH-45M2XR5-F/2/c85b9732497dd90381144c1f99832d9c.

James A. Cavender and Joseph Felsenstein. Invariants of phylogenies in a simple case with discrete states. *Journal of Classification*, 4(1):57–71, 1987.

Joseph T. Chang. Full reconstruction of Markov models on evolutionary trees: Identifiability and consistency. *Mathematical Biosciences*, 137(1): 51–73, 1996.

Joseph T. Chang and John A. Hartigan. Reconstruction of evolutionary trees from pairwise distributions on current species. In *Proceedings of the 23rd symposium on the interface*, pages 254–257, 1991.

David M. Chickering and David Heckerman. Efficient approximations for the marginal likelihood of Bayesian networks with hidden variables. *Machine Learning*, 29(2):181–212, 1997.

Myung Jin Choi, Vincent Y. F. Tan, Animashree Anandkumar, and Alan S. Willsky. Learning latent tree graphical models. *J. Mach. Learn. Res.*, 12: 1771–1812, 2011. ISSN 1532-4435.

Benny Chor, Michael D. Hendy, Barbara R. Holland, and David Penny. Multiple maxima of likelihood in phylogenetic trees: An analytic approach. *Molecular Biology and Evolution*, 17(10):1529–1541, 2000.

C. K. Chow and C. N. Liu. Approximating discrete probability distributions with dependence trees. *IEEE Trans. Inform. Theory*, 14:462–467, 1968.

Ciro Ciliberto, Angelica Cueto, Massimiliano Mela, Kristian Ranestad, and Piotr Zwiernik. Cremona linearizations of some classical varieties.

arXiv:1403.1814, 2014.

CoCoATeam. CoCoA: A system for doing Computations in Commutative Algebra. Available at `http://cocoa.dima.unige.it`.

David A. Cox, John B. Little, and Don O'Shea. *Ideals, Varieties, and Algorithms*. Springer-Verlag, NY, 3rd edition, 2007.

David A. Cox, John B. Little, and Henry K. Schenck. *Toric Varieties*, volume 124 of *Graduate Studies in Mathematics*. American Mathematical Society, Providence, RI, 2011. ISBN 978-0-8218-4819-7.

John N. Darroch and Douglas Ratcliff. Generalized iterative scaling for log-linear models. *The Annals of Mathematical Statistics*, 43(5):pp. 1470–1480, 1972. ISSN 00034851. URL `http://www.jstor.org/stable/2240069`.

Brian A. Davey and Hilary A. Priestley. *Introduction to Lattices and Order*. Cambridge University Press, New York, second edition, 2002. ISBN 0-521-78451-4.

Clintin P. Davis–Stober. Analysis of multinomial models under inequality constraints: Applications to measurement theory. *Journal of Mathematical Psychology*, 53(1):1–13, 2009. ISSN 0022-2496.

A. Philip Dawid. Conditional independence in statistical theory. *Journal of the Royal Statistical Society. Series B (Methodological)*, 41(1):1–31, 1979.

Vin de Silva and Lek-Heng Lim. Tensor rank and the ill-posedness of the best low-rank approximation problem. *SIAM J. Matrix Anal. Appl*, 30(3): 1084–1127, 2008.

Wolfram Decker, Gert-Martin Greuel, Gerhard Pfister, and Hans Schönemann. Singular 3-1-3: A computer algebra system for polynomial computations. 2011. http://www.singular.uni-kl.de.

Arthur P. Dempster, Nan M. Laird, and Donald B. Rubin. Maximum likelihood from incomplete data via the EM algorithm. *J. Roy. Statist. Soc. Ser. B*, 39(1):1–38, 1977. ISSN 0035-9246. With discussion.

Persi Diaconis. Finite forms of de Finetti's theorem on exchangeability. *Synthese*, 36(2):271–281, 1977. ISSN 0039-7857. Foundations of probability and statistics, II.

Persi Diaconis and Bernd Sturmfels. Algebraic algorithms for sampling from conditional distributions. *Ann. Statist.*, 26(1):363–397, 1998. ISSN 0090-5364.

Jan Draisma and Jochen Kuttler. On the ideals of equivariant tree models. *Mathematische Annalen*, 344(3):619–644, 2009. ISSN 0025-5831. doi: 10.1007/s00208-008-0320-6. URL `http://dx.doi.org/10.1007/s00208-008-0320-6`.

Andreas Dress, Katharina T. Huber, Jacobus Koolen, Vincent Moulton, and Andreas Spillner. *Basic Phylogenetic Combinatorics*. Cambridge University Press, Cambridge, 2012. ISBN 978-0-521-76832-0.

Suzanne Drolet and David Sankoff. Quadratic tree invariants for multivalued characters. *Journal of Theoretical Biology*, 144(1):117–129, 1990.

Mathias Drton and Martyn Plummer. A Bayesian information criterion for singular models. arXiv:1309.0911, 2013.

Mathias Drton and Seth Sullivant. Algebraic statistical models. *Statistica Sinica*, 17:1273–1297, 2007.

Mathias Drton, Bernd Sturmfels, and Seth Sullivant. Algebraic factor analysis: tetrads, pentads and beyond. *Probab. Theory Related Fields*, 138(3-4): 463–493, 2007. ISSN 0178-8051. doi: 10.1007/s00440-006-0033-2. URL http://dx.doi.org/10.1007/s00440-006-0033-2.

Mathias Drton, Bernd Sturmfels, and Seth Sullivant. *Lectures on Algebraic Statistics*. Oberwolfach Seminars Series. Birkhauser Verlag AG, 2009.

Mathias Drton, Rina Foygel, and Seth Sullivant. Global identifiability of linear structural equation models. *Ann. Statist.*, 39(2):865–886, 2011. ISSN 0090-5364. doi: 10.1214/10-AOS859. URL http://dx.doi.org/10.1214/10-AOS859.

Bradley Efron. The geometry of exponential families. *The Annals of Statistics*, 6(2):362–376, 1978.

David Eisenbud. *Commutative Algebra*, volume 150 of *Graduate Texts in Mathematics*. Springer-Verlag, New York, 1995. ISBN 0-387-94268-8; 0-387-94269-6. doi: 10.1007/978-1-4612-5350-1. URL http://dx.doi.org/10.1007/978-1-4612-5350-1. With a view toward algebraic geometry.

Alexander Engström, Patricia Hersh, and Bernd Sturmfels. Toric cubes. *Rendiconti del Circolo Matematico di Palermo*, pages 1–12, 2012.

Nicholas Eriksson. *Using Invariants for Phylogenetic Tree Construction*, volume 149 of *The IMA Volumes in Mathematics and Its Applications*, pages 89–108. Springer, 2007.

Nicholas Eriksson, Kristian Ranestad, Bernd Sturmfels, and Seth Sullivant. Phylogenetic algebraic geometry. In *Projective Varieties with Unexpected Properties*, pages 237–255. Walter de Gruyter GmbH & Co. KG, Berlin, 2005.

Steven N. Evans and Terence P. Speed. Invariants of some probability models used in phylogenetic inference. *Ann. Statist.*, 21(1):355–377, 1993. ISSN 0090-5364. doi: 10.1214/aos/1176349030. URL http://dx.doi.org/10.1214/aos/1176349030.

Jianqing Fan, Hui-Nien Hung, and Wing-Hung Wong. Geometric understanding of likelihood ratio statistics. *Journal of the American Statistical Association*, 95(451):836–841, 2000.

William Feller. *An Introduction to Probability Theory and Applications*, volume 2. John Wiley & Sons, New York, second edition, 1971.

Joseph Felsenstein. Maximum-likelihood estimation of evolutionary trees from continuous characters. *American Journal of Human Genetics*, 25 (5):471–492, September 1973. ISSN 0002-9297. URL http://www.

`ncbi.nlm.nih.gov/pmc/articles/PMC1762641/`. PMID: 4741844 PM-CID: PMC1762641.

Joseph Felsenstein. Evolutionary trees from gene frequencies and quantitative characters: Finding maximum likelihood estimates. *Evolution*, 35(6):1229–1242, November 1981. ISSN 0014-3820. doi: 10.2307/2408134. URL `http://www.jstor.org/stable/2408134`. ArticleType: research-article / Full publication date: Nov., 1981 / Copyright © 1981 Society for the Study of Evolution.

Joseph Felsenstein. *Inferring phylogenies*. Sinauer Associates Sunderland, 2004.

Jesús Fernández–Sánchez and Marta Casanellas. Invariant versus classical approaches when evolution is heterogeneous across sites and lineages. *arXiv preprint arXiv:1405.6546*, 2014.

Stephen E. Fienberg and John P. Gilbert. The geometry of a two by two contingency table. *Journal of the American Statistical Association*, 65 (330):694–701, 1970.

Ronald A. Fisher and John Wishart. The derivation of the pattern formulae of two-way partitions from those of simpler patterns. *Proceedings of the London Mathematical Society*, 2(1):195, 1932.

Cees M. Fortuin, Pieter W. Kasteleyn, and Jean Ginibre. Correlation inequalities on some partially ordered sets. *Comm. Math. Phys.*, 22:89–103, 1971. ISSN 0010-3616.

Morten Frydenberg. The chain graph Markov property. *Scand. J. Statist.*, 17(4):333–353, 1990. ISSN 0303-6898.

Satoru Fujishige. *Submodular Functions and Optimization*, volume 58 of *Annals of Discrete Mathematics*. Elsevier B. V., Amsterdam, second edition, 2005. ISBN 0-444-52086-4.

William Fulton. *Introduction to Toric Varieties*. Princeton University Press Princeton, NJ, 1993.

Luis David Garcia, Michael Stillman, and Bernd Sturmfels. Algebraic geometry of Bayesian networks. *J. Symbolic Comput*, 39(3-4):331–355, 2005.

Dan Geiger, David Heckerman, Henry King, and Christopher Meek. Stratified exponential families: Graphical models and model selection. *Ann. Statist.*, 29(2):505–529, 2001. ISSN 0090-5364.

Dan Geiger, Christopher Meek, and Bernd Sturmfels. On the toric algebra of graphical models. *Annals of Statistics*, 34:1463–1492, 2006.

Israel M. Gelfand, Mikhail M. Kapranov, and Andrei V. Zelevinsky. *Discriminants, Resultants, and Multidimensional Determinants*. Birkhäuser, 1994.

Jonna Gill, Svante Linusson, Vincent Moulton, and Mike Steel. A regular decomposition of the edge-product space of phylogenetic trees. *Adv. in Appl. Math.*, 41(2):158–176, 2008. ISSN 0196-8858. doi: 10.1016/j.aam.

2006.07.007. URL http://dx.doi.org/10.1016/j.aam.2006.07.007.

Zvi Gilula. Singular value decomposition of probability matrices: Probabilistic aspects of latent dichotomous variables. *Biometrika*, 66(2):339–344, 1979.

Daniel R. Grayson and Michael E. Stillman. Macaulay2, a software system for research in algebraic geometry. Available at http://www.math.uiuc.edu/Macaulay2/.

Geoffrey R. Grimmett and David R. Stirzaker. *Probability and Random Processes*. Oxford University Press, New York, third edition, 2001. ISBN 0-19-857223-9.

Paul Gustafson. What are the limits of posterior distributions arising from nonidentified models, and why should we care? *Journal of the American Statistical Association*, 104(488):1682–1695, 2009.

Anders Hald. The early history of the cumulants and the Gram-Charlier series. *International Statistical Review*, 68(2):137–153, 2000.

Ralf Hemmecke, Raymond Hemmecke, and Peter Malkin. 4ti2 version 1.2—computation of Hilbert bases, Graver bases, toric Gröbner bases, and more. http://www.4ti2.de/, 2005.

David G. Herr. On the history of the use of geometry in the general linear model. *The American Statistician*, 34(1):pp. 43–47, 1980. ISSN 00031305. URL http://www.jstor.org/stable/2682995.

John P. Huelsenbeck. Performance of phylogenetic methods in simulation. *Systematic Biology*, 44(1):17, 1995.

June Huh. The maximum likelihood degree of a very affine variety. *Compositio Mathematica*, pages 1–22, 2012.

June Huh and Bernd Sturmfels. Likelihood geometry. In *Combinatorial Algebraic Geometry*, pages 63–117. Springer, 2014.

Vivek Jayaswal, Lars S. Jermiin, and John Robinson. Estimation of phylogeny using a general Markov model. *Evolutionary Bioinformatics Online*, 1:62, 2005.

Vivek Jayaswal, John Robinson, and Lars S. Jermiin. Estimation of phylogeny and invariant sites under the general Markov model of nucleotide sequence evolution. *Systematic Biology*, 56(2):155–162, 2007.

Thomas H. Jukes and Charles R. Cantor. Evolution of protein molecules. *Mammalian Protein Metabolism*, 3:21–132, 1969.

Joseph B. Kadane. The role of identification in Bayesian theory. *Studies in Bayesian Econometrics and Statistics*, pages 175–191, 1974.

Thomas Kähle. *On Boundaries of Statistical Models*. PhD thesis, Leipzig University, 2010.

Robert E. Kass and Paul W. Vos. *Geometrical Foundations of Asymptotic Inference*. Wiley Series in Probability and Statistics: Probability and Statistics. John Wiley & Sons Inc., New York, 1997. ISBN 0-471-82668-5. doi: 10.1002/9781118165980. URL http://dx.doi.org/10.1002/9781118165980. A Wiley-Interscience Publication.

Junhyong Kim. Slicing hyperdimensional oranges: The geometry of phylogenetic estimation. *Molecular Phylogenetics and Evolution*, 17(1):58–75, 2000.

Motoo Kimura. A simple method for estimating evolutionary rates of base substitutions through comparative studies of nucleotide sequences. *Journal of Molecular Evolution*, 16(2):111–120, 1980.

John F. C. Kingman. The imbedding problem for finite Markov chains. *Probability Theory and Related Fields*, 1(1):14–24, 1962.

Steffen Klaere and Volkmar Liebscher. An algebraic analysis of the two state Markov model on tripod trees. *Math. Biosci.*, 237(1-2):38–48, 2012. ISSN 0025-5564. doi: 10.1016/j.mbs.2012.03.001. URL http://dx.doi.org/10.1016/j.mbs.2012.03.001.

D. Koller and N. Friedman. *Probabilistic Graphical Models: Principles and Techniques*. The MIT Press, 2009.

Chris Kottke and Richard Melrose. Generalized blow-up of corners and fiber products. In *Microlocal Methods in Mathematical Physics and Global Analysis*, pages 59–62. Springer, 2013.

Steven G. Krantz and Harold R. Parks. *A Primer of Real Analytic Functions*. Birkhäuser Advanced Texts: Basler Lehrbücher. [Birkhäuser Advanced Texts: Basel Textbooks]. Birkhäuser Boston Inc., Boston, MA, second edition, 2002. ISBN 0-8176-4264-1.

Bernadette Krawczyk and Roland Speicher. Combinatorics of free cumulants. *J. Combin. Theory Ser. A*, 90(2):267–292, 2000. ISSN 0097-3165. doi: 10.1006/jcta.1999.3032. URL http://dx.doi.org/10.1006/jcta.1999.3032.

Germain Kreweras. Sur les partitions non croisées d'un cycle. *Discrete Math.*, 1(4):333–350, 1972. ISSN 0012-365X.

Joseph B. Kruskal, Jr. On the shortest spanning subtree of a graph and the traveling salesman problem. *Proc. Amer. Math. Soc.*, 7:48–50, 1956. ISSN 0002-9939.

James A. Lake. A rate-independent technique for analysis of nucleic acid sequences: Evolutionary parsimony. *Molecular Biology and Evolution*, 4(2):167, 1987.

James A. Lake. Reconstructing evolutionary trees from DNA and protein sequences: Paralinear distances. *Proceedings of the National Academy of Sciences of the United States of America*, 91(4):1455–1459, 1994.

Joseph M. Landsberg. *Tensors: Geometry and Applications*, volume 128 of *Graduate Studies in Mathematics*. American Mathematical Society, Providence, RI, 2012. ISBN 978-0-8218-6907-9.

Steffen L. Lauritzen. *Graphical Models*, volume 17 of *Oxford Statistical Science Series*. Oxford University Press, 1996. ISBN 0-19-852219-3. Oxford Science Publications.

Steffen L. Lauritzen and Nanny Wermuth. Graphical models for associations between variables, some of which are qualitative and some quantitative. *Ann. Statist.*, 17(1):31–57, 1989. ISSN 0090-5364. doi: 10.1214/aos/ 1176347003. URL http://dx.doi.org/10.1214/aos/1176347003.

Paul F. Lazarsfeld and Neil W. Henry. *Latent Structure Analysis*. Houghton, Mifflin, New York, 1968.

Franz Lehner. Free cumulants and enumeration of connected partitions. *European Journal of Combinatorics*, 23(8):1025–1031, 2002. ISSN 0195-6698.

Shaowei Lin. Asymptotic Approximation of Marginal Likelihood Integrals. arXiv:1003.5338, November 2011. submitted.

Bruce G. Lindsay. *Mixture Models: Theory, Geometry and Applications*, volume 5 of *NSF-CBMS Regional Conference Series in Probability and Statistics*. Institute of Mathematical Statistics, Hayward, CA, 1995.

Peter J. Lockhart, Michael A. Steel, Michael D. Hendy, and David Penny. Recovering evolutionary trees under a more realistic model of sequence evolution. *Molecular Biology and Evolution*, 11(4):605–612, 1994.

E. Lukacs. Some extensions of a theorem of Marcinkiewicz. *Pacific Journal of Mathematics*, 8(3):487–501, 1958. ISSN 0030-8730.

Laurent Manivel and Mateusz Michałek. Secants of minuscule and cominuscule minimal orbits. *arXiv preprint arXiv:1401.1956*, 2014.

Józef Marcinkiewicz. Sur une propriété de la loi de Gauß. *Math. Z.*, 44(1): 612–618, 1939. ISSN 0025-5874. doi: 10.1007/BF01210677. URL http: //dx.doi.org/10.1007/BF01210677.

Frederick A Matsen. Fourier transform inequalities for phylogenetic trees. *IEEE/ACM Transactions on Computational Biology and Bioinformatics (TCBB)*, 6(1):89–95, 2009.

Peter McCullagh. *Tensor Methods in Statistics*. Monographs on Statistics and Applied Probability. Chapman & Hall, London, 1987. ISBN 0-412-27480-9.

Mateusz Michałek, Bernd Sturmfels, Caroline Uhler, and Piotr Zwiernik. Exponential varieties. arXiv:1412.6185, December 2014.

Mateusz Michałek, Luke Oeding, and Piotr Zwiernik. Secant cumulants and toric geometry. *International Mathematics Research Notices*, 2015(12): 4019–4063, 2015. doi: 10.1093/imrn/rnu056.

David Mond, Jim Q. Smith, and Duco van Straten. Stochastic factorizations, sandwiched simplices and the topology of the space of explanations. *R. Soc. Lond. Proc. Ser. A Math. Phys. Eng. Sci.*, 459(2039):2821–2845, 2003. ISSN 1364-5021.

Guido Montúfar. Mixture decompositions of exponential families using a decomposition of their sample spaces. *Kybernetika*, (1):23–39, 2013.

Vincent Moulton and Mike Steel. Peeling phylogenetic oranges. *Advances in Applied Mathematics*, 33(4):710–727, 2004.

William C. Navidi, Gary A. Churchill, and Arndt von Haeseler. Phylogenetic inference: linear invariants and maximum likelihood. *Biometrics*, 49(2): 543–555, 1993. ISSN 0006-341X. doi: 10.2307/2532566. URL http://dx.doi.org/10.2307/2532566.

Richard E. Neapolitan. *Probabilistic Reasoning in Expert Systems*. A Wiley-Interscience Publication. John Wiley & Sons Inc., New York, 1990. ISBN 0-471-61840-3. Theory and algorithms.

Jerzy Neyman. Molecular studies of evolution: A source of novel statistical problems. In *Statistical Decision Theory and Related Topics (Proc. Sympos., Purdue Univ., Lafayette, Ind., 1970)*, pages 1–27. Academic Press, New York, 1971.

Lior Pachter and Bernd Sturmfels, editors. *Algebraic Statistics for Computational Biology*. Cambridge University Press, New York, 2005. ISBN 978-0-521-85700-0; 0-521-85700-7.

Judea Pearl. Fusion, propagation, and structuring in belief networks* 1. *Artificial Intelligence*, 29(3):241–288, 1986.

Judea Pearl. *Probabilistic Reasoning in Intelligent Systems: Networks of Plausible Inference*. The Morgan Kaufmann Series in Representation and Reasoning. Morgan Kaufmann, San Mateo, CA, 1988. ISBN 0-934613-73-7.

Judea Pearl. *Causality: Models, Reasoning, and Inference*. Cambridge University Press, New York, NY, USA, 2000. ISBN 0-521-77362-8.

Judea Pearl and Rina Dechter. Learning structure from data: A survey. In *Proceedings of the Second Annual Workshop on Computational Learning Theory*, pages 230–244. Morgan Kaufmann Publishers Inc., 1989.

Judea Pearl and Michael Tarsi. Structuring causal trees. *J. Complexity*, 2 (1):60–77, 1986. ISSN 0885-064X. Complexity of approximately solved problems (Morningside Heights, N.Y., 1985).

Judea Pearl and Lei Xu. Structuring causal tree models with continuous variables. In *Proceedings of the Third Annual Conference on Uncertainty in Artificial Intelligence*, 1987.

Giovanni Pistone and Henry P. Wynn. Finitely generated cumulants. *Statist. Sinica*, 9(4):1029–1052, 1999. ISSN 1017-0405.

Giovanni Pistone and Henry P. Wynn. Cumulant varieties. *Journal of Symbolic Computation*, 41(2):210–221, 2006.

Giovanni Pistone, Eva Riccomagno, and Henry P. Wynn. *Algebraic Statistics*, volume 89 of *Monographs on Statistics and Applied Probability*. Chapman & Hall/CRC, Boca Raton, FL, 2001. ISBN 1-58488-204-2. Computational commutative algebra in statistics.

Dale J. Poirier. Revising beliefs in nonidentified models. *Econometric Theory*,

14(04):483–509, 1998.

Robert C. Prim. Shortest connection networks and some generalizations. *Bell System Technical Journal*, 36(6):1389–1401, 1957.

Claudiu Raicu. Secant varieties of segre–veronese varieties. *Algebra & Number Theory*, 6(8):1817–1868, 2012.

Johannes Rauh, Thomas Kahle, and Nihat Ay. Support Sets in Exponential Families and Oriented Matroid Theory. *Proc. WUPES'09, invited for special issue of IJAR*, 2009.

Eva Riccomagno. A short history of algebraic statistics. *Metrika*, 69(2-3): 397–418, 2009. ISSN 0026-1335. doi: 10.1007/s00184-008-0222-3. URL http://dx.doi.org/10.1007/s00184-008-0222-3.

Steven Roman. *Advanced Linear Algebra*, volume 135 of *Graduate Texts in Mathematics*. Springer, New York, third edition, 2008. ISBN 978-0-387-72828-5.

Gian-Carlo Rota. On the foundations of combinatorial theory I. Theory of Möbius Functions. *Probability Theory and Related Fields*, 2(4):340–368, 1964.

Gian-Carlo Rota and Jianhong Shen. On the combinatorics of cumulants. *J. Combin. Theory Ser. A*, 91(1-2):283–304, 2000. ISSN 0097-3165. In memory of Gian-Carlo Rota.

Gian-Carlo Rota, D. Kahaner, and A. Odlyzko. On the foundations of combinatorial theory. VIII- Finite operator calculus(Umbral/finite operator/calculus in combinatorial theory of special polynomial sequences as technique for expressing one polynomial set in terms of another). *Journal of Mathematical Analysis and Applications*, 42:684–760, 1973.

Thomas J. Rothenberg. Identification in parametric models. *Econometrica: Journal of the Econometric Society*, pages 577–591, 1971.

Alberto Roverato. A unified approach to the characterization of equivalence classes of DAGs, chain graphs with no flags and chain graphs. *Scand. J. Statist.*, 32(2):295–312, 2005. ISSN 0303-6898. doi: 10.1111/j.1467-9469. 2005.00422.x. URL http://dx.doi.org/10.1111/j.1467-9469.2005. 00422.x.

Dmitry Rusakov and Dan Geiger. Asymptotic model selection for naive Bayesian networks. *J. Mach. Learn. Res.*, 6:1–35 (electronic), 2005. ISSN 1532-4435.

Sujit K. Sahu and Alan E. Gelfand. Identifiability, Improper Priors, and Gibbs Sampling for Generalized Linear Models. *Journal of the American Statistical Association*, 94(445):247–254, 1999.

David Sankoff. Designer invariants for large phylogenies. *Molecular Biology and Evolution*, 7(3):255, 1990.

Richard Scheines, Peter Spirtes, Clark Glymour, Christopher Meek, and Thomas Richardson. The TETRAD project: Constraint based aids to causal model specification. *Multivariate Behavioral Research*, 33(1):65–117, 1998. ISSN 0027-3171. doi: 10.1207/s15327906mbr3301_3. URL http://www.tandfonline.com/doi/abs/10.1207/s15327906mbr3301_3.

Charles Semple and Mike Steel. *Phylogenetics*, volume 24 of *Oxford Lecture Series in Mathematics and Its Applications*. Oxford University Press, Oxford, 2003. ISBN 0-19-850942-1.

Raffaella Settimi and Jim Q. Smith. On the geometry of Bayesian graphical models with hidden variables. In Gregory F. Cooper and Serafín Moral, editors, *UAI*, pages 472–479. Morgan Kaufmann, 1998.

Raffaella Settimi and Jim Q. Smith. Geometry, moments and conditional independence trees with hidden variables. *Ann. Statist.*, 28(4):1179–1205, 2000. ISSN 0090-5364.

Igor R. Shafarevich. *Basic Algebraic Geometry. 1.* Springer-Verlag, Berlin, second edition, 1994. ISBN 3-540-54812-2. Varieties in projective space, Translated from the 1988 Russian edition and with notes by Miles Reid.

A. Shapiro. Towards a unified theory of inequality constrained testing in multivariate analysis. *International Statistical Review*, 56(1):49–62, 1988.

Alexander Shapiro. Asymptotic distribution of test statistics in the analysis of moment structures under inequality constraints. *Biometrika*, 72(1):133–144, 1985.

Karen E. Smith, Lauri Kahanpää, Pekka Kekäläinen, and William Traves. *An Invitation to Algebraic Geometry.* Universitext. Springer-Verlag, New York, 2000. ISBN 0-387-98980-3.

Frank Sottile. Toric ideals, real toric varieties, and the moment map. In *Topics in Algebraic Geometry and Geometric Modeling: Workshop on Algebraic Geometry and Geometric Modeling*, volume 334, page 225. Amer Mathematical Society, 2003.

Charles Spearman. The abilities of man. *Science*, 68(1750):38, Jul 1928. doi: 10.1126/science.68.1750.38-a.

Terence P. Speed. Cumulants and partition lattices. *Austral. J. Statist*, 25(2):378–388, 1983.

Roland Speicher. Multiplicative functions on the lattice of noncrossing partitions and free convolution. *Math. Ann.*, 298(4):611–628, 1994. ISSN 0025-5831. doi: 10.1007/BF01459754. URL http://dx.doi.org/10.1007/BF01459754.

Roland Speicher. Free probability theory and non-crossing partitions. *Sém. Lothar. Combin.*, 39:Art. B39c, 38 pp. (electronic), 1997. ISSN 1286-4889.

Roland Speicher and Reza Woroudi. Boolean convolution. In *Free Probability Theory (Waterloo, ON, 1995)*, volume 12 of *Fields Inst. Commun.*, pages 267–279. Amer. Math. Soc., Providence, RI, 1997.

Michael Spivak. *Calculus on Manifolds. A Modern Approach to Classical Theorems of Advanced Calculus.* W. A. Benjamin, Inc., New York-Amsterdam,

1965.

Richard P. Stanley. *Enumerative Combinatorics. Volume I.* Number 49 in Cambridge Studies in Advanced Mathematics. Cambridge University Press, 2002.

Mike Steel. Recovering a tree from the leaf colourations it generates under a Markov model. *Appl. Math. Lett.*, 7(2):19–23, 1994. ISSN 0893-9659. doi: 10.1016/0893-9659(94)90024-8. URL http://dx.doi.org/10.1016/0893-9659(94)90024-8.

Mike Steel and Beáta Faller. Markovian log-supermodularity, and its applications in phylogenetics. *Appl. Math. Lett.*, 22(7):1141–1144, 2009. ISSN 0893-9659. doi: 10.1016/j.aml.2008.10.005. URL http://dx.doi.org/10.1016/j.aml.2008.10.005.

Milan Studený. Characterization of essential graphs by means of the operation of legal merging of components. *Internat. J. Uncertain. Fuzziness Knowledge-Based Systems*, 12(January 2004, suppl.):43–62, 2004. ISSN 0218-4885. doi: 10.1142/S0218488504002576. URL http://dx.doi.org/10.1142/S0218488504002576. New trends in probabilistic graphical models.

Milan Studený. *Probabilistic Conditional Independence Structures.* Information Science and Statistics. Springer, London, 1 edition, 11 2004. ISBN 9781852338916.

Bernd Sturmfels. *Gröbner Bases and Convex Polytopes*, volume 8 of *University Lecture Series*. American Mathematical Society, Providence, RI, 1996. ISBN 0-8218-0487-1.

Bernd Sturmfels and Seth Sullivant. Toric ideals of phylogenetic invariants. *Journal of Computational Biology*, 12(2):204–228, 2005.

Bernd Sturmfels and Caroline Uhler. Multivariate Gaussians, semidefinite matrix completion, and convex algebraic geometry. *Annals of the Institute of Statistical Mathematics*, 62(4):603–638, 2010.

Bernd Sturmfels and Piotr Zwiernik. Binary cumulant varieties. *Annals of Combinatorics*, pages 1–22, 2012. ISSN 0218-0006. doi: 10.1007/s00026-012-0174-1. URL http://dx.doi.org/10.1007/s00026-012-0174-1.

Seth Sullivant. Algebraic geometry of Gaussian Bayesian networks. *Advances in Applied Mathematics*, 40(4):482–513, 2008.

Jeremy G Sumner, Jesús Fernández–Sánchez, and Peter D Jarvis. Lie Markov models. *J. Theoret. Biol.*, 298:16–31, 2012a. ISSN 0022-5193. doi: 10.1016/j.jtbi.2011.12.017. URL http://dx.doi.org/10.1016/j.jtbi.2011.12.017.

Jeremy G. Sumner, Peter D. Jarvis, Jesús Fernández–Sánchez, Bodie T. Kaine, Michael D. Woodhams, and Barbara R. Holland. Is the general time-reversible model bad for molecular phylogenetics? *Systematic Biol-

ogy, 61(6):1069–1074, 2012b.

Vincent Y. F. Y. F. Tan, Animashree Anandkumar, and Alan S. Willsky. Learning Gaussian tree models: Analysis of error exponents and extremal structures. *Signal Processing, IEEE Transactions on*, 58(5):2701–2714, May 2010. ISSN 1053-587X. doi: 10.1109/TSP.2010.2042478.

Simon Tavaré. Some probabilistic and statistical problems in the analysis of DNA sequences. In *Some Mathematical Questions in Biology—DNA Sequence Analysis (New York, 1984)*, volume 17 of *Lectures Math. Life Sci.*, pages 57–86. Amer. Math. Soc., Providence, RI, 1986.

Jozef L. Teugels. Some representations of the multivariate Bernoulli and binomial distributions. *Journal of Multivariate Analysis*, 32(2):256–268, 1990. ISSN 0047-259X. doi: http://dx.doi.org/10.1016/0047-259X(90) 90084-U. URL http://www.sciencedirect.com/science/article/pii/ 0047259X9090084U.

Thorvald Nicolai Thiele. Om iagttagelseslaxens halvinvarianter. *Kgf. damke Kdenskabernes Selskabs Forhandfinge*, (3):135–141, 1899.

Louis L. Thurstone. The vectors of mind. *Psychological Review*, 41(1):1, 1934.

Chris Tuffley and Mike Steel. Links between maximum likelihood and maximum parsimony under a simple model of site substitution. *Bulletin of Mathematical Biology*, 59(3):581–607, 1997.

Thomas S. Verma and Judea Pearl. Equivalence and Synthesis of Causal Models. In Piero P. Bonissone, Max Henrion, Laveen N. Kanal, and John F. Lemmer, editors, *UAI '90: Proceedings of the Sixth Annual Conference on Uncertainty in Artificial Intelligence, MIT, Cambridge, MA, USA, July 27-29, 1990*. Elsevier, October 1991. ISBN 0-444-89264-8.

Martin J. Wainwright and Michael I. Jordan. Graphical models, exponential families, and variational inference. *Foundations and Trends in Machine Learning*, 1(1-2):1–305, 2008.

Yi Wang and Nevin L. Zhang. Severity of local maxima for the EM algorithm: Experiences with hierarchical latent class models. In *Probabilistic Graphical Models*, pages 301–308. Citeseer, 2006.

Sumio Watanabe. *Algebraic Geometry and Statistical Learning Theory*. Number 25 in Cambridge Monographs on Applied and Computational Mathematics. Cambridge University Press, 2009. ISBN-13: 9780521864671.

Louis Weisner. Abstract theory of inversion of finite series. *Trans. Amer. Math. Soc.*, 38(3):474–484, 1935. ISSN 0002-9947. doi: 10.2307/1989808. URL http://dx.doi.org/10.2307/1989808.

Joe Whittaker. *Graphical Models in Applied Multivariate Statistics*. Wiley New York, 1990.

Sewall Wright. Correlation and causation. *J. Agric. Res.*, 20:557–585, 1921.

Ziheng Yang. *Computational Molecular Evolution (Oxford Series in Ecology and Evolution)*. Oxford University Press, USA, Oxford, USA, 12 2006. ISBN 9780198567028.

G. Alastair Young and Richard L. Smith. *Essentials of Statistical Inference*. Cambridge Series in Statistical and Probabilistic Mathematics. Cambridge University Press, Cambridge, 2005. ISBN 978-0-521-83971-6; 0-521-83971-8. doi: 10.1017/CBO9780511755392. URL http://dx.doi.org/10.1017/CBO9780511755392.

Nevin L. Zhang. Hierarchical latent class models for cluster analysis. *J. Mach. Learn. Res.*, 5:697–723, 2003/04. ISSN 1532-4435.

Günter M. Ziegler. *Lectures on Polytopes*, volume 152 of *Graduate Texts in Mathematics*. Springer-Verlag, New York, 1995. ISBN 0-387-94365-X. doi: 10.1007/978-1-4613-8431-1. URL http://dx.doi.org/10.1007/978-1-4613-8431-1.

Piotr Zwiernik. Asymptotic behaviour of the marginal likelihood for general Markov models. *J. Mach. Learn. Res.*, 12:3283–3310, 2011.

Piotr Zwiernik. L-cumulants, L-cumulant embeddings and algebraic statistics. *Journal of Algebraic Statistics*, 3:11–43, November 2012.

Piotr Zwiernik and Jim Q. Smith. Implicit inequality constraints in a binary tree model. *Electron. J. Statist.*, 5:1276–1312, 2011. ISSN 1935-7524. doi: 10.1214/11-EJS640.

Piotr Zwiernik and Jim Q. Smith. Tree cumulants and the geometry of binary tree models. *Bernoulli*, 18(1):290–321, 02 2012. doi: 10.3150/10-BEJ338. URL http://dx.doi.org/10.3150/10-BEJ338.

Index